Lecture Notes in Physics

Springer-Verlag Berlin Heidelberg GmbH

Physics and Astronomy ONLINE LIBRARY

http://www.springer.de/phys/

Editorial Policy

The series *Lecture Notes in Physics* (LNP), founded in 1969, reports new developments in physics research and teaching -- quickly, informally but with a high quality. Manuscripts to be considered for publication are topical volumes consisting of a limited number of contributions, carefully edited and closely related to each other. Each contribution should contain at least partly original and previously unpublished material, be written in a clear, pedagogical style and aimed at a broader readership, especially graduate students and nonspecialist researchers wishing to familiarize themselves with the topic concerned. For this reason, traditional proceedings cannot be considered for this series though volumes to appear in this series are often based on material presented at conferences, workshops and schools (in exceptional cases the original papers and/or those not included in the printed book may be added on an accompanying CD ROM, together with the abstracts of posters and other material suitable for publication, e.g. large tables, colour pictures, program codes, etc.).

Acceptance

A project can only be accepted tentatively for publication, by both the editorial board and the publisher, following thorough examination of the material submitted. The book proposal sent to the publisher should consist at least of a preliminary table of contents outlining the structure of the book together with abstracts of all contributions to be included.
Final acceptance is issued by the series editor in charge, in consultation with the publisher, only after receiving the complete manuscript. Final acceptance, possibly requiring minor corrections, usually follows the tentative acceptance unless the final manuscript differs significantly from expectations (project outline). In particular, the series editors are entitled to reject individual contributions if they do not meet the high quality standards of this series. The final manuscript must be camera-ready, and should include both an informative introduction and a sufficiently detailed subject index.

Contractual Aspects

Publication in LNP is free of charge. There is no formal contract, no royalties are paid, and no bulk orders are required, although special discounts are offered in this case. The volume editors receive jointly 30 free copies for their personal use and are entitled, as are the contributing authors, to purchase Springer books at a reduced rate. The publisher secures the copyright for each volume. As a rule, no reprints of individual contributions can be supplied.

Manuscript Submission

The manuscript in its final and approved version must be submitted in camera-ready form. The corresponding electronic source files are also required for the production process, in particular the online version. Technical assistance in compiling the final manuscript can be provided by the publisher's production editor(s), especially with regard to the publisher's own Latex macro package which has been specially designed for this series.

Online Version/ LNP Homepage

LNP homepage (list of available titles, aims and scope, editorial contacts etc.):
http://www.springer.de/phys/books/lnp/

LNP online (abstracts, full-texts, subscriptions etc.):
http://link.springer.de/series/lnp/

L.Fernández-Jambrina L.M.González-Romero (Eds.)

Current Trends in Relativistic Astrophysics

Theoretical, Numerical, Observational

Springer

Editors

Dr. Leonardo Fernández-Jambrina
Universidad Politécnica de Madrid
E.T.S.I. Navales
Arco de la Victoria s/n
28040 Madrid, Spain

Prof. Luis Manuel González-Romero
Universidad Complutense de Madrid
Fac. Ciencias Físicas
Avenida Complutense s/n
28040 Madrid, Spain

Cataloging-in-Publication Data applied for

A catalog record for this book is available from the Library of Congress.

Bibliographic information published by Die Deutsche Bibliothek
Die Deutsche Bibliothek lists this publication in the Deutsche Nationalbibliografie; detailed bibliographic data is available in the Internet at http://dnb.ddb.de

ISSN 0075-8450
ISBN 978-3-662-14420-6 ISBN 978-3-540-36973-8 (eBook)
DOI 10.1007/978-3-540-36973-8

http://www.springer.de

Originally published by Springer-Verlag Berlin Heidelberg New York in 2003
Softcover reprint of the hardcover 1st edition 2003

Typesetting: Camera-ready by the authors/editor
Camera-data conversion by Steingraeber Satztechnik GmbH Heidelberg
Cover design: *design & production*, Heidelberg

Printed on acid-free paper
54/3141/du - 5 4 3 2 1 0

Preface

This volume contains selected lectures presented at the Spanish Relativity Meeting on Relativistic Astrophysics held at the Escuela Técnica Superior de Ingenieros Navales of Madrid, Spain, in September 2001.

The meeting centered on the study of several aspects of Relativistic Astrophysics from various viewpoints: theoretical, numerical and observational. Lectures and contributions dealt with issues related to black holes (accretion, thermodynamics, gravitational collapse, stability), numerical and perturbative aspects of astrophysical processes (black hole evolution, relativistic stars, jet hydrodynamics...) and production and detection of gravitational waves. Other topics and animations are enclosed in the accompanying CD-ROM.

Lectures and contributions are intended to cover the gap between undergraduate courses and current research in the field, including the most recent advances in Relativistic Astrophysics.

Madrid, Spain,
July 2002

Leonardo Fernández-Jambrina
Luis Manuel González-Romero

Acknowlegements

The 24th Spanish Relativity Meeting was run by Universidad Complutense de Madrid and Universidad Politécnica de Madrid. Funds for the organization of the meeting were provided by both universities, the Spanish Ministerio de Ciencia y Tecnología, the Escuela Técnica Superior de Ingenieros Navales, the Departamento de Enseñanzas Básicas de la Ingeniería Naval and Comunidad de Madrid.

We also wish to thank the Director of the Escuela Técnica Superior de Ingenieros Navales, Dr. L.R. Núñez, for kindly offering the premises for the meeting, Dr. F. Robledo, Head of the Departamento de Enseñanzas Básicas, Prof. R. Cercós, for his support during the meeting, and Ms. C. Milans, for her generous help with the bureaucratic tasks.

We also acknowledge the work and patience of the contributors and referees of this book, specially Prof. W. Israel, who was not able to attend the meeting in person due to the sad events of September 11th, but kindly agree to contribute to this volume.

Finally, our thanks go also to the other co-organizers of the meeting, Prof. F.J. Chinea and Mr. F. Navarro-Lérida.

Table of Contents

Part III Gravitational Waves: Production and Detection

List of Contributors

Carles Bona
Universidad de las Islas Baleares
Departamento de Física
Ctra. de Valldemossa km. 7.5
Palma de Mallorca, Spain
cbona@uib.es

Leonardo Fernández-Jambrina
Universidad Politécnica de Madrid
E.T.S.I. Navales
Arco de la Victoria s/n
28040-Madrid, Spain
lfernandez@etsin.upm.es

Valeria Ferrari
Università di Roma I
Dipartimento de Fisica
Piazzale Aldo Moro 2
I-00185-Roma, Italy
valeria@roma1.infn.it

Luis Manuel González-Romero
Universidad Complutense de Madrid
Departamento de Física Teórica II
Avda. Complutense s/n
28040-Madrid, Spain
mgromero@eucmax.sim.ucm.es

Carsten Gundlach
University of Southampton
Faculty of Mathematical Studies
Highfield, Southampton
SO17 1BJ, U.K.
C.Gundlach@maths.soton.ac.uk

Sascha Husa
Max-Planck-Institut
Dept. Astrophysical Relativity
Am Muehlenberg 1
D-14476 Golm bei Potsdam, Germany
shusa@aei.mpg.de

José María Ibáñez
Universidad de Valencia
Dept. Astronomía y Astrofísica
Avda. Vicent Andrés Estelles s/n
46100-Burjassot, Valencia, Spain
jose.m.ibanez@uv.es

Werner Israel
University of Victoria
Department of Physics and Astronomy
Box 3055
Victoria B.C., V8W 3P6, Canada
israel@uvphys.phys.uvic.ca

José Alberto Lobo
Universitat de Barcelona
Departament de Física Fonamental
Martí i Franquesa 1
08028-Barcelona, Spain
lobo@hermes.ffn.ub.es

José María Martín-García
University of Southampton
Faculty of Mathematical Studies
Highfield, Southampton
SO17 1BJ, U.K.
jmm@maths.soton.ac.uk

Carlos Palenzuela
Universidad de las Islas Baleares
Departamento de Física
Ctra. de Valldemossa km. 7.5
Palma de Mallorca, Spain
vdfscpl4@uib.es

Gerhard Schäfer
Universität-Friedrich-Schiller-Jena
Theoretisch-Physikalisches Institut
Max-Wien-Platz 1
D-07743 Jena, Germany
gos@tpi.uni-jena.de

Ryoji Takahashi
Max-Planck-Institut
Dept. Astrophysical Relativity
Am Muehlenberg 1
D-14476 Golm bei Potsdam, Germany
ryoji@origin.aei-potsdam.mpg.de

Relativistic Astrophysics

Leonardo Fernández-Jambrina[1] and Luis Manuel González-Romero[2]

[1] ETSI Navales, Universidad Politécnica de Madrid,
Arco de la Victoria s/n, E-28040-Madrid, Spain
[2] Departamento de Física Teórica II, Universidad Complutense de Madrid,
Avenida Complutense s/n, E-28040-Madrid, Spain

Abstract. This chapter is devoted to the origins of relativistic astrophysics, both from the theoretical and observational point of view. Supernova explosions, pulsars, active galactic nuclei and gamma-ray bursts are some of the observed processes that are the object of this discipline. On the other hand, the intriguing features of black holes, singularities and hypotheses like the no-hair conjecture and cosmic censorship have been discussed for decades. Perturbative and numerical approaches are mandatory to tackle most of these issues.

1 Introduction

The term relativistic astrophysics was coined in the sixties but has its roots in 1931 when Chandrasekhar discovered the existence of a limiting mass for white dwarfs [1]. He arrived at the conclusion that if the mass of a star is above a value it will not evolve to a white-dwarf stage. Therefore other choices have to be found. The alternatives proposed by the works by Oppenheimer and Volkov [2] and Oppenheimer and Snyder [3] (1939) were the collapse to form a neutron star or a black hole. Even when these alternatives sound revolutionary and very interesting, it was not until the sixties that the relativistic effects began to have some influence in astrophysics, when the need to explain the features of the recently discovered quasars appeared. The immense amount of energy emitted by such a small region does not allow another explanation that a compact and very massive object requiring a relativistic treatment. The term relativistic astrophysics is used for the first time to name the first Texas Symposium (1963), developed to discuss the quasar phenomenon and related topics. After forty years there are large accumulated observational evidences (pulsars, quasars and active galactic nuclei, supernova explosion, collapsars and hipernovae, microquasars, X-Ray binaries, X-Ray burst, jets, gamma ray bursts, ...) of the existence in our universe of very compact objects (white dwarfs, neutron stars, and black holes), the description of which needs to be done in a relativistic framework. Relativistic Astrophysics tries to draw a theoretical picture where all these highly relativistic radiation phenomena fit smoothly.

The observational data are for us as some of the pieces of an incomplete puzzle which we have to solve. Some of these data correspond to several aspects of the same process. When the relation among some of the different aspects is established we begin to figure the image that the puzzle, finally, will show.

In this introduction we would like to describe roughly some of the features of these phenomena, just to develop a framework for the following chapters.

2 Some Observational Phenomena

The subject of the book is mainly theoretical. Therefore, to bridge the gap with experimental observations, in this section we describe briefly some of the phenomena treated in relativistic astrophysics.

Supernova explosions are classified in two types by the features of their spectrum; type I has no hydrogen lines and type II has hydrogen lines.

Type II supernova explosions are believed to come from gravitational collapse of supergiant stars [4]. In its evolution a star more massive than 8 $M_\odot$ after the thermonuclear stages continues to evolve until it reaches the Chandrasekhar mass (1.4 $M_\odot$). In this situation the relativistic gas has a soft equation of state and is compressed by gravity. In just a second the core of the star, after living for ten millions years burning hydrogen to produce heavier elements, implodes to something of the size of some tens of kilometers and densities of the order of 10^{14} g/cm^3. At nuclear densities matter is not compressible and the core bounces, rebounding into the infalling inner mantle and generating a strong shock wave. The shock formed at a radius of 20 - 30 kilometers has to cross many tens of solar masses of infalling material. The heavy nuclei it encounters are split into nucleons and also neutrinos of all three flavours are radiated (the initial neutrino luminosity is approximately of 10^{54} erg/s). As a consequence the shock stalls at a radius of 100-200 km forming a quasi-stationary accretion shock. The infalling shells of matter reach it, are shock compressed and heated and are deposited on the protoneutron star. The problem is to determine how the accretion shock is revived into a supernova explosion (actually Type II supernova). The solution points to the neutrinos [5]. They do not stream out immediately but must diffuse to escape. The surface of emission is named *neutrinosphere*. Instead of milliseconds they need seconds to leave the protoneutron star. The neutrinos detected from SN1987A [6] confirm this scale of time and the number of them is compatible with a neutrino total emission of $2-3\ 10^{53}$ erg. The accreted mantle is being heated by the absorption of the escaping neutrinos. The energy transfer from the core to the mantle and the accretion from the collapsing outer core are essential for the supernovae mechanism. The supernova phenomenon is a competition between the neutrino luminosity and the accretion mass process. When the stalled shock reaches the critical relation, the supernova explosion is produced, sending to the interstellar medium heavy elements and leaving behind a neutron star or a black hole [7].

Type I supernova explosions appear in a different scenario. Some white dwarfs composed by carbon and oxygen can be found in a binary with another companion star. The white dwarf can accrete enough matter from the companion to reach the Chandrasekhar mass and then the collapse continues. Due to its composition, the compression and heating lead to thermonuclear explosion of the white dwarf, leaving nothing behind except the companion star [4].

Type I supernovae have been recently proposed to study the curvature of the universe using them as standard candles [8]. Surveying the sky and considering supernovae with redshifts from 0.2 to 1.0 allows to study non-linear regions of the Hubble flow. Hence, this can be used to discern if the universe is geometrically open or closed and whether is accelerating or not. The results indicate that the universe expands forever and it seems to be accelerating. The acceleration in the framework of the inflationary cosmological model implies a non-zero cosmological constant. However, there are some open questions that have to be considered before these results can be considered as definitive.

Pulsars were first discovered in 1967 [9] as radio sources. Pulsar radio emission shows a uniform series of pulses, spaced with great precision at periods between few milliseconds and several seconds. Some pulsars have been detected by optical, X-ray and gamma-ray telescopes. Pulsars are weak radio sources. This means that the coherent addition of many pulses is required in order to produce an integrated profile. Although the individual pulses vary from pulse to pulse, the integrated profile is very stable. The pulsars were soon identified as rotating magnetized neutron stars [10]. After the discovery of neutron by Chadwick in 1932, Landau predicted the existence of neutron stars and Baade and Zwicky [11] suggested that neutron stars may be formed in supernovae explosions.

Radio pulsars are rapidly rotating magnetized neutron stars with periods P in the range of milliseconds to seconds [12]. They lose rotational energy due to the spinning down of the magnetic torque with period derivatives $\dot{P}$ of order of 10^{-12} - 10^{-6}. The magnetic field induced by the rotation spin down and the cyclotron absorption lines found in the X-ray spectra is of the order of 10^{11}-10^{13} G. The pulsar age can be estimated by $P/2\dot{P}$. Pulsars are observed to "glitch", to suddenly spin up with a relative period change in the range from 10^{-8} to 10^{-6}. In post-glitch relaxation most of the period increase decays. The glitches suggest that the neutron stars are composed by a solid crust containing superfluid neutrons.

A subclass of radio pulsars are the millisecond pulsars with periods ranging from a few milliseconds to hundreds of milliseconds. The period derivatives are small, indicating small magnetic fields, 10^8 to 10^{10} G. Most of them are in binaries. This suggests that they are old pulsars spun up by accretion from a companion star. Some of them have a white dwarf or neutron star companion. Double neutron star binaries are of special interest. The first one PSR 1913+16 was found by Hulse and Taylor [13]. Pulsar timing observations allow to calculate many parameters of the system including orbital period and period derivative, both masses, orbital distance and inclination. In these binaries general relativity can be tested accurately measuring the inward spiralling due to the emission of gravitational radiation.

Some pulsars also emit in X-rays wavelengths. These are X-ray pulsars and X-ray bursters. Around two hundred X-ray pulsars and bursters have been detected. The X-ray emission is due to the accretion of mass to the neutron star from a companion. X-ray pulses are thought to be due to strong accretion on the magnetic poles. X-ray bursts are due to slow accretion over the neutron star surface. After accumulating hydrogen on the surface, a thermonuclear explosion happens and a X-ray burst is observed, lasting for a few seconds. It is also possible to find X-ray bursters, as the low mass X-ray burster XTE J1808-369, with X-ray pulsations.

We have two different classes of X-ray binaries: the low mass X-ray binaries (LMXRB), where the companion mass is of the order of the solar mass, and the high mass X-ray binaries (HMXRB), where the companion is around 10 solar masses. For LMXRB the X-ray emission is produced through Roche-lobe overflow (Roche-lobe is the location between the two stars in a binary where the gravitational pull of the two stars is equal in modulus and with opposite direction). The angular momentum of the accreting matter tends to form a differentially rotating disk around the neutron star (also we can have LMXRB with a white dwarf or a black hole as the compact component). The material in the disk, spiralling to the compact object, heats up and emits X-rays. For HMXRB the massive companion of the neutron star (also a black hole can be found in HMXRB as the compact object. The prototype is Cyg X1) emits a stellar wind. The compact companion captures part of the emitted matter emitting X-rays.

There are also the so called anomalous X-ray pulsars, corresponding to slowly rotating, $P = 10$ s, but very rapidly spinning down due to a huge magnetic field of the order of 10^{14} G. Sometimes they are called *magnetars* [14].

In some low mass X-ray binaries the pulsation seems to be unstable exhibiting the so called quasi-periodic oscillations (QPO) [15]. These QPOs seem to arise from the interaction of the neutron star with the accretion disk, or from the instabilities in the captured material caused by X-ray emission.

Quasi-stellar objects or quasars belong to a class of galaxies named active galactic nuclei (AGN)[16]. This name is due to the large amount of energy emitted from their nuclei. Some general features of this class of galaxies are non thermal spectra, high luminosity in a non-optical region of the spectrum (radio, UV,...), strong emission lines, rapid variability. Some of them emit radio jets. It is interesting to note that some of these features are also found in normal galaxies but at lower luminosity levels.

This class of galaxies includes: Seyfert galaxies: spiral galaxies with bright starlike nuclei, strong broad emission lines implying velocities of thousand of km/s. There are two types. Type I which have very broad hydrogen emission lines and Type II, with narrower emission lines. They have also compact radio sources in their nuclei.

Radio galaxies: elliptical galaxies which have a powerful compact radio source in their center and usually also radio jets.

Blazars: they are objects with strong non-thermal radiation, no emission lines and a variability representing a large percentage of their total luminosity. They are also very rapid. It seems that blazars are radio galaxies which we are looking at right along the jet.

Quasars: They are unresolved points of light. In their spectrum there is a continuous part of the type of the synchrotron emission. There are also broad strong emission lines. They exhibit high variability in their total luminosity, with periods ranging from day to weeks or even years. Then we have an emission region of a few parsecs, emitting 10^{13} times the luminosity of the sun. Radio Quasars show a compact radio source and an extended radio synchrotron emission in the form of oppositely directed radio jets of megaparsecs extension. These jets terminate in what is called radio lobes, regions of diffuse radio emission with a final shock wave.

The only possible explanation for all these features is the release of gravitational energy by matter falling towards a black hole. These black holes need to be very massive, from 10^6 to 10^9 solar masses, to produce the luminosities observed in quasars. The falling of matter to the supermassive black hole takes place at an accretion disk. A good example of accretion disk can be found in NGC 4258. The radiation generated at its nucleus excites water molecules which produce stimulated emission of radio waves. These water masers reveal a resolved disk orbiting a 43 million solar mass black hole [17]. How the accretion produces the observed jets, or how the collimation is obtained is a matter of very interesting debates. Other evidence of a supermassive black hole can be found in the nearby AGN MCG-6-30-15. The emission line due to iron in the X-ray spectrum of the galaxy is extremely broad and skewed. This is due to relativistic effects only expected near the event horizon of a supermassive black hole [18].

An important observation related with quasars is that their number density has a peak when the universe was 2.5 10^9 years old and decreases constantly ever since. Simply many of the quasars disappear at lower redshifts. Therefore we arrive at the conclusion that the local universe is filled with supermassive black holes that have exhausted their energy supply. The evidences suggest that every galaxy has one of these supermassive black hole in its center. One of the most remarkable cases is our own galaxy. The Milky Way has a powerful radio source at its center SgrA*, with a size of the order of 3 10^9 km across. Monitoring the motion of nearby stars, the effects of a supermassive black hole of 3 10^6 solar masses are revealed [19]. A very interesting tool to measure the mass of a black hole is the stellar velocity dispersion, which is just an indicator of the typical velocity of the stars moving through a given point [20].

Other important phenomenon is galactic merging. In this situation the supermassive black holes at the center of the two galaxies will form a binary supermassive black hole of about one light year across. This could explain some effects observed in active galaxies. Some galaxies emit radio jets that twist symmetrically on either side of the nucleus, which could correspond to the precession of the spinning black hole that produces the jets. Also some outbursts observed in quasars could be interpreted as produced by a small supermassive black hole passing through the accretion disk of a larger one.

Observations in hard X-rays and radio wavelengths revealed the existence of stellar sources of relativistic jets denoted by the name of "microquasars" [21]. These are stellar mass black holes that mimic the phenomena seen in quasars. The microquasars combine accreting black holes, identified by the production of hard X-rays and gamma-rays from the surrounding accretion disk, and relativistic jets of particles detected by synchrotron emission. Comparing the microquasars with AGN we have that in microquasars the black hole has only a few solar masses instead of several million solar masses, the accretion disk has thermal temperatures of several million degrees instead of several thousand degrees and the particles ejected can travel distances of light years, instead of several million light years. Simple scaling laws are found in the physics of flows around black holes suggesting unity in the physics of accreting black holes over an enormous range of scales [22].

Gamma-ray bursts (GRB) are brief pulses of gamma-rays lasting for tens of seconds. They are detected about once a day. The current interpretation of this radiation is in terms of the so called fireball shock model and the blast wave model following the afterglow. A tremendous amount of energy is released in a very short time in a very small region and then it expands in a highly relativistic outflow, which undergoes both internal shocks producing gamma-rays, and later develops a blast wave and reverse shock, as it decelerates by interaction with the external medium. Several progenitors have been proposed for the GRB, hypernovae or collapsars, merging of two neutron stars or a neutron star and a black hole, accretion-induced collapse, ... all of them lead to a central black hole and a temporary torus of matter around it [23].

3 Some Relativistic Astrophysics Basics

In this section we will describe some theoretical basics to complement the following chapters.

3.1 Black Holes

The origin of the concept of black hole is usually dated back to 1795. In this year Pierre Simon de Laplace [24] combined both corpuscular theory of light and theory of gravity to reach the following conclusion. If a star were so dense that the velocity of escape, v_{esc}, from its surface were as large as that of light, c, not even the "photons" would be able to leave the star and this would disappear from our sight. A simple undergraduate calculation shows that the mass of the star, M, and its radius, R, are related by

$$v_{\mathrm{esc}} = \sqrt{2\frac{GM}{R}} = c\,.$$

This classical result is quite surprising since it coincides with the radius of the horizon of a black hole.

The relativistic formulation of the black hole had to wait longer than a century to Karl Schwarzschild's derivation of the metric for the static and spherically symmetric solution of Einstein's vacuum field equations [25],

$$\mathrm{d}s^2 = -\left(1-\frac{2M}{r}\right)\mathrm{d}t^2 + \left(1-\frac{2M}{r}\right)^{-1}\mathrm{d}r^2 + r^2(\mathrm{d}\theta^2 + \sin^2\theta\,\mathrm{d}\phi^2)\,,$$

which may be interpreted as the gravitational field of a point mass, M, or as the exterior vacuum field of a spherically symmetric star.

It is taken to be the first exact solution of General Relativity, since it was published in 1916, just a few months after Einstein's seminal paper. At that time Schwarzschild was a soldier in Russia during the Great War and died soon after his return to Germany and after finishing his most reknowned contribution to Physics. In a subsequent paper [26] he produced the metric for a spherically symmetric incompressible perfect fluid, thereby completing the first and simplest model for a star.

Schwarzschild's spacetime shows some intriguing features. First of all, relativity does not remove the singularity of its classical analogue, the Coulomb potential, restricting the applicability of the theory. On the contrary, quantum mechanics did get rid of the singularity of the hydrogen atom, so one would expect that a quantum theory of gravity would solve the problem. We are far from that anyway.

Second, as Birkhoff stated [27], staticity is not needed to derive Schwarzschild's solution, since it arises naturally from spherical symmetry. Therefore, Schwarzschild's metric is in some sense unique.

The first generalization of this metric did not wait long. A few months later a point charge e was added to Schwarzschild's spacetime [28]. Although just an additional e/r^2 term was needed, the spacetime acquired a richer structure. It took a little longer, nearly half a century, till Kerr included rotation [29]. The complete black hole solution, with mass, charge and angular momentum, was obtained in 1965 [30]. On the contrary, the Kerr metric has not been useful as an exterior vacuum field for a star. It has been used just as a black hole, since it has not been possible to match it to any interior perfect fluid to provide a whole model for a rotating star.

Black hole singularities have been an interesting issue for research. There is not just a region $r = 0$ out of the spacetime where the fields diverge, but at $r = 2m$ we meet a coordinate singularity. It is not a true singularity since observers travelling on geodesics do not find the end of their journey there. In fact, it can be removed by suitable changes of coordinates, like Eddington-Finkelstein [31].

However, the surface at $r = 2m$, the horizon of the black hole, though it is not singular, it does act as a one-way membrane. Geodesics starting inside the horizon cannot cross it and finally meet their fate at the singularity. Therefore the horizon splits the spacetime into two regions that are causally disconnected, that is, the interior region cannot influence the exterior. There is a further extension of the Schwarzschild spacetime due to Kruskal and Szekeres [32] that encloses two exterior regions and that is maximal in the sense that it cannot be extended.

The charged Reissner-Nordstrøm black hole exhibits new features, although its metric is rather similar to Schwarzschild's. The horizon splits into two surfaces $r_\pm = m \pm \sqrt{m^2 - e^2}$ for lightly charged black holes, $|e| < m$. The main difference with Schwarzschild is that the singularity can be avoided by infalling observers. In fact, Reissner-Nordstrøm's black hole can be also extended to include two exterior regions and constitutes a "wormhole", a sort of traversable throat between two asymptotically flat regions.

On crossing the first horizon, r_+, from the outer region, the observers appear infinitely redshifted to those that remain in the "safe" region. In fact, they would never see them cross at all, since that last ray of light would not leave the black hole. That is the reason why black holes have been also named "frozen stars". The name black hole was suggested by Wheeler depicting the fact that not even light can overcome its gravitational field.

The surface $r = r_-$ is called the Cauchy horizon since it is the boundary of the region of the spacetime where evolution can be predicted from initial data on a Cauchy hypersurface. An observer crossing the Cauchy horizon would see neighbouring objects infinitely blueshifted and the whole history of the outer region in a finite time. Since this behaviour does not seem physically sensible, it has been argued that the Cauchy horizon must be unstable under small perturbations and that it would develop a singularity. This has been studied in detail from the early 90s. It has been confirmed that a singularity appears [33], for instance, when pulses of radiation are added to the spacetime. On the other hand, this singularity is null and weak [34] in the sense that tidal distortion is finite and would not prevent an observer from crossing the event horizon.

For $|e| > m$ ($|a| > m$ for Kerr's spacetime) no horizon hides the singularity from observers in the outer region and the spacetime bears what is called a "naked singularity". It has been postulated that, although mathematically admissible, these spacetimes must be rejected in physical evolution, that is, for realistic physical configurations naked singularities would not come out from gravitational collapse or would be unstable under small perturbations, which would finally develop a black hole. This is the "cosmic censorship hypothesis". Numerical results are controversial on this issue.

A counterexample for the cosmic censorship hypothesis is namely the Choptuik spacetime [35], which describes numerically the collapse a scalar field. As one would expect, two different situations arise. Either the waves disperse or a black hole forms. What it was not to be expected is that the description were so simple. The mass of the forming black hole depends on a power of a parameter, which is the same for every set of initial data (universality). For the critical value of the parameter, a zero-mass naked singularity is formed, but it is unstable. First results showed that the critical exponent could be universal for other sorts of collapsing material. But Maison [36] found out that the exponents were different for collapsing perfect fluids. Critical collapse will be the topic of Gundlach's chapter.

The rotating Kerr black hole also incorporates new features that did not show up in static ones. Horizons subsist for slowly rotating black holes, $|a| < m$, at $r_\pm = m \pm \sqrt{m^2 - a^2}$ and the singularity is naked for $|a| > m$. In addition to them, a new exotic region appears, the ergosphere, which is limited by r_+ and the stationary limit surface, $r = m \pm \sqrt{m^2 - a^2 \cos^2 \theta}$. Inside the ergosphere an observer cannot remain at rest with respect to infinity because the generator of the stationary symmetry is not timelike in this region. Penrose suggested that this fact could be used to draw energy from a rotating black hole [37].

The fact that the rotating charged blackhole is solely determined by its mass, angular momentum and electric charge suggests that every other degree of freedom of matter is wiped out during the gravitational collapse. This lead Israel, Penrose and Wheeler to postulate the "no-hair conjecture", which claims that a black hole with a horizon with spherical topology is characterized just by those three values. The proof of this conjecture is complex and requires additional technical assumptions. It has required thirty years to be proven [38]. However, if one allows other exotic fields in the spacetime, new "hair" sources may appear. For instance, Yang-Mills black holes have non-abelian hair [39]. This issue is of course related to uniqueness of black holes.

Another intriguing issue is the entropy of a black hole. Since a black hole in equilibrium is described by a reduced set of variables, it could be considered as a sort of statistical system. Comparison with thermodynamics lead Bekenstein [40] to suggest that the area of the horizon could be assigned to an entropy, since it cannot decrease for matter with non-negative density, as Hawking had shown [41]. On the other hand, surface gravity, i.e. gravity at the horizon, could be related to a sort of temperature, since it takes the same value at every point of the horizon. The four laws of black hole dynamics appeared in [42]. This was the beginning of black hole dynamics, which is the subject of Werner Israel's chapter. There has been an intense work to establish the origin of this entropy, from theory of information to quantum theory. String theory is also considered a candidate to solve the problem.

If a black hole is to have a temperature different from zero [43], it should radiate and that this emission could be in principle detected. This was pointed out by Hawking. Therefore black holes no longer can be considered eternal. In fact, they radiate most violently the greater their mass is.

3.2 Numerical Relativity

One possible approach to study the phenomena found in relativistic astrophysics is through numerical methods. Usually this requires accurate, large scale, three-dimensional numerical simulations. The dynamics of a system is described by a coupled system of time-dependent partial differential equations, including Einstein gravitational equations and matter equations (Maxwell equations, magneto-hydrodynamical equations,...) leading usually to a non-linear hyperbolic system of equations. If we are interested in stationary solutions the system of equations transforms to a non-linear elliptic one.

Time-dependent dynamical problems draw the attention of many groups because of the importance they have for gravitational radiation.

Most approaches for general relativistic equations use spacelike foliations of the spacetime, within a 3+1 formulation (ADM). In ADM formulation spacetime is foliated into a family of non-intersecting spacelike hypersurfaces. The Einstein equations split into evolution equations for the three metric and the extrinsic curvature and constraint equations that must be satisfied at every time slice. On the other hand, the general covariant formulations, not attached to the spacetime foliation, allow techniques for fluid dynamics to be applied straightforwardly.

Numerical schemes make use mainly of finite differences, providing solutions of the discretized version of the original system of partial differential equations. Schemes using artificial viscosity are based in the idea of modifying the equations by introducing some terms providing artificial viscosity, such that the spurious oscillations near the discontinuities be damped. On the other hand, shock-capturing methods use the explicit knowledge of the characteristic fields of the equations and the corresponding eigenvectors to integrate the equations by means of Riemann solvers, computing at every interface of the numerical grid the solution of the simplest initial value problem with discontinuous initial data (Riemann problem). The main advantage of these methods is that the physical discontinuities (shock waves,...) are treated adequately.

Other techniques are used, for instance, smoothed particle hydrodynamics, where extended Lagrangian particles replace the continuum hydrodynamical variables. The extent of the particles is determined by the length scale contained in a smoothing function. Other approach is the spectral method which transforms the partial differential equations into a system of ordinary differential equations by expanding the solution in a series on a complete basis. The spectral methods are well suited for elliptic and parabolic problems. For hyperbolic systems they provide good results if there are no discontinuities in the solution.

3.3 Stability and Oscillations of Compact Objects

Practically every stellar object oscillates and although there is great difficulty in observing such oscillations, there are already results for various types of stars. For instance, observations of solar oscillation, helioseismology, have revealed a large amount of information on the Sun [44]. For normal main sequence stars the study of these oscillations can be done with a Newtonian theory and will have a small relevance for gravitational radiation. On the other hand, the oscillations of very compact objects (neutron stars and black holes), produced mainly during the formation stage, are of the main importance for gravitational wave astronomy, because they could be detected by the gravitational detectors which are under construction. These oscillations have to be studied in a relativistic theory.

In general relativity, due to gravitational radiation, there are no normal mode oscillations but instead we have "quasi-normal" modes (QNM). The frequencies become complex. The real part represents the actual frequency of the oscillation and the imaginary part corresponds to the damping produced by the emission of gravitational waves. The quasi-normal modes appear as perturbations of the spacetimes describing neutron stars or black holes.

The study of black hole perturbations began with the work of Regge and Wheeler [45] in the fifties and was followed by Zerilli [46]. It was mainly concentrated on the stability of a black hole under small perturbations. The development of relativistic star perturbations was initiated in the sixties by Thorne group [47]. They were interested in extending the known properties of Newtonian oscillation theory to general relativity, as well as to estimate the frequencies and energy radiated away as gravitational waves. Vishveshwara was the first in defining QNM [48], which latter have been found in several circumstances: particles falling into Schwarzschild and Kerr black holes, the collapse of a star to form a black hole, ...

3.4 Gravitational Radiation and Post-Newtonian Approximation

In post-Newtonian approximation it is assumed weak gravity, slow motion of the matter and small stresses and internal energies in order to make a simultaneous expansion in small parameters characterizing these quantities. Such a weak-field slow-motion expansion yields at zeroth order the empty spacetime and at first order, the Newtonian approximation. Further orders provide post-Newtonian corrections for our problem. This approach can be used, for instance, in our solar system for different metric theories.

Gravitational radiation, i.e. waves of curvature in the spacetime, can be of main importance to study many relativistic astrophysical phenomena. It is produced by dynamical time-dependent changes of the spacetime curvature in astrophysical systems. It may contribute to measure spacetime geometry around black holes, study highly nonlinear vibrations of curved spacetime in black hole collisions, check the equation of state and the structure of neutron stars, the inner part of the accretion disk and many other effects. Great effort has been devoted to develop gravitational wave detectors and to calculate theoretically the wave form for different astrophysical processes.

References

1. S. Chandrasekhar: Ap. J. **74**, 81 (1931)
2. J.R. Oppenheimer, G. Volkov: Phys. Rev. **55**, 374 (1939)
3. J.R. Oppenheimer, H. Snyder: Phys. Rev. **56**, 455 (1939)
4. H.A. Bethe: Rev. Mod. Phys. **62**, 801 (1990)
5. S.A. Colgate, R.H. White: Ap. J. **143**, 628 (1966)
   H.A. Bethe, J.R. Wilson: Ap. J. **295**, 14 (1985)
6. R.M. Bionta et al.: Phys. Rev Lett. **58**, 1494 (1987)
   K. Hirota et al.: Phys. Rev. Lett. **58**, 1490 (1987)
7. A. Burrows, J. Gosgy: Ap. J. (Lett.) **416**, L75 (1993)
   H.T. Janka, E. Muller, Astron. Astrophys. **306**, 167 (1991)
8. S. Perlmutter et al.: Ap. J. (Lett.) **517**, 565 (1999)
   A. Riess et al.: Astron. J. **116**, 1009 (1998)
9. A. Hewish, S.J. Bell, J.D.H. Ikington, P.F. Scott, R. A. Collins: Nature **217**, 709 (1968)
10. T. Gold: Nature **221**, 25 (1969)
11. W. Baade, F. Zwicky: Proc. Nat. Acad. Sci. **20**, 255 (1934)
12. A. G. Lyne, F. Graham-Smith: *Pulsar Astronomy* (Cambridge University Press, Cambridge 1998)
13. R.A. Hulse, J. H. Taylor: Ap. J. (Lett.) **195**, L51 (1975)
14. C. Thompson, R.C. Duncan: Ap. J. **408**, 194 (1994)
    C. Kouveliotou et al.: Ap. J. (Lett.) **510**, L115 (1999)

15. W. Zhang, T.E. Strohmayer, J.H. Swank: Ap. J. (Lett.) **482**, L167 (1997)
M.C. Miller, F.K. Lamb, D. Psaltis: Ap. J. **508**, 791 (1998)
T.E. Strohmayer: Ap. J. **552**, L49 (2001); Ap. J. **554**, L37 (2001)
16. J.H. Krolik: *Active Galactic Nuclei: From the Central Black Hole to the Galactic Environment* (Princeton Series in Astrophysics, Princeton University Press, Princeton 1998)
17. W.D. Watson, K. Wallin: Ap. J. (Lett.) **432**, L35 (1994)
K. Miyoshi et al.: Nature **373**, 127 (1995)
18. Y. Tanaka et al: Nature **375**, 659 (1995)
19. A. Eckart, R. Genzel: Nature **383**, 415 (1996)
20. L. Ferrarese, D. Merrit: Ap. J. (Lett.) **539**, L9 (2000)
21. I.F. Mirabel, L.F. Rodriguez: Nature **358**, 215 (1992); Nature **371**, 47 (1994)
22. B.J. Sams, A. Eckart, R. Sunyaev: Nature **382**, 47 (1996)
23. T. Piran: Phys. Rep. **314**, 575 (1999)
24. P.S. Laplace: *Le Système du Monde*, Vol. II (Paris, 1795)
25. K. Schwarzschild: Sitzber. Deut. Akad. Wiss., 189 (1916)
26. K. Schwarzschild: Sitzber. Deut. Akad. Wiss., 424 (1916)
27. G.D. Birkhoff: *Relativity and Modern Physics* (Harvard University Press, Cambridge, 1923)
28. H. Reissner: Ann. Phys. **50**, 106 (1916)
29. R.P. Kerr: Phys. Rev. Lett. **11**, 237 (1963)
30. E.T. Newman, E. Couch, K. Chinnapared, A. Exton, A. Prakash, R. Torrence: J. Math. Phys. **6**, 918 (1965)
31. A.S. Eddington: Nature **113**, 192 (1924)
D. Finkelstein: Phys. Rev. **110**, 965 (1958)
32. M.D. Kruskal: Phys. Rev. **119**, 1743 (1960)
G. Szekeres: Publ. Mat. Debrecen **7**, 285 (1960)
33. E. Poisson, W. Israel: Phys. Rev. Lett. **63**, 1663 (1989)
E. Poisson, W. Israel: Phys. Rev. **D41**, 1796 (1990)
34. A. Ori: Phys. Rev. Lett. **67**, 789 (1991)
A. Ori: Phys. Rev. Lett. **68**, 2117 (1992)
35. M.W. Choptuik: Phys. Rev. Lett. **70**, 9 (1993)
36. D. Maison: Phys. Lett. **B366**, 82 (1996)
37. R. Penrose: Nuovo Cimento **1**, 252 (1969)
38. P.T. Chruściel: Helv. Phys. Acta **69**, 529 (1996) gr-qc/9610010
39. R. Bartnik, J. McKinnon: Phys. Rev. Lett. **61**, 141 (1988)
40. J.D. Bekenstein: Phys. Rev. **D7**, 2333 (1973)
J.D. Bekenstein: Phys. Rev. **D9**, 3292 (1974)
41. S.W. Hawking: Phys. Rev. Lett. **26**, 1344 (1971)
42. J.M. Bardeen, B. Carter, S.W. Hawking: Commun. Math. Phys. **31**, 161 (1973)
43. S.W. Hawking: Commun. Math. Phys. **43**, 199 (1975)
44. D. O. Gough, J. W. Leibacher, P. H. Scherrer, and J. Toomre: Science **272**, 1281 (1996); D. O. Gough: Science **291**, 2325 (2001)
45. T. Regge and J. A. Wheeler: Phys. Rev. **108**, 1063 (1957)
46. F. J. Zerilli: Phys. Rev. D **2**, 3141 (1970)
47. K. S. Thorne and A. Campolattaro: Ap. J. **149**, 591 (1967); K. S. Thorne: Phys. Rev. Lett. **21**, 320 (1968)
48. C. V. Vishveshwara: Nature **227**, 936 (1970)

Part I

Black Holes: Theory, Evolution, Accretion and Stability

Black Hole Thermodynamics

Werner Israel

Canadian Institute for Advanced Research Cosmology Program,
Department of Physics and Astronomy, University of Victoria,
Victoria BC, Canada V8W 3P6

Abstract. This chapter reviews the conceptual developments on black hole thermodynamics and the attempts to determine the origin of black hole entropy in terms of their horizon area. The brick wall model and an operational approach are discussed. An attempt to understand at the microlevel how the quantum black hole acquires its thermal properties is included. The chapter concludes with some remarks on the extension of these techniques to describing the dynamical process of black hole evaporation.

1 Introduction: Black Holes 1930-75

Just over 70 years ago, Chandrasekhar and Landau independently discovered that there is an upper limit to the mass of a cold body in equilibrium. A cooling star heavier than this faces a crisis. A moment arrives when pressure support fails: the cooling material cracks and crumbles under its own weight and goes quickly into free fall.

What happens next in an exactly spherical implosion was described by Oppenheimer and Snyder in 1939. The star disappears within its gravitational radius and loses all ability to causally influence the outside, leaving behind only the deformation which its gravity had imprinted on the outer vacuum – the Schwarzschild geometry.

However, this idealized picture was expected to bear no more than a cursory resemblance to a real collapse, whose terminal state was imagined to be a forbiddingly complex object, bearing the imprint, probably in greatly magnified form, of the magnetic field, asymmetries and other peculiarities of the original star. During the late 1960s it gradually became clear that this is not the case. The picture that emerged instead for the endstate was of a very simple object, called a "black hole" by John Wheeler. This can be visualized as an elemental, self-sustaining gravitational field which has severed all causal connection with the material source that created it and settled, like a soap bubble, into the simplest configuration consistent with the external constraints. Only three characteristics of the collapsing star survive in this final state: mass, angular momentum and (in principle) charge. This circumstance, summed up by Wheeler's graphic phrase "A black hole has no hair", means that the gravitational field of a collapsed object is known with greater precision than anything else in astrophysics, and provides a firm foundation for modelling the (inevitably very complicated) magnetohydrodynamics of accretion and jet formation in active galactic nuclei and X-ray binaries containing black holes.

The no-hair theorems established the idea of the black hole as an object with many internal degrees of freedom whose external configuration is completely specified by just a few parameters. In this respect, the black hole resembles a thermodynamical system. Work by Christodoulou and by Bardeen, Carter and Hawking [1] in the early 1970s showed that this similarity goes much further and that there is, in fact, a detailed formal similarity between the laws of black hole mechanics and the laws of thermodynamics.

A pivotal role in this analogy is played by the second law of black hole mechanics, which arises from the character of the black hole's boundary (called the event horizon) as a one-way membrane for causal effects. The black hole can absorb matter and radiation but (at least classically) emits nothing, so one expects it to grow in some sense. However it is not correct to say that the mass of a black hole can never decrease, because there are mechanisms that can extract energy from charged or rotating black holes – for instance the process of discharging a charged hole (see (1) below), or the Penrose process of mining energy from the "ergosphere" of a spinning hole. Christodoulou and, in more generality, Hawking showed that the nondecreasing quantity is the *area*.

More precisely, Hawking's 1971 area theorem states that the area of the future event horizon can never decrease provided the stress-energy tensor of matter and fields satisfies the condition $T_{\mu\nu}l^\mu l^\nu \geq 0$ on the horizon. (The history of the horizon is a lightlike 3-space; l^μ points along the unique lightlike direction tangent to this space.) This condition requires that material accreted by the hole should have nonnegative energy density, at least while crossing the horizon. It is, of course, satisfied by "ordinary", i.e., classical materials. But it can be violated by the renormalized stress-energy tensor of quantum fields in a curved spacetime. Thus, the classical statement of the second law – "the area of the event horizon can never decrease" – must be expected to break down when quantum effects are taken into account. But we shall see that a suitably generalized form of the second law does remain valid.

As a simple illustration of the area law, consider the process of adding an infinitesimal charge dQ to a spherical hole of mass M and charge Q. In this process, the mass of the black hole must increase by at least the work done in pushing the charge dQ as far as the horizon $r = r_0$ (after which it has no alternative but to drop of its own accord:

$$dM = \frac{Q}{r_0} dQ + dE_{\text{diss}} \,. \tag{1}$$

The term dE_{diss} is nonnegative, and represents the irreversible inward transfer of energy across the horizon associated with the rest-mass and kinetic energy of the accreted charge and the accreted portion of any gravitational or electromagnetic waves which may have been generated in the process.

An uncharged spherical hole of mass M has radius $r_0 = 2GM/c^2$. For a charged hole this generalizes to

$$r_0 = 2\left(M - \frac{1}{2}\frac{Q^2}{r_0}\right)$$

in relativistic units $G = c = 1$. (The subtracted term is the electrostatic field energy excluded from a sphere of radius r_0.) The mass increment $\mathrm{d}M$ in (1) can have either sign, depending on the sign of the accreted charge. However, re-expressing (1) in the form

$$\frac{\kappa_0}{8\pi}\mathrm{d}A = \mathrm{d}M - \frac{Q}{r_0}\mathrm{d}Q = \mathrm{d}E_{\mathrm{diss}} \geq 0 \tag{2}$$

shows that the area $A = 4\pi r_0^2$ always increases in the process. Here,

$$\kappa_0 = \frac{M - Q^2/r_0}{r_0^2}$$

is called the "surface gravity" of the hole, borrowing from a formal Newtonian analogy.

The generalization of (2) to a quasi-stationary process which produces changes $\mathrm{d}Q$, $\mathrm{d}J$ in the charge and angular momentum of a spinning black hole reads [1]

$$\frac{\kappa_0}{8\pi}dA = dM - \Phi_0 dQ - \omega_0 dJ \geq 0\ , \tag{3}$$

and in this form includes both the first and the second laws of black hole mechanics. The three coefficients – the surface gravity κ_0, the electrostatic potential Φ_0 and the angular velocity ω_0 of the horizon – are necessarily constant over a stationary horizon, even if the black hole is spinning and tidally deformed by neighbouring masses. This is the content of the zeroth law. For comparison, recall the zeroth law of thermodynamics: the temperature and chemical potentials have the same value everywhere for as system in thermodynamical equilibrium.

There is an obvious resemblance between (3) and the thermodynamical relation

$$T\mathrm{d}S = \mathrm{d}E + P\mathrm{d}V - \sum_A \mu_A \mathrm{d}N_A \tag{4}$$

with S nondecreasing for a closed system. This leads one to draw a formal analogy between black hole area A and thermodynamic entropy S, and between surface gravity κ_0 and temperature T, and to interpret Φ_0, ω_0 as black hole "chemical potentials".

There is also a black hole analogue of the third law of thermodynamics in its weaker (Nernst) form – the temperature of a system cannot be reduced to absolute zero in a finite number of operations. This states that the surface gravity cannot be reduced to zero in a finite (advanced) time in any interaction with matter whose density is bounded and nonnegative [2]. (The stronger (Planck) form of the third law – $S = 0$ when $T = 0$ – does not have a classical black hole analogue. At the quantum level the situation is less clearcut, see Sect. 5)

The analogy between black hole mechanics and thermodynamics was at first considered to be strictly formal. However, in 1972 Bekenstein suggested that it should be taken seriously, and that the entropy of a black hole should be identified with its area (up to a universal constant factor). For a time this remained a one-man minority view, since it was difficult to see how one could assign any temperature other than zero to an object which emits no radiation. These preconceptions were overturned in 1974 when Hawking announced his discovery that black holes do radiate by a quantum tunnelling process. His detailed calculation [3] showed that the energy spectrum of the emitted particles is thermal and corresponds to a temperature

$$T_{\mathrm{H}} = \frac{\hbar}{2\pi}\kappa_0 \,. \tag{5}$$

The parallel between (3) and (4) then suggests that an entropy

$$S_{\mathrm{BH}} = \frac{1}{4}A/\hbar \tag{6}$$

is in some sense associated with the hole. Exactly what this might mean will be the focus of interest in the following pages.

A black hole which radiates freely into space will slowly evaporate and the event horizon will shrink, violating the classical statement of the second law. However, the decrease of black hole area (i.e., entropy) in the course of evaporation is more than compensated by the entropy of the emergent Hawking radiation. In a quantum context, the second laws of black hole mechanics and thermodynamics are subsumed and sublimated into a *generalized second law*: the sum of the entropy of a black hole (as represented by its area) and the entropy of its surroundings can never decrease.

About the same time as Hawking's discovery, work by Fulling, Davies and Unruh [4] showed that thermal effects are also associated with uniform acceleration in flat space. The ground state for an observer moving with uniform acceleration a is different from the usual Minkowski vacuum, the difference being associated with a characteristic "acceleration" or Unruh temperature

$$T_{\mathrm{U}} = \frac{\hbar}{2\pi}a \,.$$

The accelerated observer "perceives" the Minkowski vacuum as a thermal bath at temperature T_{U}. What this really means is that his ground state is depressed to negative energy density, so he views the Minkowski vacuum as excited (thermally) above his ground state. Unlike Hawking radiation, the Fulling-Davies-Unruh thermal bath is not a source of gravity even in principle, and it would be legitimate to take the view that it has no objective reality but, like centrifugal force, merely provides a convenient alternative mode of description for observations made in an accelerated frame. As such it can nevertheless shed useful light on several aspects of black hole thermodynamics, as we shall see.

In familiar units, the temperature of an uncharged black hole of mass M is

$$T_{\mathrm{H}} \sim 10^{-7}\left(\frac{M}{M_{\odot}}\right)^{-1}\ \mathrm{K}\,, \tag{7}$$

where $M_{\odot}$ is the solar mass. This is quite negligible for astrophysical purposes for a black hole formed in a stellar collapse. For a hypothetical black hole of mass 10^{15} g (about the mass of a mountain) the temperature reaches the more impressive figure of 10^{11} K, but such a black hole is so tiny (about the size of a proton) and its radiating surface so small, that it would take as long to evaporate as the present age of the universe. In general the lifetime of a black hole of mass M is

$$t \sim \left(\frac{M}{M_\odot}\right)^3 (10^{64} \text{ yr}) .$$

According to (7), the black hole gets warmer as it loses energy and contracts. In this respect, it resembles a gaseous star – both have a negative specific heat. At first, the hole radiates only massless particles – photons and neutrinos. Radiation of electron-positron pairs begins once the temperature gets high enough ($kT \sim 1$ Mev) for their creation. Baryons begin to emerge only when $kT \sim 1$ Gev, i.e, $T \sim 10^{13}$ K; by this stage the mass has been reduced to only 10^{13} g.

Black hole evaporation raises fundamental issues. Consider the evaporation of a black hole formed by the collapse of a star of mass $5\,M_\odot \sim 10^{34}$ g. The original star had a baryon number $\sim 10^{58}$. On the other hand, the evaporation product can have a baryon number $\sim 10^{13}/10^{-24} = 10^{37}$ at the very most, since the hole radiates particle-antiparticle pairs. Thus, the possibility of black hole evaporation provides a very indirect but compelling line of evidence, independent of particle physics, that conservation of baryon number cannot be an absolute law.

In relativistic units, Planck's constant $\hbar$ is an area:

$$\hbar^{1/2} = l_{\text{Pl}} \approx 10^{-33} \text{ cm}$$

which makes explicit that S_{BH} in (6) is a pure number, about 10^{78} for a $5\,M_\odot$ black hole. By contrast, the thermal entropy of the progenitor star was of the order 10^{58} (roughly the number of particles in the star). This enormous disparity makes it clear that S_{BH} bears no necessary relation, except as a (liberal) presumed upper bound, to the thermal entropy of the material that went into its formation.

What is the true nature of S_{BH}? If it represents inaccessible information, is this information only temporarily hidden, or is it permanently lost when the black hole evaporates? If Hawking radiation is truly random, then at the quantum level the collapse and subsequent evaporation of a body initially in a pure state would involve the evolution of a pure state into a mixed state, which quantum mechanics holds to be impossible. Various resolutions of this "information loss" paradox have been proposed. Those which hold most favour today revolve about the idea that the evaporation products contain subtle correlations which preserve unitarity. Here, the focus will be on thermodynamical aspects, but later sections will inevitably touch on this fundamental problem.

2 The Mystery of Black Hole Entropy

You can pick anyone off the street and say "Einstein". They will at once write $E = Mc^2$. But if you ask what this formula *means*, the response will be quite different. At best, you may get some mumbling about "atomic bomb".

It is sobering that after a quarter-century we are in a hardly better position regarding the formula

$$S_{\mathrm{BH}} = \frac{1}{4} A / l_{\mathrm{Pl}}^2 \,, \tag{8}$$

where, not to beg the question, I stipulate that "BH" stands for Bekenstein-Hawking and not necessarily for Black Hole. By now, this formula has been derived in so many ways and interpreted from so many angles that it has claims to be the most proved and least understood formula in all of theoretical physics. It is an entirely *superficial* result, in the literal sense that it refers solely to surface properties. Whether it is also deeply profound – whether it hints at some as yet undiscovered link between thermodynamics, gravity and the quantum world – is a question on which the jury is still out.

In the early speculations of Wheeler and Bekenstein no distinction was made between coarse – and fine – grained entropy; S_{BH} was supposed to keep track of the thermal entropy of objects thrown into the hole (Wheeler's "teacup experiment"). The generalized second law

$$\Delta(S_{\mathrm{BH}} + S_{\mathrm{mat}}) \geq 0$$

lends support to this interpretation. Nevertheless, as noted at the end of the previous section, it is not possible simply to identify S_{BH} with the thermal entropy of the matter which collapsed to form and feed the hole.

The view that entropy is somehow created in the process of evaporation also meets with difficulties. Black hole evaporation is very nearly – and can be made exactly – reversible. One simply encloses the hole in a perfect reflecting container, so that it comes into equilibrium with its own radiation, and then allows the radiation to leak out a little at a time. This process is reversible and cannot create entropy.

In 1977 Gibbons and Hawking [5] gave a statistical derivation of (7), using analytic continuation to the Euclidean sector and imposing a Matsubara period T_{H}^{-1} on Euclidean time – i.e., they considered a black hole in equilibrium with its own radiation at the Hawking temperature (Hartle-Hawking state). In this approach, S_{BH} appears already at zero-loop order, as a contribution to the partition function

$$Z = \mathrm{e}^{-W} \,, \qquad W = \frac{1}{8\pi} \int^{\infty} K \, \mathrm{d}\Sigma \tag{9}$$

from the boundary (extrinsic curvature) terms accompanying the Einstein-Hilbert action. (In (9), the surface integral is to be taken over the outer boundary only. There is no inner boundary; the horizon appears as a regular point of the Euclidean sector.) Unfortunately, this offers no clue to the dynamical origins of

$S_{\rm BH}$; rather, it seems to suggest that the entropy is in some sense topological in origin.

Interesting and suggestive, but still short of completely satisfying, are interpretations which refer to the hole's past or future history, or to ensembles of histories; for instance, "$S_{\rm BH}$ is the logarithm of the number of ways the hole could have been made" [6], or which attempt to relate $S_{\rm BH}$ to the entropy of the evaporation products.

Ideally, it should be possible to regard $S_{\rm BH}$ as a *state function*, defined at each moment of time in terms of the dynamical degrees of freedom existing at that moment. One also wants to understand how it comes to have the simple universal form (7), independent of the hole's internal structure and all details of the microphysics.

Perhaps the most promising view is that $S_{\rm BH}$ is *entanglement entropy*, associated with modes and correlations hidden from outside observers by the horizon [7]. Tracing over the hidden modes yields a density matrix which looks thermal. (If the black hole originates from a pure state, one could equally well trace over the external modes, since they will be perfectly correlated with the hidden internal modes.)

Remarkably, this yields an entropy proportional to area. Naively, the coefficient of proportionality is infinite – this arises from the existence of modes of arbitrarily high angular momentum close to the horizon – but reduces to the right order of magnitude when one allows for quantum fluctuations, which will prevent events closer to the horizon than about a Planck length $l_{\rm Pl}$ from being seen on the outside.

A 1985 calculation by 't Hooft [8] seems to be based, at least implicitly, on the idea of entanglement. This treats the statistical thermodynamics of hot quantum fields propagating on a Schwarzschild background. Divergences are controlled by a "brick wall", a reflecting spherical surface just outside the gravitational radius. 't Hooft found, in addition to the expected volume-proportional terms describing radiation inside a nearly flat, large cavity whose outer wall is at the Hawking temperature, additional wall terms proportional to the area. These latter terms diverge as α^{-2}, where α is the proper altitude of the wall above the gravitational radius. For a specific choice of α (which depends on the number of fields, etc., but is generally of order $l_{\rm Pl}$) one is able to recover the Bekenstein-Hawking result (8).

The brick wall model, discussed in more detail in Sect. 4, lends credence to the – at first glance rather fanciful – notion that $S_{\rm BH}$ is a property strongly localized near the wall or horizon. Further support comes from an operational approach described in Sect. 5: a massive thin shell in vacuum, slowly and reversibly compressed toward its gravitational radius, acquires a thermal entropy equal to $S_{\rm BH}$ in the limit. The focus of the next three sections is an attempt to understand at the microlevel how the quantum black hole acquires its thermal properties. Geometrically, this is rooted (at least for an eternal black hole) in the two-sheeted structure of the extended (Kruskal) vacuum manifold. This has close formal links to the twin Fock spaces in the thermofield dynamics of

Takahashi and Umezawa. A special ("thermally entangled") *pure* state on the double Fock space yields a thermal density matrix when traced over the hidden modes. We conclude with some remarks on the extension of these techniques to describing the dynamical process of black hole evaporation.

3 Ground States for Stars and Black Holes

In a general curved spacetime there is no unique choice of time co-ordinate. Different choices lead to different definitions of positive-frequency modes and different ground states. Much of the following discussion will be concerned with these different states and the Bogoliubov transformations which mediate between them. In this section we recall the most important of these states and their properties.

In a static spacetime with Killing parameter t, the Boulware state $|0\rangle_{\mathrm{B}}$ is the one empty of modes positive-frequency in t, i.e, static observers "see" no particles in this state. In an asymptotically flat space, $|0\rangle_{\mathrm{B}}$ is indistinguishable from the Minkowski vacuum near infinity. Further down, vacuum polarization due to curvature induces a nonvanishing stress-energy. The energy density is generally negative and, if a horizon were present, would actually diverge there. Hence the Boulware state would be unstable in a black hole spacetime; it is actually the zero-temperature ground state appropriate to the space in and around a static star.

The diverging terms in the Boulware stress-energy can be visualized as ingoing and outgoing radial lightlike streams of negative energy infinitely blueshifted at past and future horizons respectively. These streams can be neutralized, and the divergence mended, by introducing compensating positive-energy fluxes incident from and to infinity. The resulting state, the Hartle-Hawking state $|0\rangle_{\mathrm{H}}$, represents a black hole in thermal equilibrium with its own radiation in an enclosure. More exactly, $|0\rangle_{\mathrm{H}}$ is the state empty of modes positive-frequency in Kruskal time; free-falling observers at the horizon "see" no particles in this state.

For a black hole formed by collapse of a star, the past horizon is absent and only a positive outflux is needed to mend the divergence on the future horizon. This yields the Unruh state $|0\rangle_{\mathrm{U}}$, representing a black hole radiating freely into space. Formally, $|0\rangle_{\mathrm{U}}$ is empty of modes positive-frequency in ordinary advanced time v and Kruskal retarded time U, in the (analytically extended) black hole vacuum spacetime.

(The above definitions will serve for static geometries, but need careful reconsideration [9] for spacetimes with rotation, where the possibility of ergospheres, ambiguities in the definition of positive-frequency and gravitational analogues of the Klein paradox complicate the issues. These complications are not considered here.)

Calculation of the expectation values of the stress-energy operator for these states is generally difficult and closed-form expressions unavailable. In (1+1)-dimensions, however, the three independent components of $\langle T_a^{\ b}\rangle$ are fully determined just by the conservation laws $T_{a\ ;b}^{\ b} = 0$, the boundary conditions appropriate for this state, and the trace "anomaly" $T_a^{\ a}$, a state-independent c-number. For a massless scalar field,

$$T_a^{\ a} = \frac{\hbar}{24\pi} R \,, \tag{10}$$

where R is the two-dimensional curvature scalar. For illustrative purposes and later reference, it will be useful to record the explicit forms of $\langle T_{ab} \rangle$ for the different states of interest.

In conformal gauge

$$\mathrm{d}s^2 = -\mathrm{e}^{2\lambda(u,v)}\mathrm{d}u\,\mathrm{d}v = \mathrm{e}^{2\lambda(z,t)}(\mathrm{d}z^2 - \mathrm{d}t^2)\,, \tag{11}$$

the curvature is

$$R = -2\Box\lambda = 8\,\mathrm{e}^{-2\lambda}\partial_u\partial_v\lambda\,. \tag{12}$$

The most general conserved symmetric tensor T_{ab} with trace (10) has the form

$$T_{ab} = \Theta_{ab}[\lambda] + F^{\mathrm{out}}(u)u_{,a}u_{,b} + F^{\mathrm{in}}(v)v_{,a}v_{,b} \tag{13}$$

where the first term is defined by

$$\Theta_{ab}[\lambda] = \frac{\hbar}{12\pi}\left\{\lambda_{;ab} + \lambda_{,a}\lambda_{,b} - g_{ab}\left(\Box\lambda + \frac{1}{2}(\nabla\lambda)^2\right)\right\} \tag{14}$$

for any metric function $\lambda(u,v)$ in (11), and the functions F^{in}, F^{out} are arbitrary.

In (11), the lightlike co-ordinates u, v can be replaced by arbitrary functions of themselves. The resulting arbitrariness in λ and Θ_{ab} in (14) is precisely reflected in the arbitrariness of F^{in} and F^{out} in (13).

For definitiness, let us suppose that z, t are asymptotically Lorentzian co-ordinates in an asymptotically flat spacetime, so that $\lambda \to 0$ when $z \to +\infty$. If a horizon is present, it would be "off the map" in this asymptotically Lorentzian conformal gauge: it is characterized by $z \to -\infty$, $\lambda \to -\infty$. This can be remedied by transforming to Kruskal coordinates.

To introduce these co-ordinates we must first define the "surface gravities" of the two sheets of the (in general, nonstatic) horizon. It follows directly from the lightlike character of u and v that their second covariant derivatives must have the form

$$u_{;ab} = -\kappa^{\mathrm{out}}(u,v)\,u_{,a}u_{,b}\,; \qquad v_{;ab} = -\kappa^{\mathrm{in}}(u,v)\,v_{,a}v_{,b}\,. \tag{15}$$

It is easy to verify

$$\kappa^{\mathrm{out}}(u,v) = -2\,\partial_u\lambda\,, \qquad \kappa^{\mathrm{in}}(u,v) = 2\,\partial_v\lambda\,,$$

$$\kappa^{\mathrm{in}} - \kappa^{\mathrm{out}} = 2\,\partial_t\lambda(z,t)\,. \tag{16}$$

(In a static spacetime, $\kappa^{\mathrm{in}} = \kappa^{\mathrm{out}} = 2\,\lambda'(z)$ would give the redshifted force – calibrated for an observer at infinity – needed to hold a unit test mass stationary. In this sense κ^{in}, κ^{out} are generalized surface gravities.)

The co-ordinates u, v are not rectilinear since their second covariant derivatives do not vanish in general. However, it is possible to introduce lightlike co-ordinates $U(u)$, $V(v)$ which are locally rectilinear in the sense that their second covariant derivatives vanish along specified curves – in the case of interest to us, the past and future sheets of the horizon, $v = -\infty$ and $u = +\infty$ respectively. We define these "Kruskal co-ordinates" by

$$[\ln U'(u)]' = \kappa_0^{\text{out}}(u) \equiv \kappa^{\text{out}}(u, v = -\infty) , \tag{17a}$$

$$[\ln V'(v)]' = \kappa_0^{\text{in}}(v) \equiv \kappa^{\text{in}}(u = +\infty, v) . \tag{17b}$$

It then follows from (15) that U, V are locally rectilinear on the horizon:

$$U_{;ab}|_{v=-\infty} = V_{;ab}|_{u=\infty} = 0 . \tag{18}$$

(Obviously the formulas (15)-(18) could be carried over without essential change to nonstatic spherical black holes in (3+1)-dimensions.)

In Kruskal co-ordinates, metric (11) becomes

$$\mathrm{d}s^2 = -\mathrm{e}^{2\Lambda(U,V)}\mathrm{d}U\,\mathrm{d}V . \tag{19}$$

where now the function

$$\Lambda = \lambda - \frac{1}{2}\ln\{U'(u)\,V'(v)\} \tag{20}$$

is regular at the horizon if κ_0^{in}, κ_0^{out} are nonzero, as we shall assume. From (20) and definition (14),

$$\Theta_{ab}\,[\Lambda] = \Theta_{ab}\,[\lambda] + H^{\text{out}}(u)\,u_{,a}u_{,b} + H^{\text{in}}(v)\,v_{,a}v_{,b} , \tag{21}$$

where

$$H^{\text{out}}(u) = \frac{\hbar}{48\pi}\left\{(\kappa_0^{\text{out}})^2 - \frac{1}{2}R(u, v = -\infty)\right\} ;$$

$$H^{\text{in}}(v) = \frac{\hbar}{48\pi}\left\{(\kappa_0^{\text{in}})^2 - \frac{1}{2}R(u = +\infty, v)\right\} . \tag{22}$$

Expectation values $\langle T_{ab}\rangle$ in the various ground states of the scalar-field stress-energy must all take the general form (13), with the adjustable fluxes chosen to fit the appropriate boundary conditions. For the Boulware state – asymptotically vacuum, i.e., $\langle T_{ab}\rangle_{\text{B}} \to 0$ at infinity – we must choose $F^{\text{in}} = F^{\text{out}} = 0$. Hence

$$\langle T_{ab}\rangle_{\text{B}} = \Theta_{ab}[\lambda] . \tag{23}$$

The Boulware stress-energy is singular at the horizon, since $\lambda \to -\infty$ there.

By contrast, in the Hartle-Hawking state the stress-energy is bounded on both sheets of the horizon. Since this is also true of Λ, this boundary condition is satisfied by

$$\langle T_{ab}\rangle_{\text{H}} = \Theta_{ab}[\Lambda] . \tag{24}$$

At the horizon, this would be the stress-energy measured by a free-falling observer, using the locally Lorentzian co-ordinates to define his notion of positive frequency.

From (21),

$$\langle T_{ab}\rangle_{\rm H} = \langle T_{ab}\rangle_{\rm B} + H^{\rm out}(u)\, u_{,a}u_{,b} + H^{\rm in}(v)\, v_{,a}v_{,b}\,. \tag{25}$$

Thus, in contrast to the asymptotic vacuum of the Boulware state, the Hartle-Hawking state bathes the space around the hole with cross-streams of radiation at (in general, variable and different) effective temperatures

$$T_{\rm in} = \frac{\hbar}{2\pi}\,\kappa^{(0)}_{\rm in}(v)\,, \qquad T_{\rm out} = \frac{\hbar}{2\pi}\,\kappa^{(0)}_{\rm out}(u)\,. \tag{26}$$

For a static geometry (the case of physical interest here for this state), the radiation bath would be in thermal equilibrium with the hole at the Hawking temperature, and with asymptotic energy density $(\pi/6\hbar)T_{\rm H}^2$, the expected value for scalar radiation in $(1+1)$-dimensions.

The Unruh state is vacuous at past lightlike infinity $\mathscr{I}^-$. It therefore lacks the influx term in (13) which makes the Hartle-Hawking stress-energy (25) regular at the past horizon:

$$\langle T_{ab}\rangle_{\rm U} = \langle T_{ab}\rangle_{\rm B} + H^{\rm out}(u)\, u_{,a}u_{,b}\,. \tag{27}$$

At future lightlike infinity $\mathscr{I}^+$, the second term is all that survives; it represents the thermal outflow characteristic of an evaporating black hole. There is an accompanying inflow of *negative* energy through the future horizon:

$$\langle T_{ab}\rangle_{\rm U} = \langle T_{ab}\rangle_{\rm H} - H^{\rm in}(v)\, v_{,a}v_{,b}\,. \tag{28}$$

If the background geometry is static, these results can be made explicit and very simple. It is convenient to write the metric in the form

$$\mathrm{d}s^2 = \frac{\mathrm{d}r^2}{f(r)} - f(r)\mathrm{e}^{2\psi(r)}\,\mathrm{d}t^2\,. \tag{29}$$

A stationary observer in this geometry has redshifted outward acceleration

$$\kappa(r) = \sqrt{-g_{tt}} \times (\text{proper acceleration}) = \left(\frac{1}{2}\,f' + f\psi'\right)\mathrm{e}^{\psi}\,. \tag{30}$$

We consider a static medium with pressure $P = T^r_r$ and energy density $\varrho = -T^t_t$. The radial component of the conservation law $T_a{}^b{}_{;b} = 0$ reduces to

$$\frac{\mathrm{d}P}{\mathrm{d}r} = -(\varrho + P)\frac{\kappa}{f}\,, \text{equivalently } \frac{\mathrm{d}P}{\varrho + P} = -\mathrm{d}\ln\sqrt{-g_{tt}}\,. \tag{31}$$

If this "fluid" has entropy density $s(r)$, inverse local temperature $\beta(r) = T^{-1}$ and zero chemical potential, the local Gibbs-Duhem relations which define these quantities are

$$s = \beta(\varrho + P)\,, \quad \mathrm{d}s = \beta\,\mathrm{d}\varrho\,, \tag{32}$$

from which follows

$$\frac{\mathrm{d}P}{\varrho + P} = -\frac{\mathrm{d}\beta}{\beta} \, . \tag{33}$$

From equation (31) and (33),

$$T\sqrt{-g_{tt}} = \text{const.} \, , \tag{34}$$

which is the well-known law of Tolman (valid in this form for an arbitrary static spacetime) for the variation of temperature in a gravitating fluid in thermal equilibrium: the nether regions are hotter.

The above results apply to any static medium. If the medium is a massless scalar field, and T_a^b is the expectation value of its stress energy, then the trace anomaly

$$T_a^a = \frac{\hbar}{24\pi} R \, , \qquad R = -2\,\mathrm{e}^{\psi} \frac{\mathrm{d}\kappa}{\mathrm{d}r} \, , \tag{35}$$

determines T_a^b up to a constant of integration, to be fixed by boundary conditions. Using (35), i.e.,

$$\varrho = P - \frac{\hbar}{24\pi} R \, , \tag{36}$$

to eliminate ϱ from (31) and integrating yields

$$f\,\mathrm{e}^{2\psi} P = -\frac{\hbar}{24\pi} \{\kappa^2(r) + \text{const.}\} \, . \tag{37}$$

For the Boulware state in an asymptotically flat space, the boundary condition $P \to 0$ at infinity requires the constant to vanish:

$$P_{\mathrm{B}} = -\frac{\hbar}{24\pi} \frac{\kappa^2(r)}{f\mathrm{e}^{2\psi}} \, . \tag{38}$$

This is everywhere negative and would diverge to $-\infty$ at a horizon if one were present. The Boulware state is the zero-temperature ground state for quantum fields propagating in the spacetime of a static uncollapsed star.

Suppose now that there is a horizon at $r = r_0$ with nonvanishing surface gravity κ_0, so that

$$f(r_0) = 0, \qquad \kappa_0 = \kappa(r_0) = \frac{1}{2} f'(r_0) \mathrm{e}^{\psi(r_0)} \neq 0 \, . \tag{39}$$

The Hartle-Hawking stress-energy is bounded at this horizon. This fixes the constant in (37) at $-\kappa_0^2$ and we find

$$P_{\mathrm{H}} = \frac{\hbar}{24\pi} \frac{\kappa_0^2 - \kappa^2(r)}{f\mathrm{e}^{2\psi}} \, . \tag{40}$$

Far from the hole, (36) and (40) reduce to

$$\varrho_{\mathrm{H}} \approx P_{\mathrm{H}} \approx \frac{\pi}{6\hbar} T_{\mathrm{H}}^2 \, , \qquad T_{\mathrm{H}} = \frac{\kappa_0}{2\pi} \hbar \, , \tag{41}$$

representing thermal radiation at the Hawking temperature $T_{\rm H}$ in $(1+1)$-dimensional Minkowski space. The local temperature $T(r)$ of this radiation bath rises with depth according to Tolman's law,

$$T(r) = T_{\rm H}\,(-g_{tt})^{-1/2}\,, \tag{42}$$

but the radiation density always remains bounded and, in fact, small for a massive black hole.

The Unruh state superposes upon the Hartle-Hawking stress-energy an inward thermal flux of negative energy:

$$(T_t^r)_{\rm U} = \frac{\pi}{12\hbar} T_{\rm H}^2 {\rm e}^{-\psi}\,. \tag{43}$$

This becomes infinitely blueshifted at the past horizon, but is regular along the future horizon.

The difference between the Hartle-Hawking and Boulware stress-energies has the exactly thermal form

$$\varDelta P = \varDelta \varrho = \frac{\pi}{6\hbar} T^2(r)\,. \tag{44}$$

Thus the Hartle-Hawking state is thermally excited above the zero-temperature Boulware ground state to a local temperature $T(r)$ which grows without bound near the horizon. It is nevertheless the Hartle-Hawking state which most nearly conforms to what a gravitational theorist would call a vacuum at the horizon. (The word "vacuum" is here reserved for a condition of zero stress-energy. In curved space, because of vacuum polarization effects, no quantum state exactly fulfils this condition.)

The thermal character of the difference $(\varDelta T_{ab})_{\rm H-B}$ between the Hartle-Hawking and Boulware stress-energies was verified above for $(1+1)$-dimensions. But it remains at least qualitatively valid generally, with obvious changes arising from the dimensionality. In particular, the $(3+1)$- dimensional analogue of (44) for a massless scalar field,

$$3\varDelta P \approx \varDelta \varrho \approx \frac{\pi^2}{30\hbar^3} T^4(r)\,. \tag{45}$$

holds to a very good approximation, both far from the hole and very near the horizon. Deviations occur in the intermediate region [10], but they remain bounded and will be unimportant for our considerations.

4 Brick Wall Model

The simple relation between the Bekenstein-Hawking entropy $S_{\rm BH}$ and horizon area suggests that the horizon is actually the repository of this entropy. In 1985 't Hooft [8] gave a concrete form to this idea by interpreting this entropy as that of a "thermal atmosphere" extending a few Planck lengths above the horizon.

This section is a paraphrase [11] of 't Hooft's calculation. We consider the thermodynamics of hot quantum fields propagating outside a spherical star with a perfectly reflecting surface and radius r_1 a little larger than its gravitational radius r_0. To maintain thermal equilibrium and keep the total field energy bounded, suppose the system enclosed in a spherical container of radius L substantially larger than r_1.

For the space outside the star, assume a metric of the form

$$\mathrm{d}s^2 = \frac{\mathrm{d}r^2}{f(r)} + r^2\,\mathrm{d}\Omega^2 - f(r)\,\mathrm{d}t^2\ . \tag{46}$$

This is general enough to include the Schwarzschild, Reissner-Nordström and (anti-) de Sitter geometries (or any superposition of these) as special cases.

Into this space we introduce a set of quantum fields, raised to some temperature T_∞ at large distances, and in thermal equilibrium. The local temperature is then

$$T(r) = T_\infty\, f^{-1/2}\ , \tag{47}$$

and becomes very large when $r \to r_1 = r_0 + \Delta r$. We shall presently identify T_∞ with the Hawking temperature T_{H} of the horizon that would appear if $r_1 \to r_0$.

Characteristic wavelengths λ of this radiation are small compared to other relevant scales (curvature, size of container) in the regions of interest to us here. For instance, very near the star's surface,

$$\lambda \sim \frac{\hbar}{T} = f^{1/2}\frac{\hbar}{T_\infty} \ll r_0\ . \tag{48}$$

Elsewhere in the container, at considerable distances from the star,

$$f \approx 1\ , \qquad \lambda \sim \frac{\hbar}{T_\infty} \sim r_0 \ll L\ . \tag{49}$$

Therefore a particle description should be a good approximation to the statistical thermodynamics of the fields. (Equivalently, one can arrive at this conclusion by considering the WKB solution of the wave equation [8], [11].)

The extensive thermodynamical parameters then each receive two principal contributions for large L and small $\Delta r = r_1 - r_0$:

(a) A volume term, proportional to $4\pi L^3/3$, representing the entropy and mass-energy of a homogeneous quantum gas in a flat space (since $f \approx 1$ almost everywhere in the container if $L/r_0 \to \infty$) at a uniform temperature T_∞. This is the conventionally expected result and there is no need to consider it in detail.

(b) Of more interest is the contribution of gas near the inner wall $r = r_1$, which we now proceed to study further. We shall find that it is proportional to the wall area and at the same time diverging like $(\Delta r)^{-1}$ when $\Delta r \to 0$.

Because of the high local temperatures T near the wall for small Δr, we may use the ultra-relativistic formulae

$$\varrho = \frac{3\mathcal{N}}{\pi^2}T^4\ , \qquad s = \frac{4\mathcal{N}}{\pi^2}T^3 \tag{50}$$

for the energy and entropy densities. The numerical factor $\mathcal{N}$ takes care of helicities, the number of particle species and the factor 7/8 which differentiates fermionic from bosonic contributions.

The total entropy of the thermal excitations is given by the integral

$$S = \int_{r_1}^{L} s(r)\, 4\pi r^2 \frac{\mathrm{d}r}{\sqrt{f}} , \tag{51}$$

where we have used the proper volume element for the metric (46). On the other hand, the integral for their gravitational mass does not contain the factor $f^{-1/2}$ as is well known (intuitively, it is counterbalanced by a negative contribution from "gravitational potential energy"):

$$\Delta M_{\mathrm{therm}} = \int_{r_1}^{L} \varrho(r)\, 4\pi r^2\, \mathrm{d}r . \tag{52}$$

Substituting (50) and (47) into (51) gives for the wall contribution to the thermal entropy

$$S_{\mathrm{wall}} = \frac{4\mathcal{N}}{\pi^2} 4\pi r_1^2 T_\infty^3 \int_{r_1}^{r_1+\delta} \frac{\mathrm{d}r}{f^2(r)} , \tag{53}$$

where δ is a small radial interval subject to $\Delta r \ll \delta \ll r_1$ and otherwise arbitrary.

It is instructive to re-express this result in terms of the proper altitude

$$\alpha = \int_{r_0}^{r_1} f^{-1/2}\, \mathrm{d}r \tag{54}$$

of the inner wall above the horizon $r = r_0$ of the analytically extended exterior geometry (46). (Really, of course, the physical space contains no horizon and (46) is valid only for $r \geq r_1$.) For a non-extremal horizon we can write $f(r) \approx 2\kappa_0(r - r_0)$ in (54), obtaining

$$\Delta r = \frac{1}{2} \kappa_0 \alpha^2 , \tag{55}$$

and (53) can be written

$$S_{\mathrm{wall}} = \frac{\mathcal{N}}{90\pi\alpha^2} \left(\frac{T_\infty}{\kappa_0/2\pi}\right)^3 \frac{1}{4} A \tag{56}$$

in Planck units ($\hbar = c = G = 1$), where $A = 4\pi r_1^2$ is the wall area.

From (52) and (50) we find similarly that the thermal excitations near the wall contribute

$$\Delta M_{\mathrm{therm,wall}} = \frac{\mathcal{N}}{480\pi\alpha^2} \left(\frac{T_\infty}{\kappa_0/2\pi}\right)^3 T_\infty A \tag{57}$$

to the gravitational mass of the system.

Following 't Hooft, we introduce a crude "brick wall" cutoff to allow for quantum-gravity fluctuations by adjusting α so that we obtain the Bekenstein-Hawking entropy from (56):

$$S_{\text{wall}} = S_{\text{BH}} \text{when } T_\infty = T_{\text{H}} \ . \tag{58}$$

Notice that $S_{\text{BH}} = A/4$ and $T_{\text{H}} = \kappa_0/2\pi$ are purely *geometrical* quantities, determined solely by the metric (46). From (56) and (58) (momentarily restoring conventional units)

$$\alpha = l_{\text{Pl}}\sqrt{\mathcal{N}/90\pi} \ , \tag{59}$$

so the cutoff is of the order of the Planck length. It is significant and crucial that α turns out to be a *universal* constant, independent of the mass and other characteristics of the system, depending only on the number of physical fields in nature.

This universality permits a clean separation between the geometrical quantities A and T_{H} and the thermodynamical quantities S_{wall}, T_∞ in the free energy

$$F_{\text{wall}} = -\frac{1}{16}\left(\frac{T_\infty}{T_{\text{H}}}\right)^3 T_\infty A \ , \tag{60}$$

so that the entropy can be derived from it either via the Gibbs relation $S = -\partial F_{\text{wall}}/\partial T_\infty$ holding the geometrical variables fixed ("off-shell", i.e., breaking the equality $T_\infty = T_{\text{H}}$), or via the Gibbs-Duhem relation $F = \Delta M - T_\infty S$ (equivalent to $S = -\text{Tr}\,(\varrho \ln \varrho)$). Thus, there is no need in this formulation to maintain a distinction between "thermodynamical" and "statistical" entropy [12].

These results raise a number of questions. According to (57) "on shell" ($T_\infty = T_{\text{H}}$), thermal excitations near the wall contribute a mass-energy

$$\Delta M_{\text{therm,wall}} = \frac{3}{16} A T_{\text{H}} \ , \tag{61}$$

which amounts to a substantial fraction of the mass of the background geometry – $3M/8$ for a Schwarzschild geometry of mass M. But we assumed a fixed background and have taken no account of back-reaction. How is this justified? Secondly: our picture of hot quantum fields constituting a thermal atmosphere suggests that we are describing the Hartle-Hawking state. But according to (50) and (47), energy densities become huge near the wall, a behaviour quite unlike the Hartle-Hawking stress-energy, which remains bounded, and indeed everywhere very small for a massive black hole. Further: should not the entropy of the Gibbons-Hawking "instanton", arising from the Euclidean topology of a black hole (see (9)), be *added* to the entropy of the thermal excitations? But each by itself already accounts for the full value of the Bekenstein-Hawking entropy.

All these problems have a simple resolution. The key remark is that *'t Hooft's brick wall model correctly interpreted does not represent a black hole.* It represents the exterior of a starlike object with a perfectly reflecting surface compressed to nearly (but not quite) its gravitational radius. As noted in the previous section, the ground state for such object is not the Hartle-Hawking state but the Boulware state, corresponding to zero outside temperature and with a quite different behaviour near the gravitational radius. It has negative energy density and pressure, growing to Planck levels near the wall. Thus, the thermal energy density ϱ, given by (50), is not the only source of the wall's mass; it has to be supplemented by the ground-state energy. As in (45), we have for the total stress-energy (ground state + thermal excitations) near the wall,

$$\left(T_\mu^{\ \nu}\right)_{\rm B} + \left(T_\mu^{\ \nu}\right)_{\rm therm} = \left(T_\mu^{\ \nu}\right)_{\rm H} , \tag{62}$$

i.e., effectively the Hartle-Hawking stress-energy, which is bounded and small for large masses. The total gravitational mass of the inner wall is accordingly negligible. We are entirely justified in neglecting back-reaction.

Further, since there is no horizon in this model – it is replaced by the brick wall, an inner boundary with the quite different topology $S^1 \times S^2$ in the Euclidean sector (a horizon would be a regular point) – the Gibbons-Hawking instanton does not contribute. All of the entropy derives from the thermal contribution (56).

It thus emerges that we have two mutually exclusive and (in the Bohr sense) complementary ways of understanding $S_{\rm BH}$. In the brick wall model (which is not a black hole but it is externally indistinguishable from one), $S_{\rm BH}$ appears as entropy of thermal excitations above a zero-temperature ground state. In a real black hole, thermal energies near the horizon are negligible and $S_{\rm BH}$ has a purely geometrical origin. What are the deeper implications of this duality? This is a question it would be worth exploring.

The brick wall is treated here as a real physical barrier. To prevent misunderstanding this conception must be distinguished from a quite different one, according to which the wall is fictitious [13], merely a mathematical cutoff used to regularize the $(\Delta r)^{-1}$ and $\log \Delta r$ terms in the expression for the thermal entropy. These are then renormalized by adjusting the bare gravitational constant G and the coefficients of the one-loop (quadratic in curvature) terms in the effective gravitational Lagrangian. Formally, this scheme appears to work; its conceptual basis and relationship to the physical brick wall model remains to be clarified.

5 Operational Approach

The brick wall model still labours under the handicap that the wall's altitude above the horizon has to be adjusted by hand to give $S_{\rm BH}$ with the correct coefficient. This difficulty is not peculiar to the brick wall model. Of the various microscopic approaches, only the string-based calculations have so far been able, in special instances, to reproduce $S_{\rm BH}$ without special tailoring.

This section outlines a phenomenological approach that gives $S_{\rm BH}$ without cutoffs or ad hoc adjustments and further supports the idea that $S_{\rm BH}$ is a purely surface property.

In thermodynamics, the entropy of any state can be found by devising an idealized reversible process which arrives at that state, starting from a state of known entropy. The first law can be then used to compute the change of entropy in the process. The process considered here is the reversible quasistatic contraction of a massive thin shell towards its gravitational radius [14].

In the final stages of contraction towards a non-extremal horizon, the surface stress-energy of the shell is dominated by its surface pressure P, which grows without bound. The key result – which follows from the equations of mechanical equilibrium of an equipotential shell in the limit where the shell approaches a non-extremal horizon of the exterior geometry – is

$$\lim_{U\to 0} \frac{a}{8\pi P} = 1 , \tag{63}$$

where a is the proper acceleration of a static observer at the outer face of the shell.

Pressure-dominance means that in these final stages the shell violates the dominant-energy condition and develops other "unphysical" features. That is irrelevant in the present context, since the shell is nothing more than an idealized working substance designed to reversibly reach the final black hole state, which is expected to be independent of its mode of formation.

We now sketch the derivation of (63). Let ξ^α be the timelike Killing vector of the static exterior spacetime, and

$$u^\alpha = \frac{\mathrm{d}x^\alpha}{\mathrm{d}\tau} = U^{-1}\xi^\alpha , \qquad \xi_{(\alpha;\beta)} = 0 , \tag{64}$$

the 4-velocity of a static exterior observer. His proper acceleration a is given by

$$u_{\alpha;\beta}\, u^\beta = a\, n_\alpha = \partial_\alpha U , \tag{65}$$

where n_α is the unit principal normal to his world-line, and the second equality follows from (64).

The shell is assumed to lie on an equipotential surface $U =$ const. of the exterior geometry, so its outward normal coincides with n^α. The proper-time component of the extrinsic curvature K^+_{ab} of the shell's outer face is

$$K^+_{\tau\tau} = n_{\alpha;\beta}u^\alpha u^\beta = -u_{\alpha;\beta}u^\beta n^\alpha = -a . \tag{66}$$

The surface stress-energy S^b_a is given in terms of the jump of extrinsic curvature $[K^b_a]$ by the junction conditions

$$-8\pi S^b_a = [K^b_a - \delta^b_a K] . \tag{67}$$

In the limit $U \to 0$ of approach to a horizon, the acceleration a of a static observer on the upper face becomes infinite if the horizon is not extremal. Hence the components $K^\pm_{ab}$ are dominated by $(K^\tau_\tau)^+ = a$. The surface density

$$\sigma = -S^\tau_\tau = \frac{1}{8\pi}[K^\theta_\theta + K^\varphi_\varphi] \tag{68}$$

remains bounded, but the surface pressure

$$P = \frac{1}{2}(S_\theta^\theta + S_\varphi^\varphi) \approx \frac{1}{8\pi}(K_\tau^\tau)^+ = \frac{a}{8\pi} \qquad (U \to 0) \tag{69}$$

becomes infinite and asymptotically isotropic. This establishes (63).

The ground state for the space outside the shell is the Boulware state, whose stress-energy would diverge to large negative values in the limit $U \to 0$. To neutralize the back-reaction which would result, suppose the exterior filled with thermal radiation to produce a "topped-up" Boulware state (TUB) whose local temperature equals the acceleration temperature

$$T = \hbar \frac{a}{2\pi} \tag{70}$$

at the shell's outer face. (Near a horizon this is indistinguishable from the local Hawking temperature $T_{\rm H}\,(-g_{tt})^{-1/2}$.) To maintain thermal equilibrium (and hence applicability of the first law), the shell itself must be raised to the same temperature. This gives it a definite equation of state

$$T = T(M, A)\ , \tag{71}$$

whose specific form can be found once the exterior geometry is specified. Although the argument can be generalized, it is assumed here that the shell is spherical, with area $A = 4\pi\, R^2$.

The equation of state (71) relates the intensive variable T to two extensive variables, the shell's area A and proper mass $M = \sigma A$ as measured by a local observer. Since the shell is uniform, and we require it to be a thermodynamical system, it must be possible to rewrite (71) in a purely intensive form:

$$T = T(\sigma, n)\ . \tag{72}$$

How one chooses to interpret the second intensive parameter $n = N/A$ is quite immaterial; for convenience, it may be referred to as "particle density". Unlike (71), there is a certain amount of freedom in the functional form (72) (i.e., in the choice of n), but it is strongly constrained by the requirement of compatibility with both (71) and the Gibbs-Duhem relations [14]. One possible choice, n^* ("canonical equation of state", distinguished by an asterisk) makes

$$\mu^* n^* = \sigma \Rightarrow \mu^* N^* = M \tag{73}$$

where μ is the chemical potential associated with N.

The shell's entropy at any stage of the slow contraction is given by

$$TS = M + PA - \mu N\ . \tag{74}$$

The acceleration of a static observer on the outer face of a spherical shell with flat interior is

$$a = 4\pi(\sigma + 2P)\ . \tag{75}$$

In the limit of approach to the gravitational radius, a, P and T diverge in the non-extremal case, while M remains bounded. From (70) and (75) or (63),

$$\lim_{R\to r_0}\left(\frac{P}{T}\right)=\frac{1}{4}\hbar^{-1}\qquad (\kappa_0\neq 0)\,. \tag{76}$$

Combining this with (73) and (74), it immediately follows that, for a shell made of canonical material,

$$\lim_{R\to r_0} S=\frac{1}{4}\frac{A}{\hbar}\qquad (\kappa_0\neq 0)\,, \tag{77}$$

which is the Bekenstein-Hawking result.

Of course, one would expect the entropy of the final black hole to be independent of the material we choose for the shell, and, indeed, the limit (77) is very robust. Even for noncanonical material we still recover the limit (77) provided only that

$$\lim_{T\to\infty}\frac{1}{T}\frac{\mu N}{M}=0\,. \tag{78}$$

(Note that *some* restriction is obviously required, since the black hole limit represents a singular state for the shell ($P\to\infty$), and there would otherwise be nothing to prevent the equation of state (72) from becoming "singularly eccentric" in this limit.)

This derivation suggests an operational definition of the Bekenstein-Hawking entropy as the maximum thermodynamical entropy that could be stored in the material that goes to form the black hole, and that this maximum is attained when the material is gathered into a thin shell near the horizon. This idealized process bears no resemblance to actual black hole formation, but it is not dissimilar from a time-reverse of the evaporation process as usually conceived in terms of pair creation near the horizon.

The shell's entropy is already within 1% of its final, Bekenstein-Hawking value, when the redshift from its surface is 100, corresponding to an altitude of 10^{36} Planck lengths above the horizon for a solar-mass black hole. The shell thus occludes 't Hooft thermal atmosphere, which extends only a few Planck lengths above the horizon and is the repository of all the entropy in the brick wall model. To regard the shell as a direct phenomenological representation of the thermal atmosphere is in certain respects an oversimplification.

It is helpful to look at a concrete example. For a spherical shell of proper mass M and charge Q, the exterior metric is Reissner-Nordström, i.e., in (46)

$$f(r)=1-\frac{2m}{r}+\frac{Q^2}{r^2}\,, \tag{79}$$

and the explicit formulas are

$$m=M-\frac{1}{2}\frac{M^2-Q^2}{R}\,,\qquad P=\frac{1}{16\pi}\frac{M^2-Q^2}{R^2(R-M)}\,, \tag{80}$$

$$a = \frac{1}{2} \frac{Rf'(R)}{R - M} . \tag{81}$$

It is then easy to check explicitly that (63) and (77) follow in the non-extremal case ($Q^2 < m^2$).

The situation is quite different for an extremally charged shell ($Q^2 = m^2 \Rightarrow Q^2 = M^2$), because P is no longer dominant in the limit $R \to r_0$, but actually vanishes, while a and T remain finite. From (73) and (74) we now find that the entropy vanishes for a shell made of canonical material, so that a black hole formed out of this material would satisfy the third law in its strongest (Planck) form. However, it is clear from (74) that the limiting entropy now depends sensitively on the choice of shell material.

Thus it would appear than an extremal black hole differs from a generic one in that its entropy is not a thermodynamical state function, i.e., not independent of its mode of formation and past history. This may account for the conflicting results found in the literature. Arguments based on instanton topology and pair creation suggest that the entropy of extremal black holes is zero. On the other hand, the uncannily successful indirect derivations of S_{BH} by counting states of strings on D-branes recover the traditional value $A/4\hbar$ for extremal black holes. Perhaps the distinction between the extremal and non-extremal cases is less a matter of zero versus nonzero temperature but rather the finite versus infinite temperatures at the horizon measured by local stationary observers. There are many precedents for a situation where a simplicity and universality found at high temperatures breaks down at ordinary temperatures.

6 Conditional Entropy

Evaporation of a black hole is accompanied, as we saw in Sect. 3, by an influx of negative energy through the future horizon. It is also marked by a slow decrease in horizon area, i.e., in the entity we have referred to as "Bekenstein-Hawking entropy" S_{BH}. Two questions now present themselves.

First, if it were permissible to think of this "entropy" as a *signed* quantity, then we could describe the evaporation process as one in which the black hole acquires negative entropy as well as negative energy from its surroundings.

Secondly, and going considerably further, if this stuff – whatever its nature, fine-grained or coarse-grained – could be sensibly conceived as having properties that are at least partially localizable and capable of flowing from place to place like energy, then one could graduate to a description in which the influx of negative energy is accompanied by a flux of negative entropy, represented by an entropy current S^{μ}. In $(1+1)$-dimensions, for example, the energy flux in the Unruh state,

$$\Delta T_{ab} = \langle T_{ab} \rangle_{\mathrm{U}} - \langle T_{ab} \rangle_{\mathrm{H}} = H_{\mathrm{in}}\, v_{,a} v_{,b} \;, \quad H_{\mathrm{in}} = -\frac{\pi}{12\hbar} T_{\mathrm{H}}^2 \tag{82}$$

(cf. (28)), could be associated with an entropy flux

$$\Delta S_a = -T_{\mathrm{H}}^{-1} H_{\mathrm{in}}\, v_{,a} \;, \tag{83}$$

the relationship between the two being the standard one of covariant thermodynamics [15]:

$$\Delta S_a = -\Delta T_{ab}\,\beta^b\,, \qquad \beta^b = T_{\mathrm{H}}^{-1}\,\frac{\partial x^b(v,r)}{\partial v}\,. \tag{84}$$

This line of thought immediately runs into the difficulty that entropy, as conventionally defined – classically, as the averaged negative logarithm of a probability distribution, quantum-mechanically, as $-\mathrm{Tr}\,(\varrho \ln \varrho)$, where ϱ has eigenvalues between 0 and 1 – is an innately positive quantity. There is, however, an allied concept, "conditional entropy" [16], which applies to multi-component systems and is capable of taking negative values for quantum-entangled subsystems.

To introduce this concept, it is best to begin classically. Consider a system of two components A, B with joint probability p_{ab} that A and B occupy the states a and b respectively. The classical (Shannon) entropy of the composite system is

$$S(A,B) = -\langle \ln p_{ab}\rangle \equiv -\sum_{a,b} p_{ab}\, \ln p_{ab}\,, \tag{85}$$

and, of course, is non-negative, since

$$p_{ab} \geq 0\,, \quad \sum_{a,b} p_{ab} = 1\,. \tag{86}$$

The entropy of subsystem B is similarly

$$S(B) = -\langle \ln p_b \rangle = -\sum_{a,b} p_{ab}\, \ln p_b = -\sum_b p_b\, \ln p_b\,, \tag{87}$$

where $p_b = \sum_a p_{ab}$, and there is a similar formula for $S(A)$.

If it is known that B occupies state b, the conditional probability $p_{a|b}$ that A is in state a is given by Bayes' theorem:

$$p_{a|b} = \frac{p_{ab}}{p_b}\,. \tag{88}$$

The *conditional entropy* of A modulo B is defined by

$$S(A|B) = -\langle \ln p_{a|b}\rangle = -\sum_{a,b} p_{ab} \ln p_{a|b}\,. \tag{89}$$

Generally, the conditional entropy of a system is a measure of our ignorance of its internal state, given that we know the states of one or more specified other systems with which it has correlations. Substituting (88) into (89) gives at once

$$S(A|B) = S(A,B) - S(B)\,. \tag{90}$$

Classical conditional entropy is non-negative, since $p_{a|b}$ is a standard probability distribution satisfying the conditions (86).

Quantum-mechanically, (90) may be retained as the *definition* of $S(A|B)$, since the right-hand side is well-defined in terms of the joint and reduced density matrices ϱ_{AB} and $\varrho_B = \mathrm{Tr}\, \varrho_{AB}$:

$$S(A,B) = -\mathrm{Tr}\,(\varrho_{AB} \ln \rho_{AB}) \,, \qquad S(B) = -\mathrm{Tr}\,(\varrho_B \ln \rho_B) \,. \tag{91}$$

Now, however, a new feature enters. If the combined system is in a *pure* state, then $S(A,B) = 0$ and the conditional entropy $S(A|B) = -S(B)$ is *negative.* A and B are now perfectly correlated, and it can be shown that $S(A) = S(B)$ in this case. This common value is called "entanglement entropy":

$$S_{\text{tangle}} = S(A) = S(B) - S(A|B) = -S(B|A) \qquad (\{A,B\}\ \text{pure}) \,. \tag{92}$$

This indicates that it is not absurd to think about negative entropy in the context of quantum entanglement.

A more deepreaching analysis would attempt to define a conditional density matrix $\varrho_{A|B}$ by some analogue of Bayes' law (88). It is however, not straightforward to define the quotient of matrices whose eigenvalues can include zero. These questions will not be pursued here; they are still under active development [17].

7 Entropy from a Pure State

If the two subsystems of the previous section are identical copies of each other, and coupled in a special way to form a pure state, each of them becomes macroscopically indistinguishable from a hot body at a definite temperature T. In effect, each of them becomes a heat bath for the other.

We consider a composite system made up of a pair of identical subsystems (e.g., oscillators, fields, field modes propagating in opposite Kruskal sectors of an external black hole) labelled 1, 2 and having Hamiltonians H_1, H_2 identical in form and with common eigenvalues E_n:

$$H_1|n\rangle_1 = E_n|n\rangle_1 \,, \qquad H_2|n\rangle_2 = E_n|n\rangle_2 \,. \tag{93}$$

We now form a pure state $|T\rangle$ of the total system, characterized by a real, non-negative parameter T, which entangles the subsystems according to

$$|T\rangle = Z^{-1/2} \sum_n \mathrm{e}^{-E_n/2T} |n\rangle_1 \otimes |n\rangle_2 \,, \tag{94}$$

where $Z = \sum_n \mathrm{e}^{-E_n/T}$ to secure the normalization $\langle T|T\rangle = 1$.

Of course the entropy of this pure state is zero:

$$S = -\mathrm{Tr}\,(\varrho \ln \varrho) = 0 \,, \qquad \varrho \equiv |T\rangle\langle T| \,. \tag{95}$$

But tracing over the eigenstates $|n\rangle_2$ yields a density matrix for 1 whose eigenvalues are the Boltzmann factors for a system at temperature T:

$$\varrho_1 = \mathrm{Tr}_2\, \varrho = Z^{-1} \sum_n \mathrm{e}^{-E_n/T} |n\rangle_1 {}_1\langle n| \,. \tag{96}$$

Its entropy reduces to the usual thermodynamical expression,

$$S(\varrho_1) = -\mathrm{Tr}\,(\varrho_1 \ln \varrho_1) = \ln Z + T^{-1}\langle E\rangle_T \;. \tag{97}$$

The same result is obtained for $S(\varrho_2)$. Thus, the "entanglement entropy" $S_{\text{tangle}} = S_1 = S_2$ is positive, yet the entropy of the total system is zero.

If, further, the energy levels are evenly spaced ($E_n = n\omega$), then the "thermally entangled" state $|T\rangle$ is also a *ground state* for collective excitations on the double Fock space of the composite system. We have

$$\hat{A}|T\rangle = 0 \;, \text{where} \hat{A} = Z^{-1/2}(\hat{a}_1 \otimes 1 - \mathrm{e}^{-\omega/2T} 1 \otimes \hat{a}_2^{+}) \;, \tag{98}$$

and $\hat{a}_1$, $\hat{a}_2$ are the annihilation operators for the two subsystems.

Conversely, if one is interested solely in macroscopic effects or a coarse-grained description, then the mixed state of a system in thermal equilibrium can be mentally "purified" by the formal trick of doubling the Fock space, thus converting statistical averages into quantum expectation values on the enlarged space. Given a system 1 at temperature T, we mentally adjoin an identical copy 2 – it can be thought of as an ersatz heat bath for 1 – such that the enlarged system is in the entangled state $|T\rangle$. Macroscopically, $|T\rangle$ is indistinguishable from a thermal state for the original system 1. If $\mathcal{O}_1$ is any operator that acts only on states of 1, its thermal average is equal to its quantum expectation value in state $|T\rangle$:

$$\langle \mathcal{O}_1\rangle_T = \mathrm{Tr}\,(\varrho_1\, \mathcal{O}_1) = \sum_n \mathrm{e}^{-E_n/T}\, {}_1\langle n|\mathcal{O}_1|n\rangle_1 = \langle T|\mathcal{O}_1|T\rangle \;. \tag{99}$$

In this way statistical mechanics is reduced to quantum field theory and all of the sophisticated graphical techniques of that subject can be brought to bear. This is the idea underlying the "thermofield dynamics" of Takahashi and Umezawa [18].

Thermofield dynamics encodes a reflexive symmetry between the twin systems 1 and 2 that is reminiscent of the reflection symmetry between the left and right halves of Kruskal's picture of an eternal black hole (Fig. 1). In fact, there is an exact correspondence between the two representations. Systems 1 and 2 correspond to field modes propagating in the (causally disjoint) right and left Kruskal sectors R and L. The thermally entangled state $|T\rangle$ corresponds to the ground state on the full Kruskal manifold, i.e., the Hartle-Hawking state $|0\rangle_{\mathrm{H}}$. Static observers, whose natural ground state is the Boulware state $|0\rangle_{\mathrm{B}}$, and who are "parochial" in the sense that their world-lines are confined to a supra-horizon sector (say R), perceive the global state $|0\rangle_{\mathrm{H}}$ as thermal.

Possible implications of this correspondence hinge of course on the question whether mathematical models of eternal black holes sufficiently resemble real black holes formed in a stellar collapse. For a black hole of stellar origin, sectors L and P in Fig. 1 are obliterated and supplanted by the star. But enough remains of the common sector F to preserve a meaningful connection between the two halves of the vacuum manifold. In particular, a correlated pair of modes of

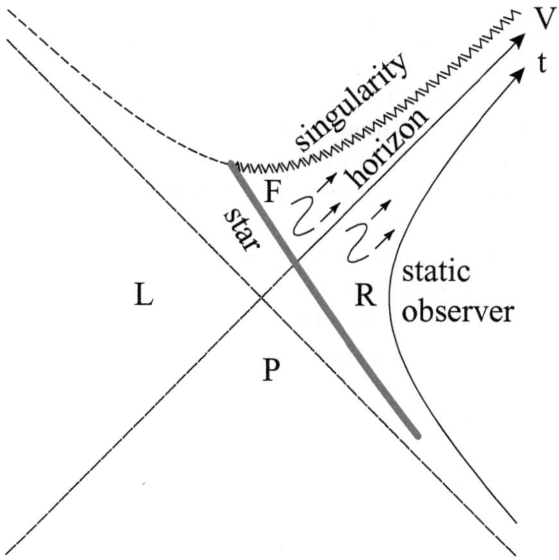

Fig. 1. Kruskal map of extended vacuum spacetime representing a Schwarzschild black hole. If the black hole was formed by stellar collapse, only the region to the right of the hatched curve (representing the star's surface) is physically meaningful. (On the left, the vacuum manifold has to be replaced by a nonsingular, more conventional geometry representing the interior field.) The figure shows a correlated pair of Boulware modes propagating alongside the horizon

opposite norm propagating in the right and left sectors of the eternal black hole should correspond in the real black hole to a virtual pair, of opposite total (i.e., including "potential"-) energies $+E$, $-E$, created near the future horizon and propagating side-by-side just above and below it [19]. (For a particle of 4-momentum p_α, $E = -p_\alpha \xi^\alpha$, where ξ^α is the Killing vector which passes from timelike to spacelike at the horizon.)

These considerations suggest that the correlations found in the thermofield-dynamical model of an eternal black hole may have a real counterpart in correlations between the inside and outside of a real black hole. Since quantum correlations grow inversely with distance, they would be strongest near the boundary – the horizon – suggesting a picture of entanglement entropy strongly localized near the horizon. Thus, it is conceivable (though very far from proven) that the thermally entangled pure state $|T\rangle$ is not merely an artefact of an artificial, stripped-down model of a heat bath, but actually represents, at least schematically, the internal state of a black hole formed by collapse.

8 Thermofield Dynamics of Black Holes

A black hole enclosed in a container will reach a state of thermal equilibrium: the Hartle-Hawking state. This state is empty of modes positive-frequency in Kruskal time: free-falling observers at the horizon register an absence of "particles". Static observers, whose notion of positive frequency is defined by the static Killing parameter t, will disagree: from the perspective of their ground state – the Boulware state, free of Killing modes – the Hartle-Hawking state is filled with a thermal distribution of "particles". This section will examine the mode structure of these two states and the relationship between them, topics treated more phenomenologically in the previous sections.

The aspects of interest here are concerned entirely with the time-dependence of field modes, with spatial dependence carried along as mere baggage. To keep the notation simple, the spatial dependence will be schematized as far as possible by working in (1+1)-dimensions. But the arguments are quite general, and apply with only cosmetic changes to arbitrary (tidally deformed) static black holes, to cosmological horizons of de Sitter type and to uniformly accelerated frames in Minkowski space.

The metric is taken in the form (29), i.e.,

$$\mathrm{d}s^2 = \frac{\mathrm{d}r^2}{f(r)} - f(r)\,\mathrm{e}^{2\psi(r)}\,\mathrm{d}t^2\,, \tag{100}$$

with a horizon at $r = r_0$, characterized by

$$f(r_0) = 0, \qquad \kappa_0 = \frac{1}{2} f'(r_0)\,\mathrm{e}^{\psi(r_0)}\,, \tag{101}$$

and Kruskal co-ordinates U, V defined by

$$\left.\begin{array}{c} \dfrac{\mathrm{d}V}{\kappa_0 V} \\[2ex] \dfrac{\mathrm{d}U}{(-\kappa_0 U)} \end{array}\right\} = \left\{\begin{array}{c} \mathrm{d}v \\ \mathrm{d}u \end{array}\right\} = \mathrm{d}t \pm \frac{\mathrm{d}r}{f(r)\mathrm{e}^{\psi(r)}}\,, \tag{102}$$

so that

$$u = -\frac{1}{\kappa_0}\ln|U|\,, \qquad v = \frac{1}{\kappa_0}\ln|V|\,, \qquad t = \frac{1}{2}(v-u)\,. \tag{103}$$

It is important to note that t, u, v run *backwards* in the left-hand Kruskal sector L (Fig. 2).

A real massless quantum field Φ, satisfying $\Box\Phi = 0$, can be decomposed as

$$\Phi(U,V) = \Phi_{\mathrm{out}}(U) + \Phi_{\mathrm{in}}(V)\,, \qquad [\Phi_{\mathrm{out}}(U), \Phi_{\mathrm{in}}(V)] = 0\,, \tag{104}$$

and we shall focus on the outgoing component $\Phi_{\mathrm{out}}(U)$, since the other piece is handled the same way.

We next introduce two sets, $F_\omega^{(\pm)}(U)$, of right-moving "Killing-Boulware" modes: $F_\omega^{(+)}(U)$ are nonzero only for $U < 0$, i.e., confined to the right-hand and white hole sectors R and P of the Kruskal geometry, see Fig. 2; $F_\omega^{(-)}(U)$ are

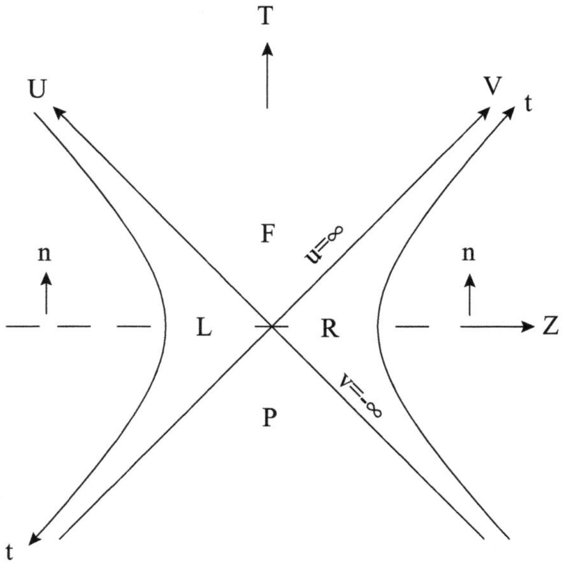

Fig. 2. Kruskal map of eternal Schwarzschild black hole. Schwarzschild time t runs backwards in the left-hand Schwarzschild sector L

nonzero for $U > 0$ and propagate in the left-hand and black hole sectors L and F. They are defined by

$$F_\omega^{(\varepsilon)}(U) = \theta(-\varepsilon U)\, f_\omega(u) \,, \qquad f_\omega(u) = \frac{1}{\sqrt{4\pi|\omega|}} \mathrm{e}^{-\mathrm{i}\omega u} \tag{105}$$

and form an orthonormal set:

$$(F_\omega^{(\varepsilon)}, F_{\omega'}^{(\varepsilon')}) = \varepsilon(\omega)\, \varepsilon\, \delta_{\varepsilon\varepsilon'}\, \delta(\omega - \omega') \,, \tag{106}$$

where $\varepsilon(\omega) \equiv \mathrm{sign}\,(\omega)$ and θ is the unit step function. The Klein-Gordon inner product for arbitrary wave modes f, g is defined as usual by

$$(f, g) = \mathrm{i} \int (f^* \overset{\leftrightarrow}{\partial}_a g)\, \mathrm{d}\Sigma^a = \mathrm{i} \int_{-\infty}^{\infty} (f^* \overset{\leftrightarrow}{\partial}_T g)\, \mathrm{d}Z \,, \tag{107}$$

where the integration is over a complete Cauchy slice with a globally consistent future normal (indicated by n in Fig. 2), and Z, T are Kruskal-like co-ordinates, i.e., $U = T - Z$, $V = T + Z$. The factor ε in (106) arises from the contraposition of n and the direction of increasing u in sector L. Thus, $F_\omega^{(\varepsilon)}$ has positive norm, and is effectively "positive-frequency" in Killing time, if $\omega\varepsilon$ (rather than simply ω) is positive.

The functions $F_\omega^{(\varepsilon)}(U)$ form a complete set of outgoing modes and define the *Boulware quantization* of the Klein-Gordon field Φ through the decomposition

$$\Phi_{\mathrm{out}}(U) = \sum_{\varepsilon=\pm} \int_{-\infty}^{\infty} \mathrm{d}\omega\, F_\omega^{(\varepsilon)}(U)\, b_\omega^{(\varepsilon)} \,, \tag{108}$$

with commutation relations

$$\left[b_\omega^{(\varepsilon)}, b_{\omega'}^{(\varepsilon')}\right] = \varepsilon(\omega)\, \varepsilon\, \delta_{\varepsilon\varepsilon'}\, \delta(\omega - \omega')\,, \quad b_{-\omega}^{(\varepsilon)} = b_\omega^{(\varepsilon)+}\,, \tag{109}$$

and a similar expansion of $\Phi_{\rm in}(V)$. (The corresponding b-operators should carry distinguishing "out" and "in" labels, omitted here for simplicity.) According to this notation, $b_\omega^{(\varepsilon)}$ is an annihilation or creation operator according as $\omega\varepsilon$ is positive or negative. This is unconventional, but permits compact and convenient expression of mode expansions and commutation relations.

An alternative quantization scheme is based on the Kruskal-Hartle-Hawking (KH^2) modes $H_\Omega(U)$, $H_\Omega(V)$, where

$$H_\Omega(U) = \frac{1}{\sqrt{4\pi|\Omega|}} \mathrm{e}^{-\mathrm{i}\Omega U}\,. \tag{110}$$

However, in this harmonic form the KH^2 modes are not related quite as simply as they can be to the Killing- Boulware (KB) modes. Instead, we resort to a device due to Unruh [4], who builds an equivalent form of KH^2 mode by fitting together a pair of KB modes of opposite norm, one from the L-sector and one from the R-sector, to make a linear combination which is analytic and positive-frequency in Kruskal time.

Recall that an analytic function $f(t)$ is positive-frequency in t (i.e., its Fourier spectrum contains only positive frequencies) if it is regular and bounded in the lower half of the complex t-plane. The function $\ln_+ t$ – defined as that branch of $\ln t$ equal to $\ln|t|$ on the lower imaginary axis and with branch cut in the upper half-plane – has the first of these two properties. Note that its values on the real axis are given by

$$\ln_+ t = \ln|t| + \mathrm{i}\frac{\pi}{2}\,\varepsilon(t).$$

It follows that $\mathrm{e}^{\mathrm{i}\omega \ln_+ t}$ (for *both* signs of ω) is positive-frequency in t. Generalizing slightly, the two functions defined for real t by

$$\ln_\varepsilon t = \ln|t| + \mathrm{i}\frac{\pi}{2}\,\varepsilon(t)\varepsilon \tag{111}$$

have imaginary exponentials

$$\mathrm{e}^{\mathrm{i}\omega \ln_\varepsilon t} = \mathrm{e}^{\mathrm{i}\omega \ln|t|}\left\{\mathrm{e}^{-\varepsilon\pi\omega/2}\theta(t) + \mathrm{e}^{\varepsilon\pi\omega/2}\theta(-t)\right\} \tag{112}$$

which are positive-frequency if $\varepsilon = +1$, negative-frequency if $\varepsilon = -1$.

The right-hand side of (112) has the form of a linear combination of two KB-modes (105), recalling that $u = -\kappa_0^{-1}\ln|U|$. Thus, the functions

$$K_\omega^{(\varepsilon)}(U) = \sqrt{\frac{\sinh\chi_\omega \cosh\chi_\omega}{4\pi\,|\omega|}}\, \exp\left\{\frac{\mathrm{i}\omega}{\kappa_0}\ln_{\varepsilon(\omega)\varepsilon} U\right\} \tag{113}$$

$$= F_\omega^{(\varepsilon)}(U)\,\cosh\chi_\omega + F_\omega^{(-\varepsilon)}(U)\,\sinh\chi_\omega \tag{114}$$

are positive-frequency in U if $\omega\varepsilon$ is positive, negative-frequency if $\omega\varepsilon$ is negative. They have the same orthonormality properties as $F_\omega^{(\varepsilon)}$:

$$\left(K_\omega^{(\varepsilon)}, K_\omega^{(\varepsilon')}\right) = \varepsilon(\omega')\,\varepsilon\,\delta_{\varepsilon\varepsilon'}\,\delta(\omega-\omega')\,. \tag{115}$$

The Bogoliubov parameter χ_ω is defined by

$$\tanh\chi_\omega = \mathrm{e}^{-\pi|\omega|/\kappa_0}\,. \tag{116}$$

The KH^2-modes $K_\omega^{(\varepsilon)}$ define *Hartle-Hawking quantization* in accordance with

$$\Phi(U) = \sum_{\varepsilon=\pm}\int_{-\infty}^{\infty} \mathrm{d}\omega\, K_\omega^{(\varepsilon)}(U)\, a_\omega^{(\varepsilon)}\,, \tag{117}$$

where the Hartle-Hawking operators $a_\omega^{(\varepsilon)}$ have the same commutation relations as the Boulware operators $b_\omega^{(\varepsilon)}$:

$$\left[a_\omega^{(\varepsilon)}, a_{\omega'}^{(\varepsilon')}\right] = \varepsilon(\omega)\,\varepsilon\,\delta_{\varepsilon\varepsilon'}\,\delta(\omega-\omega')\,,\quad a_{-\omega}^{(\varepsilon)} = a_\omega^{(\varepsilon)+}\,. \tag{118}$$

The Bogoliubov transformation which relates the two sets of operators is contragradient to the mode transformation (114):

$$a_\omega^{(\varepsilon)} = (\cosh\chi_\omega)\, b_\omega^{(\varepsilon)} - (\sinh\chi_\omega)\, b_\omega^{(-\varepsilon)}\,. \tag{119}$$

Remarkably, this has precisely the “thermal” form encountered in Sect. 6. Comparison of (119) and (116) with (98) shows that the temperature associated with this transformation is the Hawking temperature $T_{\mathrm{H}} = \kappa_0/2\pi$.

It is straightforward to check, with the aid of the commutation relations (109), that (119) can be re-expressed in the convenient exponential form

$$a_\omega^{(\varepsilon)} = \mathrm{e}^{-\mathrm{i}G}\, b_\omega^{(\varepsilon)}\, \mathrm{e}^{\mathrm{i}G}\,, \tag{120}$$

where

$$G = G^+ = \frac{\mathrm{i}}{2}\sum_{\varepsilon=\pm}\int_{-\infty}^{\infty}\mathrm{d}\omega\,\varepsilon(\omega)\,\varepsilon\,\chi_\omega\, b_{-\omega}^{(\varepsilon)} b_\omega^{(-\varepsilon)} \tag{121}$$

$$= \mathrm{i}\int_0^\infty \mathrm{d}\omega\,\chi_\omega\left(b_\omega^{(+)+} b_\omega^{(-)} - b_\omega^{(+)} b_\omega^{(-)+}\right)\,. \tag{122}$$

In accordance with our conventions, the first term in (122) involves the product of two creation operators in opposite sectors R and L, and correspondingly the second term the product of two annihilation operators.

The ground states for the two quantization schemes, defined by

$$a_\omega^{(\varepsilon)}|0\rangle_{\mathrm{H}} = 0\,,\qquad b_\omega^{(\varepsilon)}|0\rangle_{\mathrm{B}} = 0\,,\qquad (\omega\varepsilon > 0)\,, \tag{123}$$

are, accordingly to (120), formally linked by

$$|0\rangle_{\mathrm{H}} = \mathrm{e}^{-\mathrm{i}G}|0\rangle_{\mathrm{B}}\,, \tag{124}$$

although this relation needs to be treated with some caution [18]. (An infinite number of soft Boulware modes are contained in $|0\rangle$, which leads to difficulties with convergence of the partition function, i.e., the normalizing factor. Strictly speaking, $|0\rangle_{\rm H}$ and $|0\rangle_{\rm B}$ are "unitarily inequivalent".)

A more explicit and physically clearer form of (124) can be obtained by factorizing the operator $\exp(-\mathrm{i}G)$ into its creation and annihilation parts. This is accomplished by a generalization of the Baker-Campbell-Hausdorff (BCH) identity. We now state this, deferring an outline of the derivation to the Appendix.

Generalized BCH identity. If two operators A, B and their commutator C satisfy the commutation relations

$$[A,B]=C\ , \qquad [C,A]=2\,n^2 A\ , \qquad [B,C]=2\,n^2 B \tag{125}$$

for some number n (real or complex), then for any parameter χ,

$$\mathrm{e}^{\chi(A+B)}=\exp\left\{\frac{1}{n}(\tanh n\chi)A\right\}\exp\left\{\frac{1}{2n}(\sinh 2n\chi)B\right\} \cdot\exp\left\{-\frac{1}{n^2}(\ln\cosh n\chi)C\right\}\ . \tag{126}$$

(As a simple check: in the limit $n\to 0$, (126) reduces to the familiar BCH identity

$$\mathrm{e}^{\chi(A+B)}=\mathrm{e}^{\chi A}\mathrm{e}^{\chi B}\mathrm{e}^{-\chi^2 C/2}$$

as it should for the case where C commutes with A and B.)

In the application of interest here – see (122) –

$$A_\omega=b_\omega^{(+)+}b_\omega^{(-)}\ , \qquad B_\omega=-b_\omega^{(+)}b_\omega^{(-)+} \qquad (\omega>0)$$

so that A is composed of creation operators and B of annihilation operators. Then

$$C_\omega=b_\omega^{(+)+}b_\omega^{(+)}+b_\omega^{(-)+}b_\omega^{(-)}\ ,$$

and the commutation relations (125) are found to be satisfied with $n=1$, so (126) is applicable.

Discretizing the integral (122) as $G=\sum_\omega G_\omega$ then yields immediately

$$\mathrm{e}^{-\mathrm{i}G_\omega}|0\rangle_{\rm B}=\mathrm{e}^{\chi_\omega(A_\omega+B_\omega)}|0\rangle_{\rm B}$$

$$=\frac{1}{\cosh\chi_\omega}\exp\left\{(\tanh\chi_\omega)b_\omega^{(+)+}b_\omega^{(-)}\right\}|0\rangle_{\rm B}$$

$$=\frac{1}{\cosh\chi_\omega}\sum_{n=0}^{\infty}(\tanh^n\chi_\omega)\,|n_\omega^{(+)},n_\omega^{(-)}\rangle_{\rm B}\ , \tag{127}$$

where $|n_\omega^{(+)}, n_\omega^{(-)}\rangle_{\rm B}$ is the Boulware state with an equal number n of correlated Boulware modes of frequency ω in the L and R sectors.

Now, introducing the inverse Hawking temperature β by

$$\beta^{-1} = T_{\rm H} = \frac{\kappa_0}{2\pi} \tag{128}$$

and recalling (116), we can write

$$\tanh \chi_\omega = {\rm e}^{-\pi|\omega|/\kappa_0} = {\rm e}^{-\beta\omega/2} \tag{129}$$

$$\cosh^2 \chi_\omega = \frac{1}{1-{\rm e}^{-\beta\omega}} = \sum_{n=0}^{\infty} {\rm e}^{-n\beta\omega} \equiv Z_\omega\ , \tag{130}$$

so that (127) becomes

$${\rm e}^{-{\rm i}G_\omega}|0\rangle_{\rm B} = Z_\omega^{-1/2} \sum_{n_\omega=0}^{\infty} {\rm e}^{-\beta n_\omega \omega/2}\, |n_\omega^{(+)}, n_\omega^{(-)}\rangle_{\rm B}\ . \tag{131}$$

This result is for a single, fixed frequency ω. Finally, summing over all frequencies, we arrive at

$$|0\rangle_{\rm H} = \prod_\omega {\rm e}^{-{\rm i}G_\omega}|0\rangle_{\rm B} = Z^{-1/2} \sum_{\boldsymbol{n}} {\rm e}^{-\beta E_{\boldsymbol{n}}/2} |\boldsymbol{n}^{(+)}, \boldsymbol{n}^{(-)}\rangle_{\rm B}\ , \tag{132}$$

where $\boldsymbol{n}$ stands for a *set* $\{n_\omega,\ \forall\omega > 0\}$ of occupation numbers for all positive ω, and

$$E_{\boldsymbol{n}} \equiv \sum_\omega n_\omega\, \omega\ , \qquad Z \equiv \sum_{\boldsymbol{n}} {\rm e}^{-\beta E_{\boldsymbol{n}}} = \prod_\omega Z_\omega\ . \tag{133}$$

(The occupation number for ingoing modes $F_\omega^{(+)}(v)$ should also be folded into $\boldsymbol{n} = (\boldsymbol{n}^{\rm in}, \boldsymbol{n}^{\rm out})$.) Thus, we have found that $|0\rangle_{\rm H}$ has precisely the form of the thermally entangled state $|T\rangle$ encountered in (94).

The states $|\boldsymbol{n}^{(-)}\rangle_{\rm B}$ represent Boulware modes $F_\omega^{(-)}$ which never enter sector R and are unobservable by static observers confined to that sector. Tracing out these modes from the pure state $|0\rangle_{\rm H}$ leads to the thermal density matrix

$$\varrho_+ \equiv {\rm Tr}_-\big(|0\rangle_{\rm HH}\langle 0|\big) = Z^{-1} \sum_{\boldsymbol{n}} {\rm e}^{-\beta E_{\boldsymbol{n}}} |\boldsymbol{n}^{(+)}\rangle\langle \boldsymbol{n}^{(+)}|\ . \tag{134}$$

The expectation value in state $|0\rangle_{\rm H}$ of any "parochial" operator $\mathcal{O}^{(+)}$ (a function of the "+" operators $b_\omega^{(+)}$ only) is the same as its (thermal) average with respect to the parochial density matrix ϱ_+:

$${}_{\rm H}\langle 0|\mathcal{O}^{(+)}|0\rangle_{\rm H} = {\rm Tr}\,(\varrho_+ \mathcal{O}^{(+)}) = \langle \mathcal{O}^{(+)}\rangle_T\ . \tag{135}$$

We might (almost) say that the sensation of heat near a black hole is *caused* by our ignorance of modes hidden behind the horizon, echoing Pauli's words, "Was ich nicht weiss, macht mich heiss."

9 Black Hole Evaporation

If information really is recovered when a black hole evaporates, and if this is a continuous process extending through the lifetime of the hole, then closer study of slow evaporation should turn up clues to how this recovery takes place. This concluding section offers some pointers on how thermofield-dynamical techniques can be extended to describe a slowly evaporating black hole.

During evaporation the surface gravity κ_0, i.e., the temperature, changes ("adiabatic transformation"), and the horizon shrinks, i.e., shifts inward ("isothermical transformation"). Adiabatic transformations are described by the group of thermal Bogoliubov transformations (119) or (120) and are elementary. (Sect. 3 engaged the question of how κ_0 is defined for a nonstatic black hole.) We shall therefore concentrate here on the effects of horizon-displacement.

To describe a moveable horizon, we rewrite (103) as

$$u = -\frac{1}{\kappa_0} \ln |U - U_0| \, ,$$

so the future horizon is now at $U = U_0$. We consider the effect of a displacement of this horizon to $U = \bar{U}_0$, where $\bar{U}_0 = U_0 - \alpha$. To find directly the effect on the Boulware modes (105) is quite difficult. (Wave packets sandwiched in the range $U_0 - \alpha < U < U_0$ are converted from negative-norm $F^{(-)}(U)$ to positive-norm $F^{(+)}(U)$ modes.) But on Hartle-Hawking modes in their harmonic form (118),

$$H_\Omega(U - U_0) = \frac{1}{\sqrt{4\pi|\Omega|}} \mathrm{e}^{-\mathrm{i}\Omega(U-U_0)} \, ,$$

the effect of the displacement is trivial:

$$H_\Omega(U - \bar{U}_0) = \mathrm{e}^{-\mathrm{i}\Omega\alpha} H_\Omega(U - U_0) \, .$$

The relation (114) between KH^2 and KB modes then allows us to derive the effect on the latter.

The eventual result is that in the Ω-representation the Boulware operators b_Ω and the Boulware U_0-dependent ground state $|0; U_0\rangle_\mathrm{B}$ transform according to

$$\bar{b}_\Omega(\alpha) = \mathrm{e}^{-\mathrm{i}\alpha G} b_\Omega \mathrm{e}^{\mathrm{i}\alpha G} \, , \qquad |0; U_0 - \alpha\rangle_\mathrm{B} = \mathrm{e}^{-\mathrm{i}\alpha G} |0; U_0\rangle_\mathrm{B} \tag{136}$$

where

$$G = G^+ = \frac{1}{2} \int_{-\infty}^{\infty} \int_{-\infty}^{\infty} \mathrm{d}\Omega_1 \mathrm{d}\Omega_2 \, g(\Omega_1, \Omega_2) b_{\Omega_1} b_{\Omega_2} \, ,$$

and

$$g(\Omega_1, \Omega_2) \equiv -\varepsilon(\Omega_1, \Omega_2) \int_{-\infty}^{\infty} \mathrm{d}\Omega \, A(\Omega_1, -\Omega) \, |\Omega| \, A(\Omega, \Omega_2) \, ,$$

$$A(\Omega, \Omega') \equiv C(\Omega, \Omega') \, \theta(\Omega\Omega') + S(\Omega, \Omega') \, \theta(-\Omega\Omega') \, ,$$

$$\left.\begin{matrix} C \\ S \end{matrix}\right\} (\Omega, \Omega') \equiv \int_{-\infty}^{\infty} \left.\begin{matrix} \cosh \chi \\ \sinh \chi \end{matrix}\right\} \mu(\Omega, \Omega'; x)\, \mathrm{d}x$$

$$\mu(\Omega, \Omega'; x) \equiv \frac{1}{2\pi^2} |\Omega\Omega'|^{-1/2} \left|\frac{\Omega}{\Omega'}\right|^{\mathrm{i}x/\pi}, \qquad \tanh \chi \equiv \mathrm{e}^{-|x|} .$$

The Bogoliubov transformation (136) implicitly describes how shrinkage of the horizon unveils previously hidden outgoing Boulware wave packets and restores to the exterior correlations between pairs previously separated by the horizon. To unravel the details of this is a nontrivial task which may have its rewards. Understanding how the black hole loses its entropy – i.e., how information escapes from a layer near the surface – could hold clues to the seemingly unrelated question, usually considered the crux of the information-loss problem: how does information about the state of the collapsed star leak to the outside from the deep interior in apparent defiance of causality? [20]

10 Concluding Remarks

It has been widely held that the entropy contributed by thermal excitations or entanglement is a one-loop correction to the zero-loop (or "classical") Gibbons-Hawking geometrical contribution. The view advocated here is (at least superficially) quite different. One may consider these two entropy sources – (a) brick wall, no horizon, strong thermal excitations near the wall, Boulware ground state; and (b) black hole, horizon, weak (Hartle-Hawking) stress-energy near the horizon, Hartle-Hawking ground state – as equivalent but mutually exclusive (complementary in the sense of Bohr) descriptions of what is externally virtually the same physical situation. The near-vacuum experienced by free-falling observers near the horizon is eccentrically but defensibly explainable, in terms of description (a), as a delicate cancellation between a large thermal energy and an equally large and negative ground-state energy – just as the Minkowski vacuum is explainable to a uniformly accelerated observer as a thermal excitation above his negative-energy (Rindler) ground state. (This corresponds to setting $f(r) = r$, $\psi = 0$ in (29).) The artificiality of such a description is underlined by the fact that the delicate balance must extend to fluctuations: fluctuations of the Boulware ground-state energy would have to be exactly correlated with the enormous thermal fluctuations near a horizon to reproduce the relatively small fluctuations of the Hartle-Hawking state.

The brick wall model (as well as numerous other attempts to derive S_{BH} statistically by focusing on the neighbourhood of the horizon) presents us with a feature which is logically not impossible but strange and counterintuitive from a gravitational theorist's point of view. Although the wall is insubstantial (just like a horizon) – i.e., space there is practically a vacuum and the curvature low – it is nevertheless the repository of all of the Bekenstein-Hawking entropy of the model.

Frolov and Novikov [19] and others have argued that this is just what may be expected of black hole entropy in the entanglement picture. Entanglement will arise from virtual pair-creation in which one partner is "invisible" and the other "visible". Thus, on this picture, the entanglement entropy arises almost entirely from the strong correlations between the near field variables on the two sides of the partition, an effect already present in flat space.

This in turn suggests that $S_{\rm BH}$ is (in the literal sense) a *superficial property*, that it should be considered an *effective entropy* for a black hole, in the sense that 6000 K is an effective temperature for the sun. As far as their interactions with the environment are concerned, both objects are indistinguishable from shells (of the same size and mass) whose entropy and temperature have the effective values.

Acknowledgements

I am indebted to Shinji Mukohyama, my collaborator in much of the work reported here, for many stimulating discussions.

The work was supported by NSERC of Canada and by the Canadian Institute for Advanced Research.

Appendix. Derivation of Generalized BCH Identity (126)

Consider the operator

$$F(\chi) = \mathrm{e}^{xA}\mathrm{e}^{yB}\mathrm{e}^{zC} \, , \tag{137}$$

where x, y, z are undetermined functions of the parameter χ. We shall attempt to choose these functions so that F is reducible to $\exp(\chi(A+B))$.

Differentiating (137) with respect to χ,

$$F^{-1}F'(\chi) = x'\mathrm{e}^{-zC}(\mathrm{e}^{-yB}A\mathrm{e}^{yB})\mathrm{e}^{zC} + y'\mathrm{e}^{-zC}B\mathrm{e}^{zC} + z'C. \tag{138}$$

Now, from the commutation relations $[C,A] = 2m\,A$, $[B,C] = 2m\,B$ (we write m in place of n^2 for typographical convenience),

$$\mathrm{e}^{-yB}A\mathrm{e}^{yB} = A + y[A,B] + \frac{1}{2}y^2\big[[A,B],B\big] + \ldots = A + yC - y^2 mB \, ,$$

$$\mathrm{e}^{-zC}A\mathrm{e}^{zC} = \mathrm{e}^{-2mz}A \, , \qquad \mathrm{e}^{-zC}B\mathrm{e}^{zC} = \mathrm{e}^{2mz}B$$

Hence (138) reduces to

$$F^{-1}F' = x'(\mathrm{e}^{-2mz}A + yC - y^2 m\mathrm{e}^{2mz}B) + y'\mathrm{e}^{2mz}B + z'C \, . \tag{139}$$

We require this expression to equal $A+B$. Equating coefficients of A, B and C yields three equations for x, y, z:

$$x'\mathrm{e}^{-2mz} = 1 \, , \quad (y' - mx'y^2)\mathrm{e}^{2mz} = 1 \, , \quad z' + x'y = 0 \, .$$

Let $f(\chi) = \mathrm{e}^{-2mz}$. The first and third of these equations then give $x' = f^{-1}$, $y = f'/2m$. Substituting into the second equation results in a second-order linear equation for $\sqrt{f}$:

$$\frac{\mathrm{d}^2\sqrt{f}}{\mathrm{d}\chi^2} = n\sqrt{f}\,.$$

The solution, subject to the initial conditions $x \approx y \approx \chi$, $z \approx 0$ when $\chi \to 0$ is $f(\chi) = \cosh^2 n\chi$. This leads to the identity (126).

References

1. J.M. Bardeen, B. Carter, S.W. Hawking: Comm. Math. Phys. **31**, 161 (1973)
2. W. Israel: Phys. Rev. Lett. **57**, 397 (1986)
3. S.W. Hawking: Comm. Math. Phys. **43**, 199 (1975)
4. W.G. Unruh: Phys. Rev. **D14**, 870 (1976)
5. G.W. Gibbons, S.W. Hawking: Phys. Rev. **D15**, 2752 (1977)
6. W.H. Zurek, K.S. Thorne: Phys. Rev. Lett. **54**, 2171 (1985)
7. L. Bombelli, R.K. Koul, J. Lee, R.D. Sorkin: Phys. Rev. **D34**, 373 (1986)
8. G. 't Hooft: Nucl. Phys. **B256**, 727 (1985)
9. A.C. Ottewill, E. Winstanley: Phys. Rev. **D62**, 084018 (2000) gr-qc/0004012
10. B. P. Jensen, J.C. McLaughlin, A.C. Ottewill: Phys. Rev. **D45**, 3002 (1992)
    P.R. Andersson, W.A. Hiscock, D.A. Samuel: Phys. Rev. **D51**, 4337 (19795)
11. S. Mukohyama, W. Israel: Phys. Rev. **D58**, 104005 (1998) gr-qc/9806012
12. V.P. Frolov: Phys. Rev. Lett. **74**, 3319 (1995)
13. E. Winstanley: Phys. Rev. **D63**, 084013 (2001) hep-th/0011176
    V.P. Frolov, D. Fursaev: Class. Quantum. Grav. **15**, 2041 (1998) hep-th/9802085
14. F. Pretorius, D. Vollick, W. Israel: Phys. Rev. **D57**, 6311 (1998) gr-qc/9712085
15. W. Israel: Physica **106A**, 204 (1981)
    W. Israel: 'Covariant Fluid Mechanics and Thermodynamics'. In: *Lecture Notes in Mathematics #1385: Relativistic Fluid Dynamics.* ed. by A. Anile, Y. Choquet-Bruhat (Springer, Berlin, Heidelberg 1989)
16. N.J. Cerf, C. Adami: Phys. Rev. Lett. **79**, 5194 (1997)
    A. Mann, M. Rerzen, E. Santos: Phys. Lett. **A238**, 85 (1998)
17. C. Adami, N.J. Cerf: quant-ph/9904006 (1999)
18. Y. Takahashi, H. Umezawa: Collective Phenomena **2**, 55 (1975)
    H. Umezawa: *Advanced Field Theory* (AIP, New York, 1993)
19. V.P. Frolov, I.D. Novikov: Phys. Rev. **D48**, 4545 (1993)
20. J. Maldacena: hep-th/0106112 (2001)
    S.B. Giddings, A. Nudelman: hep-th/0112099 (2001)
    B.S. Kay: Class. Quantum. Grav. **15**, L89 (1998) hep-th/9810077

Perturbations of Black Holes

Valeria Ferrari

Dipartimento di Fisica "G.Marconi",
Università di Roma "La Sapienza" and Sezione INFN ROMA1,
piazzale Aldo Moro 2, I-00185 Roma, Italy

Abstract. The theory of non-charged, either static or rotating, black hole perturbations is reviewed in this chapter. Perturbations are classified into axial and polar. Their equations decouple and reduce to Schrödinger wave equations. Special attention is paid to quasi-normal modes for black holes and how they can be excited, since gravitational waves are emitted at their frequencies and they appear in many dynamical processes.

1 Introduction

The theory of perturbations is one of the most powerful tools to study how stars and black holes evolve, interact and "inform" the rest of the Universe about their changes sending gravitational waves to the outer space. The simplest formulation of the theory, known as the quadrupole formalism, considers small perturbations of a flat spacetime (weak field approximation):

$$g_{\mu\nu} = \eta_{\mu\nu} + h_{\mu\nu} \,, \qquad\qquad |h_{\mu\nu}| \ll 1 \,.$$

The Einstein equations are linearized about the flat background and solved under the assumption that the velocities of the bodies involved in the problem are much smaller than the speed of light (slow motion approximation). In this way, the perturbation can be shown to be related only to the time variation of the energy density of the source, i.e.

$$\begin{cases} \bar{h}^{\mu 0} = 0 \,, & \mu = 0,3 \\ \bar{h}^{ik}(t,r) = \dfrac{2G}{c^4 r}\left[\dfrac{\mathrm{d}^2}{\mathrm{d}t^2}\, q^{ik}\left(t - \dfrac{r}{c}\right)\right] & \end{cases} \,,$$

where $\bar{h}^{\mu\nu} = h^{\mu\nu} - \frac{1}{2}\eta^{\mu\nu}h$, and

$$q^{ik}(t) = \frac{1}{c^2}\int_V T^{00}(t, x^n) x^i x^k \mathrm{d}x^3$$

is the quadrupole moment of the source. Although this formalism is based on two very strong assumptions, i.e. that gravitational interactions do not dominate and that the velocities are small, it allows to estimate the amount of radiation emitted, for instance, by a triaxial, rotating neutron star and to predict the slowing down of its rotational period due to the energy loss; in addition, it has been successfully applied to compact binary systems to predict the variation of the orbital period. Indeed, because the energy is lost in gravitational waves, the orbit shrinks, and the period decreases by an amount which can be evaluated if the masses and the orbital parameters are known.

In 1975, Hulse and Taylor applied the quadrupole formalism to predict the slowing down of the period of the binary pulsar PSR 1913+16 [1]. They found $\mathrm{d}P/\mathrm{d}t = -2.4 \cdot 10^{-12}$, in excellent agreement with the observed value, $\mathrm{d}P/\mathrm{d}t = -(2.3 \pm 0.22) \cdot 10^{-12}$, thus providing the first indirect evidence of the existence of gravitational waves. It may be reminded that PSR 1913+16 is composed of two neutron stars, with masses $m_1 = 1.4411\ M_\odot$ and $m_2 = 1.3874\ M_\odot$, revolving on an eccentric orbit ($e = 0.617139$), with orbital separation $l_0 = 0.19 \cdot 10^{12}$ cm and keplerian frequency $\nu_\mathrm{k} = 3.583 \cdot 10^{-5}$ Hz.

However, there are cases when the weak field assumption and/or the slow motion approximation need to be released. The slow motion hypothesis implies that the wavelength of the emitted radiation must be much larger than the typical size of the source. This is certainly true for a binary system like PSR 1913+16, since in that case the radiation is emitted in several spectral lines at frequencies multiple of the orbital frequency, thus

$$\lambda_\mathrm{GW} \sim \frac{c}{\nu_\mathrm{k}} \sim 10^{15}\ \mathrm{cm}\ , \qquad \text{and} \qquad \lambda_\mathrm{GW} \gg l_0\ .$$

But if one wants to describe a neutron star pulsating in its fundamental mode of frequency $\nu_\mathrm{f} = 2 - 3$ kHz, we find that $\lambda_\mathrm{GW} \sim 10^7$ cm, and since the typical size of a neutron star is $D \sim 20$ km, $\lambda_\mathrm{GW} \sim D$.

In addition, there are cases when also the weak field assumption must be released, as for instance, when a mass is captured by a black hole and we want to find the signal which is emitted when the mass is close to the black hole horizon. In such cases, instead of considering perturbations of a flat spacetime one can consider perturbations of a given exact solution of Einstein's equations which can describe either a star or a black hole

$$g_{\mu\nu} = g^0_{\mu\nu} + h_{\mu\nu}\ , \qquad\qquad |h_{\mu\nu}| \ll |g^0_{\mu\nu}|\ ,$$

and solve the Einstein equations linearized about this background. In this chapter I will describe some of the most interesting issues related to the theory of black hole perturbations, whereas another chapter will be devoted to stellar perturbations. Since fifty years of work cannot be summarized in a few pages, I will select some topics that are, in my opinion, relevant. First of all I will explain in some detail how the equations governing the perturbations of a Schwarzschild black hole can be separated and reduced to simple wave equations of a Schrödinger type. The same procedure will be applied to non-rotating stars, and we will see that in that case a wave equation can be obtained only for the axial perturbations.

Then I will focus on the notion of quasi-normal modes (QNMs), which is central to the theory of black hole and stellar perturbations, because the frequencies of these modes are characteristic of many dynamical processes involving star and black hole oscillations, and because the gravitational radiation is emitted at these frequencies.

Finally I will show that the QNMs can be excited, discussing, as an example, the case of a point particle captured by a black hole. Only non-charged black holes will be considered in this chapter.

2 The Perturbations of a Schwarzschild Black Hole

The theory of perturbations of black holes was initiated by T. Regge and J.A. Wheeler in 1957 [2]. As they explain in their paper, the question they wanted to answer was the following:

"A sphere of water held together by gravitational forces is stable against small departures from sphericity. A sphere of water surrounded by a spherical shell of liquid mercury is also an equilibrium configuration for gravitational forces, but a situation of unstable equilibrium. Initial small departures from sphericity at the water-mercury interface will grow exponentially, and the mercury will concentrate with a rush at the center of the sphere. Which situation will more closely correspond to the behavior of a Schwarzschild singularity subjected to a small initial perturbation?"

Thus the initial motivations for studying the perturbations of a static black hole was to establish if the solution is stable. By expanding the perturbed metric tensor in tensorial spherical harmonics, Regge and Wheeler showed that the equations separate, and split into two decoupled sets, belonging to different "parities". As we shall later see explicitly, the parity is associated to the behaviour of the angular part of the perturbations under the transformation $\vartheta \to \pi - \vartheta$ and $\varphi \to \pi + \varphi$. In particular those that transform like $(-1)^{(\ell+1)}$ are said to be *odd*, or *axial*, and those that transform like $(-1)^{(\ell)}$ are said to be *even*, or *polar*. The stability of the Schwarzschild solution was studied by resolving the perturbations into proper modes, which satisfy the condition that the radial part of the perturbation is well behaved both at radial infinity and at the horizon, and finding for each mode the corresponding eigenfrequency. Since imaginary proper frequencies would correspond to an unacceptable space behaviour, they concluded that a Schwarzschild black hole is stable against axial perturbations. The stability of the polar perturbations was not considered in their paper. Regge and Wheeler also showed that the equations for the radial part of the axial perturbations of a Schwarzschild black hole can be reduced to a single Schrödinger-like wave equation with a real potential barrier.

A substantial advance in the field was done more than ten years later, in 1970, when F. Zerilli succeeded in reducing the equations of the polar perturbations to a single wave equation. He also computed the source term when the perturbation is excited by an infalling massive test-particle [3]. It is interesting to remind that during those years J. Weber had started his experiments with resonant gravitational detectors, opening the way to the detection of gravitational waves, and stimulating a new interest in the theory of perturbations of stars and black holes.

The inhomogeneous equation derived by Zerilli allowed to compute the spectrum and the waveform of the signal emitted when masses are scattered or captured by a Schwarzschild black hole, and this was a very interesting information in view of a possible detection.

I shall now show explicitly how the perturbed equations can be separated and reduced to wave equations. In the following I shall assume that the background metric is the Schwarzschild metric

$$\mathrm{d}s^2 = \mathrm{e}^{2\nu}\mathrm{d}t^2 - \mathrm{e}^{2\mu_2}\mathrm{d}r^2 - \mathrm{e}^{2\mu_3}\mathrm{d}\vartheta^2 - \mathrm{e}^{2\psi}\mathrm{d}\varphi^2 \tag{1}$$

with $\mathrm{e}^{2\nu} = \mathrm{e}^{-2\mu_2} = (1 - 2M/r)$ and $\mathrm{e}^{2\mu_3} = r^2$, $\mathrm{e}^{2\psi} = r^2 \sin^2 \vartheta$.

2.1 The Separation of the Perturbed Equations

The first step to separate the angular part of the perturbed Einstein equations is to expand the metric perturbation $h_{\mu\nu}$ and the stress-energy tensor of the source $T^{\mu\nu}$, if present, into tensorial spherical harmonics. A suitable basis to expand symmetric tensors is given by the following tensor harmonics [2,3]

$$\begin{aligned}
\mathbf{a}_{\ell m} &= [\mathbf{e}_r \mathbf{e}_r \; Y_{\ell m}] \\
\mathbf{b}_{\ell m} &= 2^{1/2} n(\ell) r \, [\mathbf{e}_r \boldsymbol{\nabla} \; Y_{\ell m}] \\
\mathbf{c}_{\ell m} &= 2^{1/2} n(\ell) \, [\mathbf{e}_r \mathbf{L} \; Y_{\ell m}] \\
\mathbf{d}_{\ell m} &= 2^{1/2} m(\ell) r \left\{ [\mathbf{L} \boldsymbol{\nabla} \; Y_{\ell m}] + \frac{1}{r} [\mathbf{e}_r \mathbf{L} \; Y_{\ell m}] \right\} \\
\mathbf{f}_{\ell m} &= 2^{-1/2} m(\ell) \, (\mathbf{e}_{\ell m} + \mathbf{h}_{\ell m}) \\
\mathbf{g}_{\ell m} &= -2^{-1/2} n(\ell)^2 \, (\mathbf{e}_{\ell m} - \mathbf{h}_{\ell m}) \\
\mathbf{a}^{(0)}_{\ell m} &= [\mathbf{e}_t \mathbf{e}_t \; Y_{\ell m}] \\
\mathbf{a}^{(1)}_{\ell m} &= 2^{1/2} \, [\mathbf{e}_t \mathbf{e}_r \; Y_{\ell m}] \\
\mathbf{b}^{(0)}_{\ell m} &= 2^{1/2} n(\ell) r \, [\mathbf{e}_t \boldsymbol{\nabla} \; Y_{\ell m}] \\
\mathbf{c}^{(0)}_{\ell m} &= 2^{1/2} n(\ell) \, [\mathbf{e}_t \mathbf{L} \; Y_{\ell m}]
\end{aligned}$$

where

$$\begin{aligned}
\mathbf{e}_{\ell m} &= r^2 \left\{ [\boldsymbol{\nabla}\boldsymbol{\nabla} \; Y_{\ell m}] + \frac{2}{r} [\mathbf{e}_r \boldsymbol{\nabla} \; Y_{\ell m}] \right\} \\
\mathbf{h}_{\ell m} &= [\mathbf{L}\mathbf{L} \; Y_{\ell m}] + r \, [\mathbf{e}_r \boldsymbol{\nabla} \; Y_{\ell m}] \; ,
\end{aligned}$$

$$\begin{aligned}
n(\ell) &= [\ell(\ell+1)]^{-1/2} \\
m(\ell) &= [\ell(\ell+1)(\ell-1)(\ell+2)]^{-1/2} \; ,
\end{aligned}$$

and $\boldsymbol{\nabla}$ is the operator of covariant derivative, $\mathbf{e}_r$ is the unit vector along the radial direction, $\mathbf{e}_t = (1, 0, 0, 0)$, and $\mathbf{L}$ is the angular momentum operator

$$\mathbf{L} = -i \mathbf{r} \wedge \boldsymbol{\nabla} \; .$$

Choosing this basis, any symmetric tensor can be expanded as follows

$$\mathbf{T} = \sum_{\ell m} \Big[A^{(0)}_{\ell m} \mathbf{a}^{(0)}_{\ell m} + A^{(1)}_{\ell m} \mathbf{a}^{(1)}_{\ell m} + A_{\ell m} \mathbf{a}_{\ell m}$$
$$+ B^{(0)}_{\ell m} \mathbf{b}^{(0)}_{\ell m} + B_{\ell m} \mathbf{b}_{\ell m} + Q^{(0)}_{\ell m} \mathbf{c}^{(0)}_{\ell m}$$
$$+ Q_{\ell m} \mathbf{c}_{\ell m} + G_{\ell m} \mathbf{g}_{\ell m} + D_{\ell m} \mathbf{d}_{\ell m} + F_{\ell m} \mathbf{f}_{\ell m} \Big] ,$$

where the coefficients of the expansion are given by the inner product between the tensor and the corresponding harmonic; for instance

$$A^{(0)}_{\ell m} = \int \mathbf{a}^{(0)\mu\nu *}_{\ell m} \, T_{\mu\nu} \, \mathrm{d}\Omega ,$$

and $\mathrm{d}\Omega$ is the solid angle element. The explicit expressions of the tensor harmonics are given in the Appendix.

Parity. If we apply the parity operator which transforms

$$\theta \to \pi - \theta \qquad \text{and} \qquad \varphi \to \pi + \varphi ,$$

$\mathbf{c}_{\ell m}$, $\mathbf{d}_{\ell m}$, $\mathbf{c}^{(0)}_{\ell m}$ transform like $(-1)^{(\ell+1)}$ and are said *axial* or *odd*, whereas $\mathbf{a}_{\ell m}$, $\mathbf{b}_{\ell m}$, $\mathbf{f}_{\ell m}$, $\mathbf{g}_{\ell m}$, $\mathbf{a}^{(0)}_{\ell m}$, $\mathbf{a}^{(1)}_{\ell m}$, $\mathbf{b}^{(0)}_{\ell m}$ transform like $(-1)^{(\ell)}$ and are said *polar* or *even*. Consequently, any symmetric tensor $\mathbf{T}$ expanded in harmonics has an axial and a polar part; for this reason when we expand the perturbed metric tensor we find

$$\mathbf{h}^{\mathrm{ax}} = \sum_{\ell m} \left[Q^{(0)}_{\ell m} \mathbf{c}^{(0)}_{\ell m} + Q_{\ell m} \mathbf{c}_{\ell m} + D_{\ell m} \mathbf{d}_{\ell m} \right] ,$$

$$\mathbf{h}^{\mathrm{pol}} = \sum_{\ell m} \Big[A^{(0)}_{\ell m} \mathbf{a}^{(0)}_{\ell m} + A^{(1)}_{\ell m} \mathbf{a}^{(1)}_{\ell m} + A_{\ell m} \mathbf{a}_{\ell m}$$
$$+ B^{(0)}_{\ell m} \mathbf{b}^{(0)}_{\ell m} + B_{\ell m} \mathbf{b}_{\ell m} + G_{\ell m} \mathbf{g}_{\ell m} + F_{\ell m} \mathbf{f}_{\ell m} \Big] ,$$

The coefficients $Q^{(0)}_{\ell m}, Q_{\ell m}, \ldots$ etc. have to be found by solving the Einstein equations, or, if we consider a star, the Einstein equations coupled to the equations of Hydrodynamics. It appears convenient to write the coefficients in the following form

$$Q^{(0)}_{\ell m} = \frac{\sqrt{2}\mathrm{i}}{n(l)r} h^{\mathrm{ax}}_{0\ell m} \qquad Q_{\ell m} = \frac{\sqrt{2}\mathrm{i}}{n(l)r} h^{\mathrm{ax}}_{1\ell m}$$

$$D_{\ell m} = -\frac{\mathrm{i}}{\sqrt{2}m(l)r^2} h_{2\ell m} \qquad A^{(0)}_{\ell m} = 2N_{\ell m}\mathrm{e}^{2\nu}$$

$$A^{(1)}_{\ell m} = -\sqrt{2}H_{1\ell m} \qquad A_{\ell m} = -2L_{\ell m}\mathrm{e}^{2\mu_2}$$

$$B^{(0)}_{\ell m} = -\frac{\sqrt{2}}{n(l)r} h_{0\ell m} \qquad B_{\ell m} = \frac{\sqrt{2}}{n(l)r} h_{1\ell m}$$

$$F_{\ell m} = -\frac{\sqrt{2}}{m(l)} V_{\ell m} \qquad G_{\ell m} = \sqrt{2}[l(l+1)V_{\ell m} - 2T_{\ell m}]$$

With this choice the perturbed metric tensor takes the form (from now on, we shall omit the subscript ℓm from the metric perturbations)

$$h^{\mathrm{ax}}_{\ell m} = \begin{pmatrix} (t) & (\varphi) & (r) & (\vartheta) \\ 0 & h^{\mathrm{ax}}_0 \sin\vartheta \frac{\partial Y_{\ell m}}{\partial\vartheta} & 0 & -h^{\mathrm{ax}}_0 \frac{1}{\sin\vartheta}\frac{\partial Y_{\ell m}}{\partial\varphi} \\ h^{\mathrm{ax}}_0 \sin\vartheta \frac{\partial Y_{\ell m}}{\partial\vartheta} & -\frac{1}{2} h_2 \sin\vartheta X_{\ell m} & h^{\mathrm{ax}}_1 \sin\vartheta \frac{\partial Y_{\ell m}}{\partial\vartheta} & -\frac{1}{2} h_2 \sin\vartheta W_{\ell m} \\ 0 & h^{\mathrm{ax}}_1 \sin\vartheta \frac{\partial Y_{\ell m}}{\partial\vartheta} & 0 & -h^{\mathrm{ax}}_1 \frac{1}{\sin\vartheta}\frac{\partial Y_{\ell m}}{\partial\varphi} \\ -h^{\mathrm{ax}}_0 \frac{1}{\sin\vartheta}\frac{\partial Y_{\ell m}}{\partial\varphi} & -\frac{1}{2} h_2 \sin\vartheta W_{\ell m} & -h^{\mathrm{ax}}_1 \frac{1}{\sin\vartheta}\frac{\partial Y_{\ell m}}{\partial\varphi} & \frac{1}{2} h_2 \frac{1}{\sin\vartheta} X_{\ell m} \end{pmatrix}, \tag{2}$$

$$h^{\mathrm{pol}}_{\ell m} = \begin{pmatrix} (t) & (\varphi) & (r) & (\vartheta) \\ 2\mathrm{e}^{2\nu} N Y_{\ell m} & -h_0 \frac{\partial Y_{\ell m}}{\partial\varphi} & -H_1 Y_{\ell m} & -h_0 \frac{\partial Y_{\ell m}}{\partial\vartheta} \\ -h_0 \frac{\partial Y_{\ell m}}{\partial\varphi} & -2\mathrm{e}^{2\psi} H_{11} & h_1 \frac{\partial Y_{\ell m}}{\partial\varphi} & -r^2 V X_{\ell m} \\ -H_1 Y_{\ell m} & h_1 \frac{\partial Y_{\ell m}}{\partial\varphi} & -2\mathrm{e}^{2\mu_2} L Y_{\ell m} & h_1 \frac{\partial Y_{\ell m}}{\partial\vartheta} \\ -h_0 \frac{\partial Y_{\ell m}}{\partial\vartheta} & -r^2 V X_{\ell m} & h_1 \frac{\partial Y_{\ell m}}{\partial\vartheta} & -2\mathrm{e}^{2\mu_3} H_{33} \end{pmatrix}, \tag{3}$$

where

$$H_{11} = \left[T Y_{\ell m} + V \left(\frac{1}{\sin^2\vartheta}\frac{\partial^2}{\partial\varphi^2} + \cot\vartheta \frac{\partial}{\partial\vartheta} \right) Y_{\ell m} \right],$$
$$H_{33} = \left[T Y_{\ell m} + V \frac{\partial^2}{\partial\vartheta^2} Y_{\ell m} \right].$$

At this point we can fix the gauge: we make an infinitesimal coordinate transformation

$$x'^i = x^i + \xi^i, \tag{4}$$

expand the vector $\boldsymbol{xi}$ into vector harmonics:

$$\xi(t,\varphi,r,\vartheta) = \sum_{\ell m} \Big\{ \xi^{(0)}_{\ell m} [\boldsymbol{\nabla}\, Y_{\ell m}] + \xi^{(1)}_{\ell m} [\mathbf{L}\, Y_{\ell m}] + \xi^{(2)}_{\ell m} [\mathbf{e}_r\, Y_{\ell m}] + \xi_{\ell m} [\mathbf{e}_t\, Y_{\ell m}] \Big\}, \tag{5}$$

where the vector harmonic $[\mathbf{L}\ Y_{\ell m}]$ is axial, and the remaining three $[\boldsymbol{\nabla}\ Y_{\ell m}]$, $[\mathbf{e}_r\ Y_{\ell m}]$, $[\mathbf{e}_t\ Y_{\ell m}]$, are polar. Thus, the components of $\boldsymbol{\xi}$ can be used to eliminate four components of $h_{\mu\nu}$. For instance, we can set

$$h_2^{\mathrm{ax}} = H_1^{\mathrm{pol}} = h_0^{\mathrm{pol}} = h_1^{\mathrm{pol}} = 0,$$

and with this choice the perturbed metric becomes

$$h_{\ell m}^{\mathrm{ax}} = \begin{pmatrix} (t) & (\varphi) & (r) & (\vartheta) \\ 0 & h_0^{\mathrm{ax}} \sin\vartheta \dfrac{\partial Y_{\ell m}}{\partial \vartheta} & 0 & -h_0^{\mathrm{ax}} \dfrac{1}{\sin\vartheta} \dfrac{\partial Y_{\ell m}}{\partial \varphi} \\ h_0^{\mathrm{ax}} \sin\vartheta \dfrac{\partial Y_{\ell m}}{\partial \vartheta} & 0 & h_1^{\mathrm{ax}} \sin\vartheta \dfrac{\partial Y_{\ell m}}{\partial \vartheta} & 0 \\ 0 & h_1^{\mathrm{ax}} \sin\vartheta \dfrac{\partial Y_{\ell m}}{\partial \vartheta} & 0 & -h_1^{\mathrm{ax}} \dfrac{1}{\sin\vartheta} \dfrac{\partial Y_{\ell m}}{\partial \varphi} \\ -h_0^{\mathrm{ax}} \dfrac{1}{\sin\vartheta} \dfrac{\partial Y_{\ell m}}{\partial \varphi} & 0 & -h_1^{\mathrm{ax}} \dfrac{1}{\sin\vartheta} \dfrac{\partial Y_{\ell m}}{\partial \varphi} & 0 \end{pmatrix}, \tag{6}$$

$$h_{\ell m}^{\mathrm{pol}} = \begin{pmatrix} (t) & (\varphi) & (r) & (\vartheta) \\ 2\mathrm{e}^{2\nu} N Y_{\ell m} & 0 & 0 & 0 \\ 0 & -2\mathrm{e}^{2\psi} H_{11} & 0 & -r^2 V X_{\ell m} \\ 0 & 0 & -2\mathrm{e}^{2\mu_2} L Y_{\ell m} & 0 \\ 0 & -r^2 V X_{\ell m} & 0 & -2\mathrm{e}^{2\mu_3} H_{33} \end{pmatrix}. \tag{7}$$

We are now in a position to compute the components of the Einstein equations for the metric written above. At the time Regge, Wheeler and Zerilli did it, it was a quite formidable task; today we can do it very easily by using any program of symbolic manipulation on a PC.

2.2 The Axial and Polar Equations

The relevant equations for the axial perturbations are:

$$\begin{aligned} \delta G_{\vartheta\varphi} = 0 : \quad & \frac{\sin\vartheta}{2} \left(\partial_\vartheta^2 - \cot\vartheta \partial_\vartheta - \frac{1}{\sin\vartheta^2} \partial_{\varphi^2} \right) Y_{\ell m}(\vartheta, \varphi) \\ & \cdot \left\{ -\mathrm{i}\omega \mathrm{e}^{-2\nu} h_0^{\mathrm{ax}} - \mathrm{e}^{-2\mu_2} \left[h_{1,r}^{\mathrm{ax}} + (\nu - \mu_2)_{,r} h_1^{\mathrm{ax}} \right] \right\} = 0 \end{aligned} \tag{8}$$

$$\delta G_{r\vartheta}=0: \quad -\frac{1}{2\sin\vartheta}\partial_\varphi Y_{\ell m}(\vartheta,\varphi)\left\{e^{-2\nu}\left[\omega^2 h_0^{\rm ax}-i\omega\left(h_{0,r}^{\rm ax}-\frac{2}{r}h_0^{\rm ax}\right)\right]\right.$$
$$\left.-\frac{2n}{r^2}h_1^{\rm ax}-2e^{-2\mu_2}h_1^{\rm ax}\left[\nu_{,rr}+\left(\frac{1}{r}+\nu_{,r}\right)(\nu-\mu_2)_{,r}\right]\right\}=0\,.$$

In addition the equilibrium equation $G_{\vartheta\vartheta}=0$ gives

$$\nu_{,rr}+\left(\frac{1}{r}+\nu_{,r}\right)(\nu-\mu_2)_{,r}=0\,,$$

which allows to write the perturbed equations as

$$-i\omega e^{-2\nu}h_0^{\rm ax}-e^{-2\mu_2}\left[h_{1,r}^{\rm ax}+(\nu-\mu_2)_{,r}h_1^{\rm ax}\right]=0\,,$$
$$e^{-2\nu}\left[-i\omega\left(h_{0,r}^{\rm ax}-\frac{2}{r}h_0^{\rm ax}\right)+\omega^2 h_1^{\rm ax}\right]-\frac{2n}{r^2}h_1^{\rm ax}=0\,. \tag{9}$$

By operating in a similar way on the polar equations we find

$$\delta G_{tr}: \quad \left[\frac{d}{dr}+\left(\frac{1}{r}-\nu_{,r}\right)\right](2T-kV)-\frac{2}{r}L=0$$
$$\delta G_{t\vartheta}: \quad T-V+L=0$$
$$\delta G_{r\vartheta}: \quad (T-V+N)_r-\left(\frac{1}{r}-\nu_{,r}\right)N-\left(\frac{1}{r}+\nu_{,r}\right)L=0$$
$$\delta G_{rr}: \quad \frac{2}{r}N_{,r}+\left(\frac{1}{r}+\nu_{,r}\right)(2T-kV)_{,r}-\frac{2}{r}\left(\frac{1}{r}+2\nu_{,r}\right)L$$
$$-\frac{1}{r^2}(2nT+kN)e^{2\mu_2}+\omega^2 e^{-2\nu+2\mu_2}(2T-kV)=0$$
$$\frac{\delta G_{\vartheta\vartheta}-\delta G_{\varphi\varphi}}{\sin^2\vartheta}: \quad V_{,rr}+\left(\frac{2}{r}+\nu_{,r}-\mu_{2,r}\right)V_r+\frac{e^{2\mu_2}}{r^2}(N+L)$$
$$+\omega^2 e^{2\mu_2-2\nu}V=0 \tag{10}$$

where $k=\ell(\ell+1)$, and $2n=k-2$. In writing these equations, we have Fourier-expanded all perturbed variables

$$f(t)=\int_{-\infty}^{+\infty} f(\omega)e^{-i\omega t}d\omega$$

2.3 A Wave Equation for the Axial and Polar Perturbations of a Schwarzschild Black Hole

The two equations for the axial perturbations (9) can be reduced to a single wave equation by eliminating $h_0^{\rm ax}$ and by introducing the function

$$Z^{\rm ax}(\omega,r)=e^{\nu-\mu_2}\frac{h_1^{\rm ax}(\omega,r)}{r}\,.$$

The equation for $Z^{\rm ax}$ is

$$\frac{{\rm d}^2 Z^{\rm ax}}{{\rm d}r_*^2} + [\omega^2 - V_\ell^{\rm ax}(r)]Z^{\rm ax} = 0 \; , \tag{11}$$

which is known as the Regge-Wheeler equation [2], where $r_* = r + 2M \log(r/2M - 1)$, and

$$V^{\rm ax}(r) = \frac{{\rm e}^{2\nu}}{r^3}[\ell(\ell+1)r + r^3 - 6M] \; .$$

The reduction of the polar equations to a single wave equation is much more difficult, and requires a repeated use of the equilibrium equation. However also in that case it is possible to show that, by introducing a function $Z^{\rm pol}$ defined as

$$Z^{\rm pol}(\omega, r) = \frac{r}{nr + 3M}\Big(3MV(\omega, r) - rL(\omega, r)\Big) \; ,$$

the polar equations reduce to the Zerilli equation [3]

$$\frac{{\rm d}^2 Z^{\rm pol}}{{\rm d}r_*^2} + [\omega^2 - V_\ell^{\rm pol}(r)]Z^{\rm pol} = 0 \; , \tag{12}$$

where

$$V^{\rm pol}(r) = \frac{2(r-2M)}{r^4(nr+3M)^2}[n^2(n+1)r^3 + 3Mn^2r^2 + 9M^2nr + 9M^3] \; .$$

It must be stressed that the two potential barriers depend only on the black hole mass. They have a different analytic form, but similar shape; indeed they vanish at $\pm\infty$ like

$$V \to {\rm e}^{r_*/2M} \qquad (r_* \to -\infty) \; , \qquad\qquad V \to \frac{1}{r^2} \qquad (r_* \to +\infty) \; , \tag{13}$$

as it is shown in Fig. 1.

The curvature generated by a pointlike mass appears in the perturbed equations as a one-dimensional potential barrier and consequently the response of a black hole to generic perturbations can be studied by investigating the manner in which a gravitational wave incident on that barrier is transmitted, absorbed and reflected, a phenomenon which is familiar in elementary quantum theory.

Thus, the theory of black hole perturbations can be formulated as a scattering theory, which has been beautifully illustrated by S. Chandrasekhar in the book *The mathematical theory of black holes* [4].
We shall now introduce the important notion of *quasi-normal modes.*

3 The Quasi-normal Modes

In 1970 Vishveshwara [5] pointed out that the equations governing the perturbations of a Schwarzschild black hole should allow complex frequency solutions which satisfy the boundary condition of a pure outgoing wave at infinity. This idea was confirmed by Press [6] who found that an arbitrary initial perturbation decays as a pure frequency mode. However only in 1975 Chandrasekhar and Detweiler [7] actually computed the discrete eigenfrequencies of these modes and since then the concept of quasi-normal modes (QNMs) has been central to the theory of perturbations of stars and black holes.

In the case of black holes these modes are defined to be solutions of the axial and polar wave equations that satisfy the boundary conditions of a pure outgoing wave at infinity and of a pure ingoing wave at the black holes horizon, the latter corresponding to the requirement that nothing can escape from the horizon:

$$\begin{aligned} Z &\to \mathrm{e}^{\mathrm{i}\omega r_*}\,, \qquad r_* \to -\infty\,, \\ Z &\to \mathrm{e}^{-\mathrm{i}\omega r_*}\,, \qquad r_* \to +\infty\,, \end{aligned}$$

(as it is shown in Fig. 1).

In the scattering theory these boundary conditions identify the singularities of the reflection amplitude associated to the potential barrier. Since the oscillations are damped by the emission of gravitational waves, the eigenfrequencies of the QNMs are complex $\omega = \omega_0 + \mathrm{i}\omega_i$. The real part is the pulsation rate, the imaginary part is the inverse of the damping time.

They depend exclusively on the parameters that identify the spacetime geometry, i.e. the mass, and, if the black hole rotates or is charged, the angular momentum and the charge. Consequently these frequencies will be characteristic of many processes involving the dynamical perturbations of black holes.

The potential barriers of the Regge-Wheeler and of the Zerilli equations admit the same transmission and reflection coefficients, and consequently the axial and polar perturbations are isospectral [4]. As we shall see in the chapter on stellar perturbations, this is not true for stars! It follows that the spectrum of the axial and polar gravitational waves emitted by a black hole will be peaked at exactly the same frequencies, providing a clear signature of the nature of the source [8].

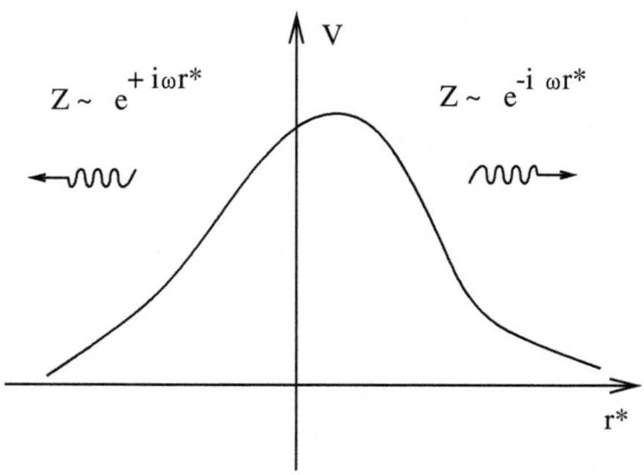

Fig. 1. The potential barrier of the wave equations

The frequency of the lowest QNMs were found by Chandrasekhar and Detweiler by transforming the wave equations into a Riccati equation [7] as follows

$$\frac{\mathrm{d}^2 Z}{\mathrm{d}r_*^2} + [\omega^2 - V_\ell(r)]Z = 0 \,, \text{ putting } Z = \exp\left\{\mathrm{i}\int^{r*} \Phi \,\mathrm{d}r*\right\} \,,$$

the equation becomes

$$\mathrm{i}\Phi_{,r} - \Phi^2 + \omega^2 - V = 0 \,; \tag{14}$$

the QNMs correspond to the solutions of (14) that satisfy the boundary conditions

$$r* \to +\infty \qquad \Phi \to -\omega \,, \qquad r* \to -\infty \qquad \Phi \to +\omega \,, \qquad \omega = \omega_0 + \mathrm{i}\omega_i \,.$$

The values of the first few eigenfrequencies for $\ell = 2,\ 3$ are given in Table 1 in units of the black hole mass.

Table 1. The values of the lowest quasi-normal mode frequencies for $\ell = 2$, 3, are given in units of the black hole mass

	$M\omega + \mathrm{i}M\omega$		$M\omega + \mathrm{i}M\omega$
$\ell = 2$	0.3737+i0.0890	$\ell = 3$	0.5994+i0.0927
	0.3467+i0.2739		0.5826+i0.2813
	0.3011+i0.4783		0.5517+i0.4791
	0.2515+i0.7051		0.5120+i0.6903

For example, if a black hole has a mass of 1 $M_\odot$ or if it is a supermassive black hole with $M = 10^6\ M_\odot$, the frequencies and damping times of the lowest QNMs are, respectively,

$$\begin{aligned} M &= 1M_\odot \,, & \nu_0 &= 12.06 \text{ kHz} \,, & \tau &= 5.55 \cdot 10^{-5} \text{ s} \,, \\ M &= 10^6 M_\odot \,, & \nu_0 &= 1.21 \cdot 10^{-2} \text{ Hz} \,, & \tau &= 55.5 \text{ s} \,. \end{aligned}$$

The gravitational signal emitted by a perturbed black hole will, during its last stages, decay as a superposition of the first few quasi-normal modes. Thus, the frequencies of the lowest modes are utmost significant physically and they are rather easy to compute. Conversely, great care must be used to determine the entire spectrum. For instance, a WKB approximation has also been used to solve for the lowest modes [9–12] and a higher order WKB approach has been developed to find the frequencies of the higher order modes [13]. To complete the study of the QNM spectrum, the method of continued fraction, the phase-integral method and the theory of the Regge poles have also been applied; a complete bibliography on the subject can be found in [14].

The results of this extensive study can be summarized shortly as follows

- For any value of the harmonic index ℓ the real part of the frequency, $\omega_{0n}(\ell)$, approaches a non zero limiting value as the order n of the mode increases.
- The imaginary part, $\omega_{in}(\ell)$, increases linearly as $n/4$.
- The asymptotic behaviour is, to a high accuracy, independent of ℓ.

4 Perturbations of a Kerr Black Hole

The perturbations of a Schwarzschild black hole described in Sect. 2 have been studied in terms of the perturbed metric tensor. An alternative approach considers the perturbations of the Weyl, Maxwell and Ricci scalars through the equations of the Newman-Penrose formalism. Using this latter approach in 1972 Teukolsky [15], [16] was able to decouple and separate the equations governing the perturbations of a Kerr black hole and to reduce them to a single master equation for the radial part of the perturbation R_{lm}:

$$\Delta R_{lm,rr} + 2(s+1)(r-M)R_{lm,r} + V(\omega,r)R_{lm} = 0$$
$$\Delta = r^2 - 2Mr + a^2 \ .$$

The potential $V(\omega,r)$ is

$$V(\omega,r) = \frac{1}{\Delta}\left[(r^2+a^2)^2\omega^2 - 4aMrm\omega + a^2m^2 \right.$$
$$\left. + 2\mathrm{i}s(am(r-M) - M\omega(r^2-a^2))\right] + \left[2\mathrm{i}s\omega r - a^2\omega^2 - A_{lm}\right] \ .$$

The angular part of the perturbations, S_{lm}, satisfies the equations of the oblate spheroidal harmonics

$$\begin{cases} [(1-u^2)S_{lm,u}]_{,u} + \left[a^2\omega^2u^2 - 2am\omega su + s + A_{lm} - \dfrac{(m+su)^2}{1-u^2}\right] S_{lm} = 0 \ , \\ u = \cos\theta \ . \end{cases}$$

The complete perturbation is

$$\psi_s(t,r,\theta\varphi) = \frac{1}{2\pi}\int \mathrm{e}^{-\mathrm{i}\omega t} \sum_{l=|s|}^{\infty} \sum_{m=-l}^{l} \mathrm{e}^{\mathrm{i}m\varphi} S_{lm}(u)R_{lm}(r)\,\mathrm{d}\omega \ , \tag{15}$$

where $s =$ is the spin-weight parameter: $s = 0, \pm 1, \pm 2$, for scalar, electromagnetic and gravitational perturbations, and A_{lm} is a separation constant.

It should be stressed that, unlike the potential barrier of a Schwarzschild black hole which is real and independent of the harmonic index m, the potential barrier of a Kerr black hole is complex and depends on the frequency and on m.

As a consequence, an interesting phenomenon occurs when electromagnetic or gravitational waves are scattered on the complex potential barrier of a rotating black hole; if the incident wave has a frequency in the range

$$0 < \omega < \omega_c \ , \quad \text{where} \quad \omega_c = \frac{am}{2Mr_+} \ , \quad m > 0 \ , \tag{16}$$

the reflection coefficient associated to the potential barrier exceeds unity [17]. This phenomenon is called *superradiance*, and it is the analogue, in the domain of wave propagation, of the Penrose process in the domain of particle creation.

4.1 The Quasi-normal Modes of a Kerr Black Hole

The quasi-normal frequencies of a Kerr black hole have been first determined by Detweiler [18], and subsequently by Leaver [19], Seidel and Iyer [20] and Kokkotas [21].

Since the rotation removes the degeneracy presented by the Schwarzschild modes, different eigenfrequencies are expected for different values of the harmonic index m. The calculations show that when a increases, $\omega_{0n}(a, \ell, m)$ is bounded but the imaginary part is not. Moreover, when $a \to M$ the more highly damped frequencies coalesce to the purely real value of the critical frequency for superradiant scattering $\omega_c = m/2M$.

In this context, an interesting result was obtained by Detweiler in 1977. He found that when a Kerr black hole becomes "extreme", i.e. when $a \to M$, the imaginary part of the frequency of its quasi-normal modes tends to zero. If excited, this mode would set the black hole into an oscillation that would never decay, suggesting that extreme Kerr black holes are "marginally unstable". It was subsequently shown by Mashoon and the author [22] that when $a \to M$, the amplitude of these modes tends to zero, and consequently quasi-normal modes with a real frequency cannot exist in the ordinary regime.

5 Can the Quasi-normal Modes be Excited?

In the preceding sections we have considered source-free perturbations. We shall now assume that the source of the perturbations is, for example, a mass falling into a Schwarzschild black hole. We shall impose that the infalling mass is much smaller than the black hole mass, $m_0 \ll M$.

Under this assumption m_0 can be assumed to move on a geodesic of the unperturbed spacetime and its stress-energy tensor

$$T^{\mu\nu} = m_0 \frac{\mathrm{d}T}{\mathrm{d}\tau}\frac{\mathrm{d}z^\mu}{\mathrm{d}\tau}\frac{\mathrm{d}z^\nu}{\mathrm{d}\tau}\frac{\delta(r - R(t))}{r^2}\,\delta^{(2)}[\Omega - \Omega(t)] \tag{17}$$

will act as a source of the first order perturbed Einstein equations

$$\delta G_{\mu\nu} = \frac{8\pi G}{c^4} T_{\mu\nu} \; .$$

In (17) $\delta^{(2)}[\Omega - \Omega(t)] = \delta[\cos\theta - \cos\Theta]\delta[\varphi - \Phi]$, and τ is the proper time of m_0 along the geodesic

$$z^\mu = [T(\tau), R(\tau), \Phi(\tau), \Theta(\tau)] \; .$$

As an example, we shall consider the case of a test-particle falling radially into the black hole. In this case, it will not excite the axial modes since the source

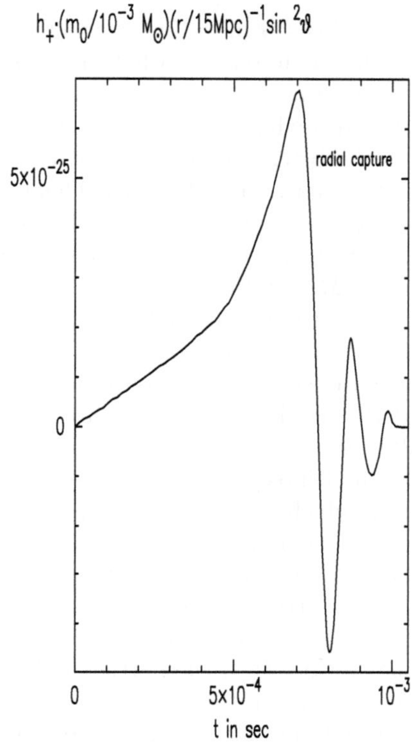

Fig. 2. The $\ell = 2$ gravitational signal emitted when a mass $m_0 = 10^{-3}\ M_\odot$ falls radially into a black hole of mass $M = 1.5\ M_\odot$, located at a distance of 15 Mpc

term of the Regge-Wheeler equation vanishes. After expanding the stress-energy tensor in spherical harmonics and Fourier-transforming all time dependent quantities, the equations for the polar perturbations can be written as

$$\frac{\mathrm{d}^2 Z^{\mathrm{pol}}(\omega, r_*)}{\mathrm{d}r_*^2} + [\omega^2 - V^{\mathrm{pol}}] Z_\ell = S^{\mathrm{pol}}((\omega, r_*),$$

where

$$S^{\mathrm{pol}} = \frac{4m_0(\ell + 1/2)^{1/2}}{nr + 3M}\left(1 - \frac{2M}{r}\right) \cdot \left[\left(\gamma_0^2 - 1 + \frac{2M}{r}\right)^{-1/2} - \frac{2\mathrm{i}n\gamma_0}{\omega(nr + 3M)}\right] \mathrm{e}^{\mathrm{i}\omega T(r)},$$

$\gamma_0 = (1 - v_\infty^2)^{-1/2}$, and v_∞^2 is the velocity of the particle at infinity [23]. These equations can be numerically integrated by imposing the boundary conditions of pure outgoing wave at radial infinity and of pure ingoing wave at the black hole horizon.

The structure of the signal is shown in Fig. 2, where the waveform is plotted for a mass $m_0 = 10^{-3}\ M_\odot$ falling into a black hole of mass $M = 1.5\ M_\odot$, located at a distance of 15 Mpc from Earth.

The waveform exhibits an oscillating tail, emitted when $2 < r/M < 4.5$, i.e. when m_0 is very close to the black hole horizon, which can be fitted with the superposition of the first two quasi-normal modes. This means that these modes are excited by the infalling mass and that the perturbed black hole oscillates until its mechanical energy is radiated away in gravitational waves. In this case the total radiated energy is $\Delta E \sim 0.01\ (m_0^2/M)$.

It should be stressed that the quadrupole formalism would not be able to reproduce the ringing part of the signal shown in Fig. 2.

Similar results can be obtained if m_0 falls spiralling into the black hole. In this case both the polar and the axial perturbations are excited (though the axial perturbations are usually less energetic than the polar ones), and due to the contribution of the higher multipoles, the total energy radiated in gravitational waves increases by a factor which, depending on the angular momentum of the infalling mass, can be as high as 50 [24].

If the black hole rotates, [25] the situation can be summarized as follows. For radial capture along the symmetry axis, the energy emitted is larger than that emitted in the non-rotating case. For example, if $a = 0.99M$ the energy is about 1.65 times larger than that for $a = 0$. If the particle falls in the equatorial plane with a vanishing angular momentum, the energy is about 4.27 times larger than that for $a = 0$. The particle will, in general, excite the quasi-normal modes of the black hole and more energy is emitted if the particle is co-rotating than if it is counter-rotating. Quasi-normal modes are barely excited when a particle, starting at rest at infinity, is scattered by a rotating black hole, unless the periastron is located very close to the black hole horizon.

6 Concluding Remarks

In this chapter we have discussed only a few of the many interesting issues which arise when the dynamical behaviour of black holes is studied by using a perturbative approach. We have seen that, both for rotating and nonrotating black holes, the perturbed equations reduce to a Schrödinger-type equation, and it is possible to analyze the perturbation in terms of the scattering of gravitational waves by the potential barrier generated by the spacetime curvature. This approach proved extremely powerful in the study of how a black hole reacts to an external perturbation, allowed to clarify the manner in which incident gravitational, electromagnetic and scalar waves are absorbed and reflected, and brought light to the superradiant nature of the scattering of waves of appropriate frequency by rotating black holes [4].

The existence of the quasi-normal modes shows that the gravitational field possesses its own modes of vibration and we have seen that the gravitational radiation emitted when a black hole is perturbed by a source exhibits an oscillating tail, which is a superposition of these modes. Since the characteristic frequencies of the QNMs depend only on the mass and on the angular momentum of the black hole, this part of the signal carries direct information on the physical parameters of the emitting source.

In the next part, we shall show how the theory of stellar perturbations can be constructed in analogy with the theory of black holes perturbations and we shall see that this approach has proven likewise fruitful.

Appendix: The Explicit Expressions of Tensor Harmonics

$$\mathbf{a}^{(0)}_{\ell m} = \begin{pmatrix} (t) & (\varphi) & (r) & (\vartheta) \\ Y_{\ell m}(\varphi,\vartheta) & 0 & 0 & 0 \\ 0 & 0 & 0 & 0 \\ 0 & 0 & 0 & 0 \\ 0 & 0 & 0 & 0 \end{pmatrix}, \quad \mathbf{a}^{(1)}_{\ell m} = \frac{1}{\sqrt{2}} \begin{pmatrix} (t) & (\varphi) & (r) & (\vartheta) \\ 0 & 0 & Y_{\ell m}(\varphi,\vartheta) & 0 \\ 0 & 0 & 0 & 0 \\ Y_{\ell m}(\varphi,\vartheta) & 0 & 0 & 0 \\ 0 & 0 & 0 & 0 \end{pmatrix}$$

$$\mathbf{a}_{\ell m} = \begin{pmatrix} (t) & (\varphi) & (r) & (\vartheta) \\ 0 & 0 & 0 & 0 \\ 0 & 0 & 0 & 0 \\ 0 & 0 & Y_{\ell m}(\varphi,\vartheta) & 0 \\ 0 & 0 & 0 & 0 \end{pmatrix}$$

$$\mathbf{b}^{(0)}_{\ell m} = \frac{n(\ell) r}{\sqrt{2}} \begin{pmatrix} (t) & (\varphi) & (r) & (\vartheta) \\ 0 & \frac{\partial Y_{\ell m}}{\partial \varphi} & 0 & \frac{\partial Y_{\ell m}}{\partial \vartheta} \\ \frac{\partial Y_{\ell m}}{\partial \varphi} & 0 & 0 & 0 \\ 0 & 0 & 0 & 0 \\ \frac{\partial Y_{\ell m}}{\partial \vartheta} & 0 & 0 & 0 \end{pmatrix}, \quad \mathbf{b}_{\ell m} = \frac{n(\ell) r}{\sqrt{2}} \begin{pmatrix} (t) & (\varphi) & (r) & (\vartheta) \\ 0 & 0 & 0 & 0 \\ 0 & 0 & \frac{\partial Y_{\ell m}}{\partial \varphi} & 0 \\ 0 & \frac{\partial Y_{\ell m}}{\partial \varphi} & 0 & \frac{\partial Y_{\ell m}}{\partial \vartheta} \\ 0 & 0 & \frac{\partial Y_{\ell m}}{\partial \vartheta} & 0 \end{pmatrix}$$

$$\mathbf{f}_{\ell m} = \frac{m(\ell) r^2}{\sqrt{2}} \begin{pmatrix} (t) & (\varphi) & (r) & (\vartheta) \\ 0 & 0 & 0 & 0 \\ 0 & -\sin^2\vartheta W_{\ell m} & 0 & X_{\ell m} \\ 0 & 0 & 0 & 0 \\ 0 & X_{\ell m} & 0 & W_{\ell m} \end{pmatrix}, \quad \mathbf{g}_{\ell m} = \frac{r^2}{\sqrt{2}} \begin{pmatrix} (t) & (\varphi) & (r) & (\vartheta) \\ 0 & 0 & 0 & 0 \\ 0 & \sin^2\vartheta Y_{\ell m} & 0 & 0 \\ 0 & 0 & 0 & 0 \\ 0 & 0 & 0 & Y_{\ell m} \end{pmatrix}$$

$$\mathbf{c}^{(0)}_{\ell m} = \frac{\imath n(\ell) r}{\sqrt{2}} \begin{pmatrix} (t) & (\varphi) & (r) & (\vartheta) \\ 0 & -\sin\vartheta \frac{\partial Y_{\ell m}}{\partial \vartheta} & 0 & \frac{1}{\sin\vartheta}\frac{\partial Y_{\ell m}}{\partial \varphi} \\ -\sin\vartheta \frac{\partial Y_{\ell m}}{\partial \vartheta} & 0 & 0 & 0 \\ 0 & 0 & 0 & 0 \\ \frac{1}{\sin\vartheta}\frac{\partial Y_{\ell m}}{\partial \varphi} & 0 & 0 & 0 \end{pmatrix}$$

$$\mathbf{c}_{\ell m} = \frac{\imath n(\ell) r}{\sqrt{2}} \begin{pmatrix} (t) & (\varphi) & (r) & (\vartheta) \\ 0 & 0 & 0 & 0 \\ 0 & 0 & -\sin\vartheta \frac{\partial Y_{\ell m}}{\partial \vartheta} & 0 \\ 0 & -\sin\vartheta \frac{\partial Y_{\ell m}}{\partial \vartheta} & 0 & \frac{1}{\sin\vartheta}\frac{\partial Y_{\ell m}}{\partial \varphi} \\ 0 & 0 & \frac{1}{\sin\vartheta}\frac{\partial Y_{\ell m}}{\partial \varphi} & 0 \end{pmatrix}$$

$$\mathbf{d}_{\ell m} = \frac{\imath m(\ell) r^2}{\sqrt{2}} \begin{pmatrix} (t) & (\varphi) & (r) & (\vartheta) \\ 0 & 0 & 0 & 0 \\ 0 & -\sin\vartheta X_{\ell m} & 0 & -\sin\vartheta W_{\ell m} \\ 0 & 0 & 0 & 0 \\ 0 & -\sin\vartheta W_{\ell m} & 0 & \frac{1}{\sin\vartheta} X_{\ell m} \end{pmatrix}$$

$$X_{\ell m}(\vartheta, \varphi) = 2\frac{\partial}{\partial \varphi}\left[\frac{\partial}{\partial \vartheta} - \cot\vartheta\right] Y_{\ell m}(\vartheta, \varphi)$$

$$W_{\ell m}(\vartheta, \varphi) = \left[\frac{\partial^2}{\partial^2 \vartheta} - \cot\vartheta \frac{\partial}{\partial \vartheta} - \frac{1}{\sin^2\vartheta}\frac{\partial^2}{\partial^2 \varphi}\right] Y_{\ell m}(\vartheta, \varphi) .$$

References

1. R.A. Hulse, J.H. Taylor: Astrophys. J. **195**, L51 (1975)
2. T. Regge, J.A. Wheeler: Phys. Rev. **108**, 1063 (1957)
3. F. Zerilli Phys. Rev. **D2**, 2141 (1970)
4. S. Chandrasekhar: *The mathematical theory of black holes* (Clarendon Press, Oxford 1984)
5. C.V. Vishveshwara: Phys. Rev. **D1**, 2870 (1970)
6. W.H. Press: Ap. J. **170**, 215 (1971)
7. S. Chandrasekhar, S.L. Detweiler: Proc. R. Soc. Lond. **A344**, 441 (1975)
8. V. Ferrari: Phys. Lett. **A171**, 271 (1992)
9. B.F. Schutz, C.M. Will: Ap. J. Lett. **291**, L33 (1985)
10. S. Iyer, C.M. Will: Phys. Rev. **D35**, 3621 (1987)
11. S. Iyer: Phys. Rev. **D35**, 3632 (1987)
12. K.D. Kokkotas, B.F. Schutz: Phys. Rev. **D37**, 12 (1988)
13. J.W. Guinn, C.M. Will, Y. Kojima, B.F. Schutz: Class. Quantum Grav. **7**, L47 (1990)
14. N. Andersson: Proc. R. Soc. Lond. **A442**, 427 (1993)
15. S. Teukolsky: Phys. Rev. Lett. **29**, 1114 (1972)
16. S. Teukolsky: Ap. J. **185**, 635 (1973)
17. A.A. Starobinski, S.M. Churilov: Soviet JEPT, **38**, 1 (1973)
    W.H. Press, S. Teukolsky: Ap. J. **185**, 649 (1973)
18. S.L. Detweiler: Proc. R. Soc. Lond. **A352**, 381 (1977)
    S.L. Detweiler: Ap. J. **239**, 292 (1980)
19. E.W. Leaver: Proc. R. Soc. Lond. **A402**, 285 (1985)
20. E. Seidel, S. Iyer: Phys. Rev. **D41**, 374 (1990)
21. K.D. Kokkotas: Class.Quantum Grav. **8**, 2217 (1991)

22. V. Ferrari, B. Mashoon: Phys. Rev. Lett. **52**, 1361 (1984)
V. Ferrari, B. Mashoon: Phys. Rev. **D30**, 295 (1984)
23. M. Davis, R. Ruffini, W.H. Press, R.H. Price: Phys. Rev. Lett. **27** (1971)
M. Davis, R. Ruffini, J. Tiomno: Phys. Rev. **D5**, 2932 (1972)
V. Ferrari, R. Ruffini: Phys. Lett. **B98**, 381 (1984)
S.L. Detweiler, E. Szedenits: Ap. J. **231**, 211 (1979)
K. Oohara, T. Nakamura: Phys. Lett **94A**, 349 (1983)
K. Oohara, T. Nakamura: Prog. Theor. Phys. **70**, 757 (1983)
K. Oohara, T. Nakamura: Phys. Lett. **98A**, 407 (1983)
K. Oohara, T. Nakamura: Prog. Theor. Phys. **71**, 91 (1984)
T. Nakamura, M. Sasaki: Phys. Lett. **106B**, 1627 (1981)
S. Shapiro, I. Wasserman: Ap. J. **260**, 838 (1982)
24. T. Nakamura, K. Oohara, Y. Kojima: Prog. Theor. Phys. Suppl. **90**, 1 (1987)
25. T. Nakamura, M. Sasaki: Phys. Lett. **89A**, 68 (1981)
T. Nakamura, M. Sasaki: Prog. Theor. Phys. **67**, 1788 (1982)
T. Nakamura, M. Haugan: Ap. J. **269**, 292 (1983)
Y. Kojima, T. Nakamura: Phys. Lett. **96A**, 335 (1983)
Y. Kojima, T. Nakamura: Prog. Theor. Phys. **71**, 79 (1984)
Y. Kojima, T. Nakamura: Prog. Theor. Phys. **72**, 494 (1984)
S.L. Detweiler: Ap. J. **225**, 687 (1978)
E. Poisson: Phys. Rev. **D47**, 1497 (1993)

Critical Phenomena in Gravitational Collapse: The Role of Angular Momentum

José M. Martín-García and Carsten Gundlach

Faculty of Mathematical Studies, University of Southampton, Southampton SO17 1BJ, UK

Abstract. After reviewing the basics of Critical Phenomena in Gravitational Collapse of a spherically symmetric perfect fluid system, we address the relevance of adding angular momentum to the process. We study two different examples: the same perfect fluid but now with angular momentum, and Vlasov matter (collisionless particles, each with angular momentum). Using linear perturbation theory we show that in the former case there are still critical phenomena, explicitly predicting the associated scaling laws. We show that, on the contrary, critical phenomena are not generic for Vlasov matter.

1 Introduction

In 1993 Choptuik [1] discovered that the threshold of black hole formation in massless scalar field collapse was much simpler than one could expect. Although this is a regime with very strong gravitational coupling, which must be studied with sophisticated numerical techniques, it can be easily described within the theory of general dynamical systems. In fact, borrowing concepts from that theory, what we now call *Critical Phenomena in Gravitational Collapse* is just the study of the boundaries between different basins of attraction in the phase space of General Relativity. It has become both a discipline by itself in General Relativity and an active area of research, and continues to be considered the most important new result in Numerical Relativity.

After Choptuik's discovery in scalar field collapse, the same phenomenology was found in many other matter models, including perfect fluids, nonabelian gauge fields, or even gravitational waves. Most of these investigations assumed spherical symmetry (actually all of them except for the collapse of axisymmetric waves, which does not have a spherical counterpart), and dealt with electrically neutral matter models. But because generic black holes have electric charge and angular momentum, apart from mass, it is important to know what happens in the generic case.

We have predicted [2], and it has been subsequently verified in nonlinear simulations [3], that the addition of electric charge to the scalar field problem does not prevent critical phenomena. On the contrary, new scaling phenomena are found, confirming their relevance in General Relativity.

In this work we present some current ideas and results about the effects of adding angular momentum to critical systems. We start with a review of Critical Phenomena in spherical symmetry, restricting ourselves to a perfect fluid matter model, and then we study the effects of adding angular momentum. We do it in two different ways: we can keep the perfect fluid description for the matter but consider nonspherical situations, or we can work with a spherical distribution of particles, each one having angular momentum. We finish with the conclusions section.

2 Critical Phenomena

2.1 Self-gravitating Perfect Fluid

In this and the following sections our matter model will be a self-gravitating perfect fluid, with stress-energy tensor

$$T^{\mu\nu} = \varrho\, u^\mu u^\nu + p\,(g^{\mu\nu} + u^\mu u^\nu)\;, \tag{1}$$

where $u^\mu(x)$ is the 4-velocity field of the fluid, and $\varrho(x), p(x)$ are its total energy density and pressure, respectively, as measured by comoving observers at x. The evolution of the metric field is given by the Einstein equations

$$G_{\mu\nu} = 8\pi T_{\mu\nu}\;, \tag{2}$$

while the evolution of the fluid is just given, for this simple matter model, by stress-energy conservation. The system of equations is closed by adding an equation of state of the form $p = p(\varrho)$. Note that because we have chosen a barotropic equation of state with ϱ as independent variable we do not need to include the particle-number density in the problem, nor particle-number conservation as an equation of motion. This simplifies the description of the system.

We will be interested in self-similar solutions of the system and therefore we restrict ourselves to the only family of equations of state which are compatible with self-similarity [4]:

$$p(\varrho) = \kappa\varrho\;, \tag{3}$$

where κ is a dimensionless constant, because this is the only way of avoiding introducing a dimensionful scale in the problem which would prevent scale invariance. The speed of sound c_s of the fluid can be calculated as

$$c_\mathrm{s}^2 = \frac{\partial p}{\partial \varrho} = \kappa\;, \tag{4}$$

and therefore, for causality, κ must be a number between 0 and 1. $\kappa = 0$ corresponds to a pressureless fluid (dust), while $\kappa = 1$ is known as the "stiff" fluid and can be shown to be equivalent to a scalar field when the fluid is irrotational. The only truly physical case in this family is the $\kappa = 1/3$ radiation fluid.

Now suppose that the system is spherically symmetric. The spacetime (M^4, g) can be given as $M^4 = M^2 \times S^2$, where S^2 is the unit two-sphere and M^2 is a 1+1 manifold with boundary (the "reduced spacetime"). The metric can be 2+2 decomposed as

$$\mathrm{d}s^2 = g_{\mu\nu}(x)\mathrm{d}x^\mu \mathrm{d}x^\nu = g_{AB}(x^D)\mathrm{d}x^A\mathrm{d}x^B + r^2(x^D)\gamma_{ab}\mathrm{d}x^a\mathrm{d}x^b\;. \tag{5}$$

Capital letters $A, B, \ldots$ are indices on M^2 while lower case letters $a, b, \ldots$ denote the usual θ, ϕ coordinates on S^2. g_{AB} is a metric and r is a scalar on M^2, while γ_{ab} is the unit metric on S^2. A spherically symmetric fluid is given by $u^\mu = (u^A, u^a = 0)$, being u^A and ϱ both functions of x^D.

We can choose coordinates in M^2 in several different ways, but it has proved to be very convenient in critical phenomena theory to choose a radial gauge, so that the scalar field r becomes a radial coordinate $x^1 \equiv r$, and the polar slicing, which here means that the time coordinate $x^0 \equiv t$ is orthogonal to the radial coordinate. The metric is then

$$\mathrm{d}s^2 = -\alpha^2(t,r)\mathrm{d}t^2 + a^2(t,r)\mathrm{d}r^2 + r^2\mathrm{d}\Omega^2 \,. \tag{6}$$

In order to have a regular center we need the functions α and a to be even in r, with $a(t, r = 0) = 1$. Finally, there is a residual gauge freedom in α related to changes $t \to t'(t)$, which is commonly fixed in Critical Phenomena theory by imposing $\alpha(t, r = 0) = 1$, so that t is the proper time of a central observer.

A complete set of equations of motion for the system is finally, in polar-radial coordinates:

$$\frac{a_{,r}}{a} + \frac{a^2-1}{2r} = 4\pi a^2 r \varrho \frac{1+\kappa V^2}{1-V^2} \,, \tag{7a}$$

$$\frac{\alpha_{,r}}{\alpha} - \frac{a^2-1}{2r} = 4\pi a^2 r \varrho \frac{\kappa + V^2}{1-V^2} \,, \tag{7b}$$

$$\frac{1-\kappa V^2}{1-\kappa}\frac{\varrho_{,t}}{\alpha} + V\frac{\varrho_{,r}}{a} = \frac{1+\kappa}{a(1-\kappa)}\left[4\pi a^2 r \varrho^2 (1+\kappa)V - r^{-1}\varrho(2V + rV_{,r})\right] \tag{7c}$$

$$\frac{1-\kappa V^2}{1-\kappa}\frac{V_{,t}}{\alpha} + V\frac{V_{,r}}{a} = \frac{1-V^2}{a(1-\kappa)}\Big[-4\pi a^2 r \varrho \kappa (1+V^2) + 2r^{-1}(1-a^2) + 2r^{-1}(3+a^2)\kappa V^2\Big] \tag{7d}$$

We have defined the velocity $V = au^r/\alpha u^t$ with respect to constant r observers.

2.2 Criticality

Following Choptuik's strategy of bisection searching, Evans and Coleman [5] evolved initial conditions of the form

$$V(0,r) = 0, \qquad \varrho(0,r) = \frac{\eta}{2\pi^{3/2} r_0^2} \exp(-r^2/r_0^2) \,, \tag{8}$$

parametrized by η and r_0. The total gravitational mass of that initial Gaussian profile is $M = \eta r_0/2$, so that $\eta = 2M/r_0$ is a dimensionless measure of the strength of the initial gravitational field, while r_0 is the typical initial length scale.

Being initially at rest, the fluid ball of size r_0 tends to fall towards the center due to its own gravitation. The energy density increases, and so does the pressure, which tries to halt the collapse. Evans and Coleman numerically evolved the system described in the previous subsection with $\kappa = 1/3$ and found the following situation: For $\eta < \eta_*$, with $\eta_* \simeq 1.0188$, pressure effects are able to stop and reverse the collapse, while for $\eta > \eta_*$ a black hole is always formed. For values close to η_* a strong, ingoing rarefaction wave with radius $R(t)$ is formed, separating an interior region where matter is still falling in from an exterior region where matter is expanding. The rarefaction wave chases the ingoing matter, so that the total mass $M[R(t)]$ in the interior region decreases in time. For $\eta \simeq \eta_*$ evolutions the dimensionless combination $M[R(t)]/R(t)$ is almost constant and the system becomes self-similar near the center, at very small scales compared with the initial r_0.

Self-similarity means that both the metric functions a, α and the fluid variables $r^2\varrho, V$ are functions of r/t, with a suitable choice of origin of time. This has important consequences for the system, as we will see later. This self-similar regime is only an intermediate state and eventually the rarefaction wave removes all the matter, leaving an almost empty region at the center ($\eta < \eta_*$, *subcritical evolution*) or, on the contrary, it is unable to do it and a small black hole is formed at the center with the matter which is still falling in ($\eta > \eta_*$, *supercritical evolution*). Only for $\eta = \eta_*$ (*critical evolution*) the self-similar regime continues until the rarefaction wave hits the center forming a zero mass singularity there, which can be shown to be naked.

Evans and Coleman [5] found that, for slightly supercritical evolutions, the mass $M(\eta)$ of the black hole formed starting from the initial condition given by η obeys the following power law

$$M(\eta) = C(\eta - \eta_*)^\beta , \qquad \beta \simeq 0.36 . \tag{9}$$

It is possible to show that this power law holds for *any* family of initial conditions interpolating between fluid dispersion and black hole formation, and the most striking fact is that the exponent β is the *same* for every family of initial conditions (while C depends on the family). This phenomenon is now referred to as *universality.* The value 0.36 was surprisingly close to the 0.37 measured by Choptuik in the collapse of massless scalar field, or the 0.36 measured by Abrahams and Evans [6] in the collapse of axisymmetric gravitational waves. Maison [7] has shown that for a perfect fluid with $p = \kappa\varrho$ the exponent β depends strongly on κ, so that the conclusion is that universality holds within a given matter model, but not among different matter models.

2.3 Self-similarity and the Critical Spacetime

Universality was not only found in the critical exponent β. The profile adopted by the metric and fluid variables in the self-similar regime was always the same, which strongly suggested the existence of an exact solution of the equations with the exact symmetry of self-similarity. The intermediate self-similar regime could then be easily understood as an approach in phase space to that exact solution with an eventual departure from it. Evans and Coleman were able to construct that exact solution and showed that the profiles observed in the nonlinear time evolutions coincided remarkably well with it, confirming the relevance of that solution in the process of critical collapse. In this subsection we construct numerically that exact self-similar spacetime.

Geometrically a spherically symmetric spacetime is self-similar (technically *continuously* self-similar) if it has a homothetic Killing vector field ξ, such that

$$\mathcal{L}_\xi g_{\mu\nu} = -2 g_{\mu\nu} \ . \tag{10}$$

For nontrivial spacetimes, this implies the existence of a singularity at the center at a certain moment of time, the point where all the integral curves of ξ converge. Using polar-radial coordinates, this vector will be

$$\xi = -t\partial_t - r\partial_r \ , \tag{11}$$

assuming that the singularity is at $t = r = 0$.

It is convenient to define the following new set of coordinates adapted to self-similarity:

$$\tau \equiv -\log\left(\frac{-t}{t_0}\right) \ , \qquad x \equiv \log\left(\frac{r}{-t}\right) \ , \tag{12}$$

valid in the region $r \geq 0$, $t < 0$, such that the homothetic vector is now $\xi = \partial_\tau$. The signs have been chosen so that both t and τ increase to the future, and t_0 is an arbitrary positive constant setting a global scale in the system. Note that the coordinate x is invariant under scale changes in spacetime, while τ is not. In other words, x labels the integral curves of ξ, while τ is a parameter on them. Using these coordinates, the self-similarity condition of the metric translates into the requirement of the metric functions α, a being independent of τ. Using the Einstein equations, the fluid variables $r^2\varrho$ and V are then also functions of x.

Following Koike, Hara and Adachi [8], we define the following variables

$$N \equiv \frac{\alpha}{a e^x} \ , \qquad A \equiv a^2 \ , \qquad \omega \equiv 4\pi a^2 r^2 \varrho \tag{13}$$

such that the evolution equations in self-similar coordinates are autonomous:

$$\frac{N_{,x}}{N} - (\kappa - 1)\omega + 2 - A = 0 \ , \tag{14a}$$

$$\frac{A_{,x}}{A} - \frac{1+\kappa V^2}{1-V^2} 2\omega - 1 + A = 0 \ , \tag{14b}$$

$$\frac{\omega_{,\tau}}{\omega} + (1+NV)\frac{\omega_{,x}}{\omega} + \frac{1+\kappa}{1-V^2}[V V_{,\tau} + (N+V)V_{,x}] \\ + \frac{NV}{2}\left[(\kappa+3)A + (\kappa-1)(3+2\omega)\right] = 0 \ , \tag{14c}$$

$$\frac{1+\kappa}{1-V^2}\left[V_{,\tau} + (1+NV)V_{,x}\right] + \kappa V \frac{\omega_{,\tau}}{\omega} + \kappa(N+V)\frac{\omega_{,x}}{\omega} \\ + \frac{N}{2}\left[-1 - 7\kappa + (1+3\kappa)A + 2(\kappa-1)\kappa\omega\right] = 0 \ . \tag{14d}$$

The imposition of self-similarity eliminates the τ-derivative terms and we are left with a system of ODEs that can be easily integrated using the following

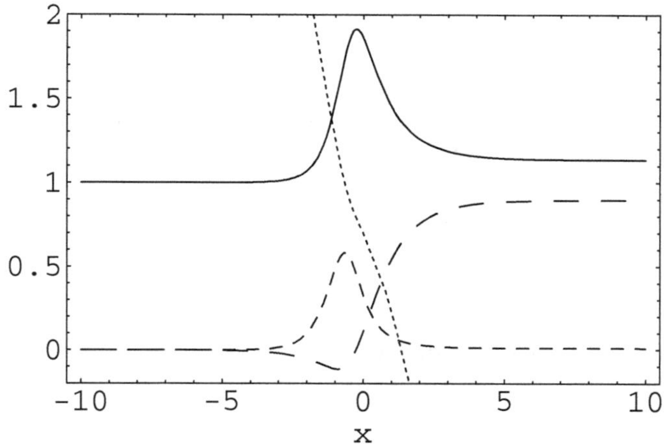

Fig. 1. Evans and Coleman spacetime. $A(x)$ (continuous line), $\log N(x)$ (dotted line), ω (short-dashed line) and V (long-dashed line). Note that $A(x \to \infty) > 1$ because this spacetime is not asymptotically flat

initial/boundary conditions. Given that we are looking for a solution which is arbitrarily approached by smooth initial conditions, we must require analyticity on every spacelike slice, before the formation of the singularity. In particular we will have a regular center ($A = 1$ and $V = 0$ at $r = 0$). On the other hand the equations contain a regular singular point x_0, physically representing a sonic point: particles at x_0 move outwards with velocity c_s with respect to constant x observers and therefore cannot send sound signals to those observers with $x < x_0$. In other words, x_0 is the past sound cone of the singularity. We use the gauge freedom in α, or equivalently in N, to fix $x_0 = 0$. Expanding in power series around $x = 0$ we see that there is a single free parameter, say $V(x = 0) \equiv V_0$, in the initial condition. Finally the regularity conditions at the center give a discrete set of values for V_0.

Using a bisection search for V_0 and shooting from the sound cone to the center it is possible to find that the critical spacetime is that with $V_0 \simeq 0.112\,439$. Integrating then outwards from the sound cone we get the rest of the spacetime. The profiles are those given in Fig. 1. In Fig. 2 the corresponding fluid worldlines are represented in t, r coordinates.

2.4 Perturbation Theory and the Critical Exponent

Evans and Coleman suggested that the critical exponent β could be understood in terms of the linearized evolution around the critical spacetime. The actual calculation was carried out by Koike, Hara and Adachi [8] and is reviewed in this subsection.

The important point is the realization that the whole phenomenology of critical collapse can be qualitatively understood in Fig. 3, representing a schematic picture of the phase space of the system. In that infinite-dimensional space a

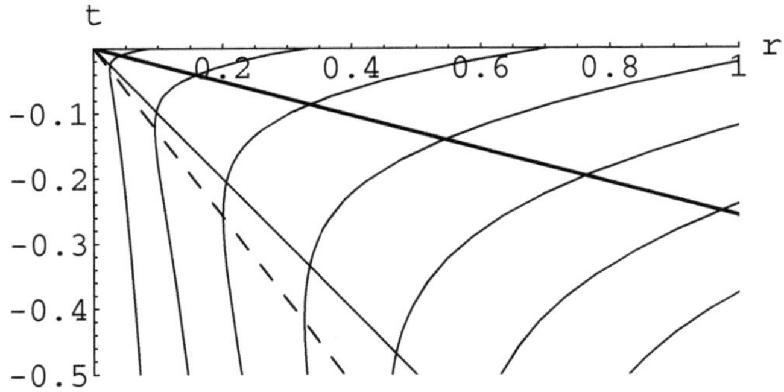

Fig. 2. Fluid world lines in coordinates (t, r) in the Evans and Coleman spacetime. The dashed line gives the points $x \simeq -0.252\,787$ where $V = 0$. The continuous thin straight line $x = 0$ gives the sonic point (the past sound cone of the singularity). The continuous thick line $x \simeq 1.357\,941$ gives the past light cone of the singularity

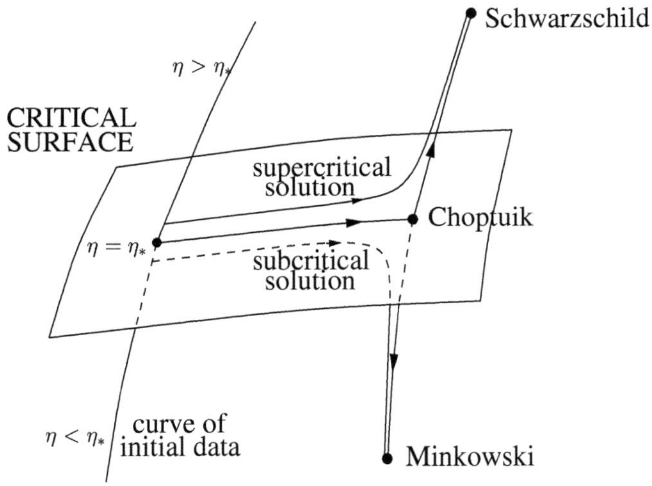

Fig. 3. Schematic picture of phase space around the critical solution, here called Choptuik spacetime. Very close initial conditions bifurcate and give very different final results: a black hole (supercritical evolution), flat space (subcritical evolution) or a naked singularity (critical evolution)

point represents an instantaneous state of the system, that is the set of functions $Z \equiv \{N, A, \omega, V\}$ of x at a certain time τ. A generic full spacetime is given by a trajectory on phase space.

There are only three possible stable end states for the system: a static star, a Schwarzschild black hole or vacuum spacetime, the only three global attractors in the system. Each of them has its own basin of attraction, with codimension-1 boundaries among them (the so-called "critical surfaces"), and here we concentrate on the boundary between black hole collapse and dispersion. The main idea is that the critical spacetime Z_* sits on that boundary and acts as an attractor within it, and therefore as a codimension-1 attractor in phase space.

We expand the solution with initial condition given by η, to linear order around Z_*, as:

$$Z(\tau, x; \eta) \simeq Z_*(x) + \sum_{n=0}^{\infty} C_n(\eta) \mathrm{e}^{\lambda_n \tau} Z_n(x) \tag{15}$$

where the Z_n are the eigenstates of the linearized evolution equations, with λ_n as eigenvalues, and the C_n are free constants which depend in a complicated way on the initial data. Codimension 1 means that there is a single mode (say Z_0) with positive real part eigenvalue:

$$\Re(\lambda_0) > 0 , \qquad \Re(\lambda_{n\geq 1}) \leq 0 . \tag{16}$$

Actually, because the eigenvalues must form complex conjugate pairs, uniqueness of the unstable mode implies that λ_0 must be real. As τ increases all the perturbations vanish, except for the unstable mode, so that the information in the $C_{n\geq 1}$ is washed out. In the following we consider this limit, and retain only the growing perturbation. By definition the critical solution corresponds to $\eta = \eta_*$, so we must have $C_0(\eta_*) = 0$. Linearizing around η_* we obtain

$$\lim_{\tau\to\infty} Z(\tau, x; \eta) \simeq Z_*(x) + \left.\frac{\mathrm{d}C_0}{\mathrm{d}\eta}\right|_{\eta_*} (\eta - \eta_*) \mathrm{e}^{\lambda_0 \tau} Z_0(x) . \tag{17}$$

The solution has this approximate form over a range of τ (infinite if perfect fine-tuning of η were possible). In this range, all that is left of the information contained in the family of initial conditions $\{C_n(\eta)\}_{n=0}^{\infty}$ is the single number $(\eta - \eta_*)K$, with $K \equiv C_0'(\eta_*)$. The later evolution of the system is determined by that number.

At sufficiently large τ, the linear perturbation has grown so much that the linear approximation breaks down. Later on a black hole forms. The crucial point is that we need not follow this evolution in detail to get the mass of the black hole: we can find it with a simple dimensional argument. Choose an arbitrary time τ_* within the linear regime, that is such that

$$\varepsilon \equiv (\eta - \eta_*) K \mathrm{e}^{\lambda_0 \tau_*} \ll 1 \tag{18}$$

for those values of η that we are interested in. The spacetime is at that time

$$Z(\tau_*, x) \simeq Z_*(x) + \varepsilon\, Z_0(x) = Z_* \left(\log \frac{r}{-t_*} \right) + \varepsilon\, Z_0 \left(\log \frac{r}{-t_*} \right) , \tag{19}$$

with

$$t_* \equiv -t_0\, \mathrm{e}^{-\tau_*} . \tag{20}$$

There are two free parameters, ε and $(-t_*)$. Again, we know that a Schwarzschild black hole forms later, with a single free parameter: its mass M, with dimension of length. Clearly M must be given by ε and $(-t_*)$, and because the first is dimensionless and the second has length dimension, we must have:

$$M(\varepsilon, t_*) = (-t_*) f(\varepsilon) = (-t_*) f \left[(\eta - \eta_*) K \left(\frac{t_0}{-t_*} \right)^{\lambda_0} \right] . \tag{21}$$

Finally, using the fact that t_* is arbitrary and therefore M must be independent of it, we get that the function f must be a power law:

$$f(y) = c\, y^{1/\lambda_0} , \tag{22}$$

and then

$$M(\eta) = c t_0 K^{1/\lambda_0} (\eta - \eta_*)^{1/\lambda_0} , \tag{23}$$

which is (9) with $\beta = 1/\lambda_0$.

The actual calculation of the exponent λ_0 for $\kappa = 1/3$ was carried out by Koike, Hara and Adachi [8] by linearizing (7a-7d) around the Evans and Coleman spacetime and solving a linear eigenvalue problem for the mode Z_0. The associated mode is $\lambda_0 \simeq 2.810\,553$, giving $\beta \simeq 0.355\,802$, in good agreement with the observed 0.36.

3 Angular Momentum in Critical Fluid Collapse

After studying critical phenomena in spherical symmetry it is obvious to ask whether they are just a consequence of the high degree of symmetry of the system or, on the contrary, they are a generic feature of any relativistic self-gravitating system.

We know for example that typical nonspherical processes of collapse in Astrophysics usually produce rapidly rotating compact objects, with large centrifugal forces acting on them. Is that the case as well for near-critical collapse? Would those centrifugal forces prevent the formation of naked singularities?

The numerical evolution in time of the nonlinear equations of motion of a self-gravitating fluid is a nontrivial task, and without the restriction to spherical symmetry it becomes highly demanding of computer resources. Several clever and powerful methods have been devised to tackle the problem (mainly concerning the fluid part), but none of them is currently able to cope with a generic situation.

Critical phenomena are a result of the strong gravitational coupling regime (we generate arbitrarily high curvatures near the singularity) and involve ultra-relativistic states for the matter. Furthermore they develop structure at arbitrarily small scales in the system, which must be resolved with some kind of dynamic regridding process. Only a code able to deal with those three kinds of problems can properly study the relevance of angular momentum for critical collapse.

Fortunately, we have seen that Critical Phenomena can be studied (at least locally in phase space) using linear perturbation theory. That is, we have to check whether a potential candidate to be a critical spacetime is a codimension-1 attractor in phase space, but now including all nonspherical perturbations in the game. This cannot be an exhaustive way of studying the problem because we must know in advance the candidate spacetime, instead of letting the nonlinear code to find it. Therefore, we can only hope to find critical solutions among those spacetimes which are already critical in spherical symmetry.

In this section we show that the Evans and Coleman spacetime is actually a critical solution in the general situation for a certain range of values of κ.

We start by reviewing the Gerlach and Sengupta [9] formalism of nonspherical perturbations and the equations derived from it. We then present the results, and finish with the description of angular momentum scaling and the prediction [10] of interesting new phenomena that will happen for small values of κ.

3.1 Nonspherical Perturbations

Given a spherical spacetime with its metric $g_{\mu\nu}(x)$ expressed in the form (5), we introduce a perturbed spacetime with metric

$$g_{\mu\nu}(x) + h_{\mu\nu}(x) , \tag{24}$$

where h is arbitrary but assumed to be small (in some sense) in comparison with g, so that we can neglect quadratic terms in h. Under these circumstances h can be considered as a tensor field living on the background spacetime. That is, $h_{\mu\nu}$ is a nonspherical tensor on a spherically symmetric background.

In 1979 Gerlach and Sengupta [9] presented a new formalism for nonspherical perturbations combining three key ideas:

- Following Regge and Wheeler [11], the angular dependence of the 10 functions in $h_{\mu\nu}$ can be decomposed in series of spherical tensor harmonics. The coefficients (labelled by l, m) are functions of the coordinates in M^2. Due to the additional symmetry of parity, those 10 coefficients can be separated into two groups: 7 of them are polar and the other 3 are axial.
- This expansion can be done in geometrical terms on the reduced spacetime M^2, that is, without imposing particular coordinates on M^2. For a given l, m the polar variables are a symmetric 2-tensor h_{AB}, a vector $h_B^{\rm polar}$ and two scalars K and G. The axial variables are a vector $h_B^{\rm axial}$ and a scalar h.
- The coordinate freedom on the perturbed spacetime (which is different from and independent of the coordinate freedom on the background) yields 4 gauge freedoms among the 10 functions in h. These gauge freedoms can be used to impose 4 conditions on the metric. However, following Moncrief [12], Gerlach and Sengupta construct 6 linear combinations of the 10 variables which are invariant under the 4 coordinate changes in the perturbed spacetime. Finally we have 4 polar variables describing the polar gravitational wave: a tensor k_{AB} and a scalar k, and 2 axial variables describing the axial gravitational wave: a vector k_A.

The same construction can be done for the fluid variables, and there we get 3 perturbations of the fluid velocity: γ (polar radial perturbation), α (polar tangential) and β (axial tangential); and the perturbation $\delta\varrho$ of the energy density field. γ and $\delta\varrho$ describe the polar sound wave degree of the fluid.

By linearizing the Einstein equations and the fluid equations of motion we can get the equations for the perturbations. As usual, the detailed equations of linear perturbation theory in general relativity are straightforward but lengthy. We just sketch the form of the equations [13,14]:

In the axial case β is just transported along the fluid worldline, decoupled from the gravitational wave k_A. However, the latter obeys a wave equation with β as a source, so that the gravitational wave is driven by the fluid perturbation:

$$\dot{\beta} + (...)\beta' = 0 , \tag{25}$$

$$-\ddot{\Pi} + \Pi'' + (...)\Pi' + (...)\Pi = (...)\beta' + (...)\beta , \tag{26}$$

where Π is the rotational of k_A and the (...) terms denote background quantities. Dots and primes denote time and radial derivatives in the comoving frame. For $l = 1$ the wave equation degenerates to an algebraic equation, reflecting the fact that we cannot have an $l = 1$ gravitational wave.

In the $l \geq 2$ polar case we can express the matter perturbations $\alpha, \gamma, \delta\varrho$ as radial derivatives of the metric perturbations (rewritten in terms of three variables k, χ, ψ), which in turn obey a closed set of linear evolution equations: a wave equation for the gravitational wave χ, a wave equation for the 'sound wave' k and a transport equation for the tangential perturbation ψ:

$$-\ddot{\chi} + \chi'' + (...)\psi' = ... , \tag{27a}$$

$$-\ddot{k} + c_s^2 k'' + (...)\psi' = ... , \tag{27b}$$

$$\dot{\psi} = ... \tag{27c}$$

Again, the cases $l = 0$ and $l = 1$ must be treated separately because they do not contain gravitational freedom.

3.2 Results

This is a brief summary of [15].

The transport equation (25) is simple enough so that we can analytically find the spectrum of β perturbations on a self-similar background: The n-th perturbation mode for the problem corresponding to the harmonic l is

$$\lambda_n^l = \frac{2(1-3\kappa) - (1+3\kappa)(l+2n)}{3(1+\kappa)} , \tag{28}$$

which is linear in $l + 2n$. From this formula we can read off that all $l \geq 2$ modes decay for all κ in the range $0 < \kappa < 1$. All $l = 1$ modes also decay for $\kappa > 1/9$, but for $\kappa < 1/9$ there is exactly one growing $l = 1$ mode (the $n = 0$ mode). As we will see, that mode is relevant for angular momentum scaling. Its associated eigenvalue is:

$$\lambda_0^1 = \frac{1-9\kappa}{3(1+\kappa)} . \tag{29}$$

For $l = 1$ the metric perturbation Π is obtained as a quadrature over β, and therefore it has the same spectrum. But for $l \geq 2$ it obeys a wave equation and therefore the spectrum of Π is given as the union of the spectrum of β and that obtained from the homogeneous wave equation. It is possible to show numerically that in that second part of the spectrum there is an unstable $l = 2$ mode in the range $0.58 < \kappa < 0.87$, while every $l \geq 3$ mode decays.

The polar perturbations must be analyzed numerically as well, and this turns out to be a nontrivial problem because of the coupling of the two wave equations (27a) and (27b) with very different characteristics (the speed of light and the speed of sound, respectively). The $l = 2$ perturbations become unstable for $\kappa > 0.49$, but all $l = 1$ and $l \geq 3$ perturbations decay. Of course there is always a polar unstable $l = 0$ mode, responsible for the critical phenomenology.

Therefore we conclude that in the range $1/9 < \kappa < 0.49$ (an interval remarkably well centered on the physical value 1/3) the Evans and Coleman spacetime for that κ value is a codimension-1 solution, and therefore critical phenomena will be present in that range, at least for slightly nonspherical, near critical collapse processes. In the following subsection we will review the consequences that the instability of the axial dipole mode $l = 1$ brings for $\kappa < 1/9$. The meaning of the $l = 2$ instabilities for $\kappa > 0.49$ is not known to us.

3.3 Angular Momentum Scaling and Scaling Functions

A slowly rotating Kerr black hole can be approximately considered as an axial $l = 1$ perturbation of a Schwarzschild black hole. Therefore, in order to find the scaling law of the angular momentum, we must include the $l = 1$ modes in the argument of Sect. 2.4.

Suppose that now we start from a 4-parameter $(\eta, \boldsymbol{q})$ family of initial conditions, such that if $\boldsymbol{q} = 0$, they are spherically symmetric and again, for $\boldsymbol{q} = 0$ the value η_* signals the black hole threshold. We assume that $\boldsymbol{q}$ introduces a small amount of angular momentum in the initial condition, such that if a black hole is formed, it will have some angular momentum $\boldsymbol{L}(\eta, \boldsymbol{q})$, with $\boldsymbol{L}(\eta, -\boldsymbol{q}) = -\boldsymbol{L}(\eta, \boldsymbol{q})$. We are interested in the scaling properties of $\boldsymbol{L}$ with respect to both η and $\boldsymbol{q}$ near criticality ($\eta = \eta_*$, $\boldsymbol{q} = \boldsymbol{0}$).

If we restrict ourselves to $\kappa < 0.49$, the axial $l = 1$ modes are either unstable or the most slowly decaying modes, apart from the unstable $l = 0$ mode. Therefore, taking into account that those modes are threefold degenerate, we can expand around criticality by neglecting the $l \geq 2$ terms:

$$Z(\tau, x, \Omega; \eta, \boldsymbol{q}) \simeq Z_*(x) + \sum_{n=0}^{\infty} C_n(\eta) Z_n(x) \mathrm{e}^{\lambda_n \tau}$$
$$+ \sum_{n=0}^{\infty} \boldsymbol{D}_n(\eta, \boldsymbol{q}) \cdot \boldsymbol{Z}_n(x, \Omega) \mathrm{e}^{\lambda_n^1 \tau} , \tag{30}$$

where the second expansion is associated with axial $l = 1$ perturbations. Ω denotes the angular dependence. Again, for high enough τ we can neglect all the nondominant terms in the summations and expand both in $\eta - \eta_*$ and $\boldsymbol{q}$:

$$\lim_{\tau \to \infty} Z(\tau, x, \Omega; \eta, \boldsymbol{q}) \simeq Z_*(x) + K(\eta - \eta_*) Z_0(x) \mathrm{e}^{\lambda_0 \tau}$$
$$+ \boldsymbol{q} \cdot \boldsymbol{K} \cdot \boldsymbol{Z}_0(x, \Omega) \mathrm{e}^{\lambda_0^1 \tau} . \tag{31}$$

Note that $\boldsymbol{K} \equiv \partial_{\boldsymbol{q}} \boldsymbol{D}_0(\eta_*, \boldsymbol{0})$ is a Jacobian matrix and again $K \equiv C_0'(\eta_*)$. We assume that if λ_0^1 is complex, the $l = 1$ term in the previous equation also contains the contribution of the complex conjugated mode. We choose an arbitrary time τ_* within the linear regime, that is such that

$$\varepsilon \equiv (\eta - \eta_*) K \mathrm{e}^{\lambda_0 \tau_*} , \tag{32}$$

$$\boldsymbol{\delta} \equiv \boldsymbol{q} \cdot \boldsymbol{K} \mathrm{e}^{\lambda_0^1 \tau_*} \tag{33}$$

are very small. The spacetime is at time τ_*:

$$Z(\tau_*, x, \Omega) \simeq Z_*(x) + \varepsilon Z_0(x) + \boldsymbol{\delta} \cdot \boldsymbol{Z}_0(x, \Omega) , \tag{34}$$

and therefore the mass and angular momentum of the black hole produced later on must be given as functions of $(t_*, \varepsilon, \boldsymbol{\delta})$, where t_* was defined in (20). By dimensional analysis

$$M = (-t_*) f(\varepsilon, \boldsymbol{\delta}) , \qquad \boldsymbol{L} = {t_*}^2 \boldsymbol{g}(\varepsilon, \boldsymbol{\delta}) , \tag{35}$$

and requiring the results to be independent of t_* we get that the functions f and g are power-laws in ε times functions of the combination $\boldsymbol{\delta}/\varepsilon^{\lambda_0^1/\lambda_0}$:

$$M = (\eta - \eta_*)^{1/\lambda_0} F \left[\frac{\boldsymbol{q}}{(\eta - \eta_*)^{\lambda_0^1/\lambda_0}} \right] , \tag{36}$$

$$\boldsymbol{L} = (\eta - \eta_*)^{2/\lambda_0} \boldsymbol{G} \left[\frac{\boldsymbol{q}}{(\eta - \eta_*)^{\lambda_0^1/\lambda_0}} \right] . \tag{37}$$

with F and $\boldsymbol{G}$ universal *scaling functions* within the model under study. Note that $F(-\boldsymbol{y}) = F(\boldsymbol{y})$, and then $F(\boldsymbol{0})$ is a constant, while $\boldsymbol{G}(-\boldsymbol{y}) = -\boldsymbol{G}(\boldsymbol{y})$, so that $\boldsymbol{G}(\boldsymbol{0}) = \boldsymbol{0}$.

If $\Re(\lambda_0^1) < 0$ (stable $l = 1$ modes) the critical regime always corresponds to very small arguments of the functions F and $\boldsymbol{G}$. We get the scalings

$$M = c(\eta - \eta_*)^{\beta} , \qquad \beta \equiv \frac{1}{\lambda_0} , \tag{38}$$

$$\boldsymbol{L} = \boldsymbol{c}(\eta - \eta_*)^{\mu} , \qquad \mu \equiv \frac{2 - \Re(\lambda_0^1)}{\lambda_0} , \tag{39}$$

irrespectively of the (nonzero) values of $\boldsymbol{q}$. From (29) we have for $\kappa = 1/3$ that $\lambda_0^1 = -1/2$ and then

$$\mu = \frac{5}{2} \beta = 0.889\,505 . \tag{40}$$

If $\Re(\lambda_0^1) > 0$ (unstable $l = 1$ modes) the critical regime covers the whole range of values for the argument of the scaling functions F and $\boldsymbol{G}$, which can be used for predictive purposes. It is particularly interesting to study the regime of very large angular momentum. For example the question "do the addition of angular momentum to a critical evolution produce a black hole or a vacuum region?" has a universal answer, independent of the initial condition. It is likely that the answer will be a vacuum region, as one would expect centrifugal forces to disrupt data that already hover between collapse and dispersion. If that is the case and centrifugal forces can destroy slightly supercritical black holes, can we produce very small black holes just by keeping fixed η and increasing the modulus of $\boldsymbol{q}$? Again the answer is universal for a given matter model and just depends on the behaviour of the scaling functions for large values of their argument. See [10] for a complete discussion of the possibilities.

4 The Einstein-Vlasov System

A very different way of studying the influence of angular momentum in critical phenomena is to replace the fluid field as a matter model by an averaged ensemble of particles. In this way it is possible to consider a situation where the individual particles have angular momentum, but their global distribution and therefore the spacetime are spherically symmetric. In order to simplify the problem we will assume collisionless evolution (Vlasov matter), and therefore there is no reason to expect obtaining the same results that hold for a fluid, even if we compare massless particles with a radiation fluid, as we will do here.

4.1 Collisionless Matter

As in classical statistical mechanics we describe the state of a many-body system with a positive distribution function over the phase space of the system. For equivalent particles that do not directly interact with each other we can use a distribution function $f(x^\mu, p^\nu)$ on the one-particle phase space, and in fact, if they all have the same mass m, we can consider the distribution f as a function $f(x^\mu, p^i)$. Greek indices denote the range 0-3 and Latin indices the range 1-3.

Noninteracting particles follow the geodesics of the spacetime, so that the distribution function is just Lie-dragged along them, obeying the Vlasov equation:

$$p^\mu \frac{\partial f}{\partial x^\mu} - \Gamma^i_{\nu\lambda} p^\nu p^\lambda \frac{\partial f}{\partial p^i} = 0 \,. \tag{41}$$

The spacetime is coupled to the matter through the Einstein equations, with

$$T^{\mu\nu}(x) = \int_{P(x)} \frac{\mathrm{d}p^3}{-p_0} \sqrt{-g}\, f(x,p) p^\mu p^\nu \,, \tag{42}$$

where $P(x)$ is the 3-momentum space at the point x^μ and p_0 is determined from p^i and the metric. The Vlasov equation is a sufficient condition for stress-energy conservation.

If we impose spherical symmetry and use polar-radial coordinates to write the metric as (6), the distribution function is just a function $f(t, r, p^t, p^r)$, and the Vlasov and Einstein equation are much simpler, but still a coupled system of integro-differential equations in four independent variables.

The solutions of the Vlasov equation are formally arbitrary functions of the constants of motion along particle trajectories. It is customary to introduce additional symmetries in the system to find simple expressions for those constants of motion. For instance, the static solutions of the system have been intensively studied, and several interesting theorems have been proved.

It is interesting to compare the Einstein-Vlasov system with its Newtonian counterpart, the Poisson-Vlasov system, where the Einstein equations are replaced by the Poisson equation. It is then possible to show that, even though there are always formally 4 different constants of motion along particle trajectories, any static solution of the Poisson-Vlasov system is just a function $f(E,F)$, where E and F are the energy and squared angular momentum of the particles, respectively. This is the so-called *Jeans theorem* [16]. There are two types of counterexamples to this theorem in General Relativity [17,18], as a reminder that all the constants of motion can be important in the construction of solutions of the Einstein-Vlasov system.

Because our goal is the study of criticality in the system, we now introduce a new symmetry in the problem: self-similarity, instead of staticity, but always keeping spherical symmetry.

4.2 Self-similar Solutions with Massless Particles

In this subsection we show numerically that there are self-similar solutions of the spherically symmetric Einstein-Vlasov system with massless particles (potential candidates for critical spacetimes in this system). Again, we only introduce the main ideas because the equations are long and uninteresting. See [19] for a complete exposition.

As we said previously, a self-similar spacetime has a homothetic Killing vector ξ^μ. The *homothetic energy*

$$J \equiv -\xi^\mu p_\mu = tp_t + rp_r = -p_\tau \tag{43}$$

is a constant of motion for massless particles. One can try to find solutions where the distribution function is of the form $f(J,F)$, imitating the Jeans solutions $f(E,F)$ of the static case. However this leads to a divergent stress-energy tensor. The difference is that both E and F are explicitly independent of time in the static case and therefore any function f is a valid solution. However J and F are scale dependent, so that only a function of the dimensionless combination $J/\sqrt{F}$ is a valid solution and one of the multiple integrals in the energy-stress tensor becomes divergent. If there are well defined solutions they must involve the additional constants of motion [19].

We introduce the remaining constant of motion, called $t_0(t,r,p^t,p^r)$, representing the time when the particle at t,r with p^t,p^r was at a certain canonical position. Of course this depends on the trajectory joining both points, so that t_0 does not have a local expression in terms of its variables. This will introduce another integration, in a system which is already integro-differential. The interesting point is that, due to the high symmetry of the system, we can classify in advance the possible trajectories of the particles, which gives us enough information about t_0 to solve the equations numerically.

Using dimensional analysis and requiring the right scaling for the $T_{\mu\nu}$, a self-similar solution will have the form

$$f(t,r,p^t,p^r) = g(t_0,J,F) = \frac{1}{F} k\left(\frac{J}{\sqrt{F}}, \frac{F}{t_0^2}\right) , \tag{44}$$

where the arguments of the function k are scale invariant. Defining the function

$$\bar{k}(Y) \equiv \int_0^\infty k(Y,Z)\mathrm{d}Z , \tag{45}$$

it is possible to show that the energy-stress tensor $t^2 T_{\mu\nu}(x)$, where x is the self-similar variable $r/(-t)$, is an integral transform of $\bar{k}(Y)$ with some complicated kernel $K(x,Y)$. It is then clear that if we had a function k independent of Z, the function $\bar{k}$ would diverge for every Y.

Using the constraint Einstein equations we end up with a set of integro-differential equations of the form

$$D_x g(x) = \int_{-\infty}^{\infty} \mathrm{d}Y K(x,Y;g)\bar{k}(Y) , \tag{46}$$

where D_x is a differential operator, g represents both metric functions a, α, and the kernel K depends nonlocally on g. For a given fluid distribution $\bar{k}(Y)$ they can be solved numerically by iteration

$$D_x g^{(n+1)}(x) \equiv \int_{-\infty}^{\infty} \mathrm{d}Y K\left(x,Y;g^{(n)}\right)\bar{k}(Y) , \tag{47}$$

starting from a given initial metric $g^{(0)}$, typically a flat spacetime. If the exact solution is very far from Minkowski it is more convenient to start from the metric of Evans and Coleman solution. Figures 4 and 5 show the convergence process for a Gaussian $\bar{k}$ centered around $Y = 6$ with width $\sigma = 1$ and maximum value of 10^{-4}.

4.3 Relevance for Critical Phenomena

We have explicitly constructed a family of spherically symmetric, self-similar solutions of the massless Einstein-Vlasov system that is parameterized by an arbitrary function of two conserved quantities $k(Y,Z)$. There is plenty of room for potential candidates to be critical solutions. However, we have seen that the spacetime only depends on the averaged function $\bar{k}(Y)$. This is due to a symmetry related to the fact that our particles are massless: we can change the particle-number distribution and the energy-momentum of each particle without changing the total energy-momentum distribution, and therefore without changing the spacetime. That explains why we can have several matter distributions giving rise to the same spacetime. The same result is expected to hold for the perturbations and therefore it is impossible to have a single unstable mode. We

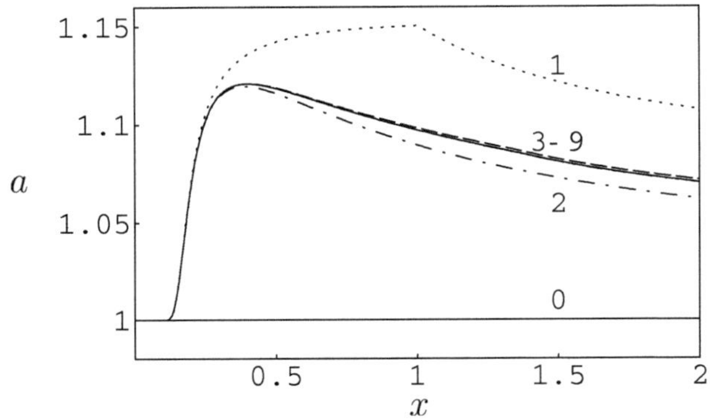

Fig. 4. Different iterations of the metric function $a(x)$, starting with the flat case in the iteration $i = 0$. The convergence is fast and starting from $i = 3$ or $i = 4$ it is not possible to resolve different iterations in the figure

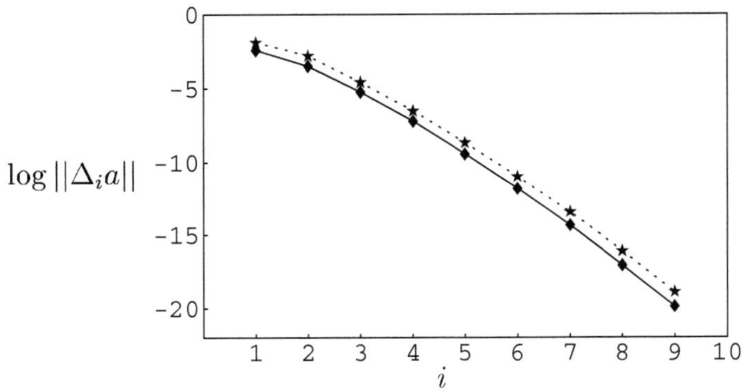

Fig. 5. Decay of differences between successive iterations in Fig. 4. We have defined $\Delta_i a \equiv a_i - a_{i-1}$. The continuous line represents the 2-norm and the dotted line the ∞-norm, both integrated between $x = 0$ and $x = 2$

either have none or an infinite family. Critical phenomena are not possible within the massless Einstein-Vlasov system.

There are numerical simulations of the evolution of a self-gravitating Einstein-Vlasov system, [20] and [21], but they evolve massive particles. The former did not found any sign of critical phenomena, and in particular it was impossible to form black holes of arbitrarily small mass. The latter reference found some evidence of a different kind of critical phenomena, where a metastable static (instead of self-similar) solution acts as an intermediate attractor at the black hole threshold.

The mass of the particles introduces a scale in momentum space. Therefore, we can not extend our arguments to the massive case because that symmetry between different matter distributions is not valid anymore.

5 Conclusions

The threshold of black hole formation is a very interesting area of research that remains mostly unexplored apart from its restriction to spherical symmetry. It involves many different disciplines: gravitational collapse and theory of singularities, PDE asymptotics, general theory of dynamical systems, self-similarity and scaling, numerical relativity, etc, and therefore benefits from the exchange of ideas and methods among those disciplines. An example of this is the prediction of scaling functions in the process of collapse, the same kind of scaling functions that are so important in condensed matter physics.

We have reviewed some current ideas and results on Critical Phenomena with matter systems involving angular momentum. Within the perfect fluid model with equation of state $p = \kappa\varrho$, and using linear perturbation theory, we predict that there will be a critical regime for generic slightly nonspherical collapse, dominated by the Evans and Coleman solution for $0 < \kappa < 0.49$. In particular we predict the critical exponent associated to the angular momentum of the black holes formed in near critical evolutions: $\mu \simeq 0.889$ for $\kappa = 1/3$. This result, together with a parallel result for the scalar field system [13] (with exponent $\mu \simeq 0.762$) and the work by Abraham and Evans on collapse of axisymmetric waves, strongly suggest that critical phenomena are a generic feature in General Relativity. Those predictions must of course be confirmed in nonlinear numerical evolutions.

If we include angular momentum using a statistical description for the matter, instead of a usual local field theory, we find that critical phenomena are not always allowed, or at least not the kind of critical phenomena that we have studied here. Nonlinear evolutions seem to support that view. It is not clear whether this is an intrinsic problem associated to the statistical description (which contains far more degrees of freedom than a field theory), or whether it is associated with the collisionless approximation, which certainly must fail in the last stages of a collapse process.

References

1. M.W. Choptuik: Phys. Rev. Lett. **70**, 9 (1993)
2. C. Gundlach, J.M. Martín-García: Phys. Rev. **D54**, 7353 (1996)
3. S. Hod, T. Piran: Phys. Rev. **D55**, 3485 (1997)
4. A.H. Cahill, M.E. Taub: Comm. Math. Phys. **21**, 1 (1971)
5. C.R. Evans, J.S. Coleman: Phys. Rev. Lett. **72**, 1782 (1994)
6. A.M. Abrahams, C.R. Evans: Phys. Rev. Lett. **70**, 2980 (1993)
7. D. Maison: Phys. Lett. **B366**, 82 (1996)
8. T. Koike, T. Hara, S. Adachi: Phys. Rev. Lett. **74**, 5170 (1995)
9. U.H. Gerlach, U.K. Sengupta: Phys. Rev. **D19**, 2268 (1979)
10. C. Gundlach: Phys. Rev. **D65**, 064019 (2002)
11. T. Regge, J.A. Wheeler: Phys. Rev. **108**, 1063 (1957)
12. V. Moncrief: Annals of Phys. **88**, 323 (1974)
13. J.M. Martín-García, C. Gundlach: Phys. Rev. **D59**, 064031 (1999)

14. C. Gundlach, J.M. Martín-García: Phys. Rev. **D61**, 084024 (2000)
15. C. Gundlach: 'Critical gravitational collapse of a perfect fluid with $p = \kappa\varrho$: Non-spherical perturbations', to be published in Phys. Rev. **D65** (2002)
16. J. Batt, W. Faltenbacher, E. Horst: Arch. Rational Mech. Anal. **83**, 159 (1986)
17. G. Rein: Math. Proc. Cambridge Philos. Soc. **115**, 559 (1994)
18. J. Schaeffer: Comm. Math. Phys. **204**, 313 (1999)
19. J.M. Martín-García, C. Gundlach: 'Self-similar spherically symmetric solutions of the massless Einstein-Vlasov system', to be published in Phys. Rev. **D65** (2002)
20. G. Rein, A. D. Rendall, J. Schaeffer: Phys. Rev. **D58**, 044007 (1998)
21. I. Olabarrieta, M.W. Choptuik: Phys. Rev. **D65**, 024007 (2002)
22. C. Gundlach: Phys. Rev. Lett. **75**, 3214 (1995)
23. C. Gundlach: Phys. Rev. **D55**, 695 (1998)
24. C. Gundlach: Phys. Rev. **D57**, R7075 (1998)
25. C. Gundlach: Phys. Rev. **D57**, R7080 (1998)
26. C. Gundlach: Living Reviews in Relativity **2**, 1 (1999)
27. J.M. Martín-García, C. Gundlach: Phys. Rev. **D64**, 024012 (2001)

Part II

Numerical and Perturbative Analysis of Astrophysical Processes

Stellar Perturbations

Valeria Ferrari

Dipartimento di Fisica “G.Marconi”,
Università di Roma “La Sapienza” and Sezione INFN ROMA1,
piazzale Aldo Moro 2, I-00185 Roma, Italy

Abstract. In this chapter the theory of non-rotating perfect fluid stellar adiabatic perturbations is derived in a similar way to black hole perturbations. In that case a wave equation can be obtained only for the axial perturbations. Whereas for black holes quasi-normal modes are purely gravitational, the modes of oscillation of stars are due to the coupling between the fluid and the gravitational field. This means that the emitted radiation carries information on the structure of the source.

1 Introduction

In this chapter I will show how the theory of non radial perturbations of non rotating stars can be developed along the same lines of the theory of black hole perturbations. Black holes are the simplest astrophysical objects, a curvature singularity hidden by a horizon, and are completely described by the metric tensor. Stars are much more complex objects, since they are composed of a fluid which obeys an assigned equation of state (EOS).

Thus, the structure of the spacetime generated by a star and its dynamical behaviour under a small perturbation has to be found by solving the Einstein equations coupled to the equations of Hydrodynamics. Due to the variety of models of stars existing in nature or proposed by theoretical investigations, the field is so wide that we need to limit our analysis to a restricted domain. In this chapter I will focus on the study of non-rotating stars that are composed by perfect fluids and I will consider only adiabatic perturbations, in which changes in the pressure and in the energy density arise without dissipation.

When stars or black holes are perturbed, they are set into non-radial oscillations and emit gravitational waves at the characteristic frequencies and damping times of the quasi-normal modes (QNMs). For black holes, these modes are purely gravitational, because the information on the detailed structure of the original star is radiated away during the gravitational collapse and only essential parameters like mass and angular momentum eventually identify the newborn collapsed object.

Conversely, the modes of oscillation of stars are due to the coupling between the fluid and the gravitational field and consequently the emitted radiation carry information on the structure of the source and on the manner in which the gravitational field couples to matter.

In addition, we will see that the QNM-spectrum of a star in General Relativity has a very rich structure and that there exist modes of the radiative field that do not have a newtonian counterpart.

Stellar pulsations have a central role in astrophysics. For instance, they are observed in the Sun and the corresponding frequencies, measured with very high accuracy, are used in modern heliosysmology to investigate the internal structure of our star.

Non-radial pulsations are thought to be at the origin of the drifting subpulses and micropulses detected in some radio sources and of the quasi-periodic variability seen in some X-ray burst sources and in a number of bright X-ray sources [1]. For this reason, much effort has been put in computing the frequencies of the quasi-normal modes for different stellar models.

A first formulation of the theory of stellar perturbations in General Relativity was developed in 1967 by K. Thorne and his collaborators [2] and the first mode calculations date back to 1969 [3]; in 1983 Detweiler and Lindblom tabulated the real and the imaginary part of the first characteristic frequencies for $\ell = 2$ for a wide range of stellar models (13 equations of state in the supranuclear density regime) [4]. This work has been extended recently by Andersson and Kokkotas [5]. I will briefly summarize the results in Sect. 3.

A new interest in the subject of stellar perturbations was stimulated by some work that S. Chandrasekhar and I did in the early nineties; we proposed a formulation of the theory, which allowed to shed a new light onto the problem and allowed the discovery of new sets of modes [6–12]. In the following of this chapter, I will describe the theory of stellar perturbations in the formulation that I contributed to develop. I shall make use of many equations already written in describing the perturbations of non-rotating black holes.

2 The Perturbations of a Non-rotating Star

The metric $g^0_{\mu\nu}$ which describes the unperturbed star, is assumed to be static and spherically symmetric

$$ds^2 = \mathrm{e}^{2\nu}(\mathrm{d}t)^2 - \mathrm{e}^{2\mu_2}(\mathrm{d}r)^2 - r^2(\mathrm{d}\theta^2 + \sin^2\theta \mathrm{d}\varphi^2) \; . \tag{1}$$

Inside the star, the functions $\nu(r)$ and $\mu_2(r)$ can be determined by solving the equations of hydrostatic equilibrium. We shall consider a star composed by a perfect fluid, whose energy-momentum tensor is given by

$$T^{\alpha\beta} = (p + \varepsilon)u^\alpha u^\beta - pg^{\alpha\beta} \; , \tag{2}$$

where p and ε are respectively the pressure and the energy density, that are assumed to have an isotropical distribution, and u^α is the four-velocity of the fluid. By defining the mass contained inside a sphere of radius r as

$$m(r) = 4\pi \int_0^r \varepsilon r^2 \mathrm{d}r \; , \tag{3}$$

the relevant equations which describe the equilibrium configuration are

$$\nu_{,r} = -\frac{p_{,r}}{p+\varepsilon} \; , \tag{4}$$

$$\left[1 - \frac{2m(r)}{r}\right] p_{,r} = -(\varepsilon + p)\left[pr + \frac{m(r)}{r^2}\right] \; , \tag{5}$$

and $e^{2\mu_2} = (1 - 2m(r)/r)^{-1}$. When the equation of state (EOS) of the fluid is specified, these equations can be solved numerically and the distribution of pressure and energy-density through the star can be determined. Once ε and p are known, (4) can be integrated

$$\nu = -\int_0^r \frac{p_{,r}}{(\varepsilon + p)} \mathrm{d}r + \nu_0 \, . \tag{6}$$

The constant ν_0 is determined by the condition that at the boundary of the star the metric reduces to the Schwarzschild metric, i.e.

$$(\mathrm{e}^{2\nu})_{r=R} = (\mathrm{e}^{-2\mu_2})_{r=R} = 1 - \frac{2M}{R} \, , \tag{7}$$

where $M = m(R)$ is the total mass. We shall now consider small perturbations of $g^0_{\mu\nu}$

$$g_{\mu\nu} = g^0_{\mu\nu} + h_{\mu\nu} \, , \qquad |h_{\mu\nu}| \ll |g^0_{\mu\nu}| \, .$$

Inside the star we shall perturb the Einstein equations coupled to the equations of Hydrodynamics

$$\delta G_{\mu\nu}(h_{\mu\nu}) = \frac{8\pi G}{c^4} \delta T_{\mu\nu} \, , \qquad \delta \left[T^{\mu\nu}{}_{;\nu}\right] = 0 \, .$$

The perturbed stress-energy tensor of the fluid is given by

$$\delta T_{ik} = (\varepsilon + p) u_i \delta u_k + (\varepsilon + p) u_k \delta u_i + (\delta\varepsilon + \delta p) u_i u_k - \delta p g_{ik} - p h_{ik} \, , \tag{8}$$

where $\delta \boldsymbol{u} = \boldsymbol{\xi}_{,t}$ and $\boldsymbol{\xi}$ is the Lagrangian displacement of the perturbed fluid element. δp and $\delta\varepsilon$ are the perturbations of the pressure and of the energy-density.

Both $h_{\mu\nu}$ and $\delta T_{\mu\nu}$ will be expanded in tensor harmonics as described in Chap. 2, Sect. 2. We shall make the same choice of gauge, i.e. the perturbed metric tensor will be written as in Chap. 2, (6) and (7). The explicit expression of $\delta T_{\mu\nu}$ expanded in harmonics is given in the Appendix.

2.1 The Axial Equations

The equations we need to solve (to be compared to (8) in Chap. 2) are

$$\begin{aligned}
\delta G_{\vartheta\varphi} = \frac{8\pi G}{c^4} \delta T_{\vartheta\varphi} : \; & \frac{\sin\vartheta}{2} \left(\partial^2_\vartheta - \cot\vartheta \partial_\vartheta - \frac{1}{\sin\vartheta^2} \partial_{\varphi^2} \right) Y_{\ell m}(\vartheta, \varphi) \\
& \left\{ -\mathrm{i}\omega \mathrm{e}^{-2\nu} h_0^{\mathrm{ax}} - \mathrm{e}^{-2\mu_2} \left[h^{\mathrm{ax}}_{1,r} + (\nu - \mu_2)_{,r} h_1^{\mathrm{ax}} \right] \right\} = \frac{8\pi G}{c^4} \delta T_{\vartheta\varphi}
\end{aligned}$$

$$\begin{aligned}
\delta G_{r\vartheta} = \frac{8\pi G}{c^4} \delta T_{r\vartheta} : \; & -\frac{1}{2\sin\vartheta} \partial_\varphi Y_{\ell m}(\vartheta, \varphi) \left\{ \mathrm{e}^{-2\nu} \left[\omega^2 h_0^{\mathrm{ax}} - \mathrm{i}\omega \left(h^{\mathrm{ax}}_{0,r} - \frac{2}{r} h_0^{\mathrm{ax}} \right) \right] \right. \\
& \left. - \frac{2n}{r^2} h_1^{\mathrm{ax}} - 2\mathrm{e}^{-2\mu_2} h_1^{\mathrm{ax}} \left[\nu_{,rr} + \left(\frac{1}{r} + \nu_{,r} \right) (\nu - \mu_2)_{,r} \right] \right\} \\
& = \frac{8\pi G}{c^4} \delta T_{r\vartheta} \, .
\end{aligned}$$

It is easy to see that

$$\delta T_{\vartheta\varphi} = \delta[(\varepsilon + p)u_\vartheta u_\varphi - p g_{\vartheta\varphi}] = 0 \,,$$

because in the unperturbed regime only $u^t \neq 0$; in addition

$$\frac{8\pi G}{c^4}\delta T_{r\vartheta} = \frac{1}{2\sin\vartheta}\partial_\varphi Y_{\ell m}(\vartheta,\varphi)\frac{8\pi G}{c^4}\,(2p\,h_1) \ .$$

The equilibrium equations $G_{\vartheta\vartheta} = 8\pi G T_{\vartheta\vartheta}/c^4$ gives

$$\nu_{,rr} + \left(\frac{1}{r} + \nu_{,r}\right)(\nu - \mu_2)_{,r} = \frac{8\pi G}{c^4} p\, \mathrm{e}^{2\mu_2} \,,$$

and by direct substitution into the axial equations, we find

$$\begin{aligned} &-\mathrm{i}\omega \mathrm{e}^{-2\nu} h_0^{\mathrm{ax}} - \mathrm{e}^{-2\mu_2}\left[h_{1,r}^{\mathrm{ax}} + (\nu - \mu_2)_{,r} h_1^{\mathrm{ax}}\right] = 0 \,, \\ &\mathrm{e}^{-2\nu}\left[-\mathrm{i}\omega\left(h_{0,r}^{\mathrm{ax}} - \frac{2}{r}h_0^{\mathrm{ax}}\right) + \omega^2 h_1^{\mathrm{ax}}\right] - \frac{2n}{r^2}h_1^{\mathrm{ax}} = 0 \,, \end{aligned} \tag{9}$$

that are formally the same as (9), which we derived in Chap. 2 for black holes. Thus, the axial perturbations do not excite fluid pulsations and, as for black holes, they are perturbations of the gravitational field only. But there is a difference: now the unperturbed metric functions ν and μ_2 have to be found by solving the equations of hydrostatic equilibrium for the star, for an assigned EOS. This means that ν and μ_2 will be related to ε and p.

2.2 A Schrödinger Equation for the Axial Perturbations

By using the same procedure as in the case of black holes, and introducing the same function Z^{ax}

$$Z^{\mathrm{ax}}(\omega, r) = \mathrm{e}^{\nu-\mu_2}\frac{h_1^{\mathrm{ax}}(\omega, r)}{r} \,,$$

it is easy to reduce the axial equations (9) to a single wave equation with a potential barrier [7]

$$\begin{aligned} &\frac{\mathrm{d}^2 Z^{\mathrm{ax}}}{\mathrm{d}r_*^2} + [\omega^2 - V_\ell^{\mathrm{ax}}(r)]Z^{\mathrm{ax}} = 0 \,, \\ &V^{\mathrm{ax}}(r) = \frac{\mathrm{e}^{2\nu}}{r^3}[\ell(\ell+1)r + r^3(\varepsilon - p) - 6m(r)] \,, \end{aligned} \tag{10}$$

where $r_* = \int_0^r \mathrm{e}^{-\nu+\mu_2}\,\mathrm{d}r$. It is remarkable that the potential barrier, which depends only on the mass in the case of a Schwarzschild black holes, now depends on how the energy-density and the pressure are distributed inside the star in the equilibrium configuration. The asymptotic behaviour of the solution of (10) at radial infinity is

$$\begin{aligned} Z^{\mathrm{ax}} \to &+ \left\{\alpha - \beta\frac{n+1}{\omega r} - \frac{1}{2\omega^2}[n(n+1)\alpha - 3M\omega\beta]\frac{1}{r^2} + ...\right\}\cos\omega r_* \\ &- \left\{\beta + \alpha\frac{n+1}{\omega r} - \frac{1}{2\omega^2}[n(n+1)\alpha + 3M\omega\beta]\frac{1}{r^2} + ...\right\}\sin\omega r_* \,. \end{aligned} \tag{11}$$

In conclusion, the axial perturbations of stars and black holes are both described by a Schrödinger like equation with a potential barrier, and therefore in both cases an incident axial wave will experience a potential scattering; however, there is a basic difference: the equation for a Schwarzschild black hole describes a problem of scattering by a one-dimensional potential barrier defined in the interval $-\infty < r_* < +\infty$, whereas in the case of a star it describes the scattering by a central potential defined within $0 < r < +\infty$.

2.3 The Polar Equations

In analogy with the polar equations derived for a Schwarzschild black hole, the relevant equations that describe the polar perturbations of a star can be written in the following form

$$\begin{aligned}
&\delta G_{tr}: && \left[\frac{\mathrm{d}}{\mathrm{d}r} + \left(\frac{1}{r} - \nu_{,r}\right)\right](2T - kV) - \frac{2}{r}L = -U \\
&\delta G_{t\vartheta}: && T - V + L = -W \\
&\delta G_{r\vartheta}: && (T - V + N)_{r} - \left(\frac{1}{r} - \nu_{,r}\right)N - \left(\frac{1}{r} + \nu_{,r}\right)L = 0 \\
&\delta G_{rr}: && \frac{2}{r}N_{,r} + \left(\frac{1}{r} + \nu_{,r}\right)(2T - kV)_{,r} - \frac{2}{r}\left(\frac{1}{r} + 2\nu_{,r}\right)L \\
& && -\frac{1}{r^2}(2nT + kN)\mathrm{e}^{2\mu_2} + \omega^2\mathrm{e}^{-2\nu+2\mu_2}(2T - kV) = 2\mathrm{e}^{2\mu_2}\Pi \\
&\frac{\delta G_{\vartheta\vartheta} - \delta G_{\varphi\varphi}}{\sin^2\vartheta}: && V_{,rr} + \left(\frac{2}{r} + \nu_{,r} - \mu_{2,r}\right)V_{r} + \frac{\mathrm{e}^{2\mu_2}}{r^2}(N + L) \\
& && +\omega^2\mathrm{e}^{2\mu_2-2\nu}V = 0 \\
& && k = \ell(\ell+1)\,, \qquad 2n = k - 2
\end{aligned} \tag{12}$$

This set of equations is the same as that for black holes (Chap. 2, (10)) but on the right hand side now there is the radial part of the variables that describe the perturbed fluid; indeed, W, U are respectively the radial part of the $\vartheta-$ and $r-$ components of the lagrangian displacement, and Π refers to the variation of the pressure (see Appendix). By a suitable manipulation of the hydrodynamical equations the fluid variables can be written as

$$\begin{aligned}
\Pi &= -\frac{1}{2}\omega^2\mathrm{e}^{-2\nu}W - (\varepsilon + p)N\,, \quad E = Q\Pi + \frac{\mathrm{e}^{-2\mu_2}}{2(\varepsilon + p)}(\varepsilon_{,r} - Qp_{,r})U\,, \\
U &= \frac{[(\omega^2\mathrm{e}^{-2\nu}W)_{,r} + (Q+1)\nu_{,r}(\omega^2\mathrm{e}^{-2\nu}W) + 2(\varepsilon_{,r} - Qp_{,r})N](\varepsilon + p)}{[\omega^2\mathrm{e}^{-2\nu}(\varepsilon + p) + \mathrm{e}^{-2\mu_2}\nu_{,r}(\varepsilon_{,r} - Qp_{,r})]}\,,
\end{aligned} \tag{13}$$

where

$$Q = \frac{(\varepsilon + p)}{\gamma p}\,, \qquad \gamma = \frac{(\varepsilon + p)}{p}\left(\frac{\partial p}{\partial \varepsilon}\right)_{\text{entropy=const}}$$

γ is the adiabatic exponent, and E is the radial part of $\delta\varepsilon$. Equations (13) show that the fluid variables $[W, U, E, \Pi]$ can be expressed as a combination of the metric perturbations $[T, V, L, N]$.

Therefore, if we replace the expressions of U and Π on the right-hand side of (12), we obtain a *system of equations which involves only the perturbations of the metric functions, with no reference to the motion of the fluid.* This fact is remarkable and it should be stressed that this decoupling requires no assumptions on the equation of state of the fluid. After eliminating the fluid variables and introducing the new functions X and G instead of V and T, the final set of equations involving only gravitational variables is

$$X_{,r,r} + \left(\frac{2}{r} + \nu_{,r} - \mu_{2,r}\right) X_{,r} + \frac{n}{r^2}\mathrm{e}^{2\mu_2}(N+L) + \omega^2 \mathrm{e}^{2(\mu_2-\nu)} X = 0\,,$$
$$(r^2 G)_{,r} = n\nu_{,r}(N-L) + \frac{n}{r}(\mathrm{e}^{2\mu_2}-1)(N+L) + r(\nu_{,r}-\mu_{2,r})X_{,r}$$
$$+\omega^2 \mathrm{e}^{2(\mu_2-\nu)} rX\,,$$
$$-\nu_{,r}N_{,r} = -G + \nu_{,r}[X_{,r} + \nu_{,r}(N-L)] + \frac{1}{r^2}(\mathrm{e}^{2\mu_2}-1)(N - rX_{,r} - r^2 G)$$
$$-\mathrm{e}^{2\mu_2}(\varepsilon+p)N + \frac{1}{2}\omega^2 \mathrm{e}^{2(\mu_2-\nu)}\left\{N + L + \frac{r^2}{n}G + \frac{1}{n}[rX_{,r} + (2n+1)X]\right\}\,,$$
$$L_{,r}(1-D) + L\left[\left(\frac{2}{r} - \nu_{,r}\right) - \left(\frac{1}{r} + \nu_{,r}\right)D\right] + X_{,r} + X\left(\frac{1}{r} - \nu_{,r}\right) + DN_{,r} +$$
$$+N\left(D\nu_{,r} - \frac{D}{r} - F\right) + \left(\frac{1}{r} + E\nu_{,r}\right)\left[N - L + \frac{r^2}{n}G + \frac{1}{n}(rX_{,r} + X)\right] = 0 \quad (14)$$

where

$$A = \frac{1}{2}\omega^2 \mathrm{e}^{-2\nu}\,, \qquad B = \frac{\mathrm{e}^{-2\mu_2}\nu_{,r}}{2(\varepsilon+p)}(\varepsilon_{,r} - Qp_{,r})\,,$$
$$D = 1 - \frac{A}{2(A+B)}\,,\; E = D(Q-1) - Q\,, \qquad F = \frac{\varepsilon_{,r} - Qp_{,r}}{2(A+B)}\,,$$

and the definitions of X_ℓ and G_ℓ are

$$X = nV\,, \quad (15)$$
$$G = \nu_{,r}\left[\frac{n+1}{n}X - T\right]_{,r} + \frac{1}{r^2}(\mathrm{e}^{2\mu_2}-1)[n(N+T)+N] + \frac{\nu_{,r}}{r}(N+L) -$$
$$-\mathrm{e}^{2\mu_2}(\varepsilon+p)N + \frac{1}{2}\omega^2 \mathrm{e}^{2(\mu_2-\nu)}\left[L - T + \frac{2n+1}{n}X\right]\,. \quad (16)$$

In order to integrate these equations numerically, we need to impose some boundary conditions:

1. The lagrangian perturbation of the pressure must vanish at the boundary of the star.
2. All functions must be regular at the center of the star where they behave as

$$(G, V, L, N) \sim (G_0, V_0, L_0, N_0)r^x + (G_2, V_2, L_2, N_2)r^{(x+2)} + O\left(r^{(x+4)}\right)\,. \quad (17)$$

The coefficients of the expansion and the exponent x can be found by direct substitution of (17) into (14).

The system of equations (14) is a fifth order linear system, thus in principle we expect five independent solutions for x, and five corresponding sets of values for the coefficients (G_0, V_0, L_0, N_0) and (G_2, V_2, L_2, N_2). However, only for two coincident values of x the solution is regular at the origin. This value is $x = \ell$, where ℓ is the harmonic index expressing the order of the multipole we are considering. Consequently there are only two independent solutions to be numerically integrated, and the corresponding coefficients can easily be found. The procedure of integration is the following: for a fixed value of the frequency ω, integrate equations (14) for the two independent, regular solutions that behave as in (17) near the origin. At the boundary, superimpose the solutions in such a way that the lagrangian perturbation of the pressure vanishes. Outside the star the perturbed equations reduce to the Schwarzschild equations, therefore, from the values of the functions $X = nV$ and L obtained in the interior (and their derivatives) we compute the Zerilli function Z^{pol}:

$$Z^{\mathrm{pol}}(\omega, r) = \frac{r}{nr + 3M}\Big[3MV(\omega, r) - rL(\omega, r)\Big] , \tag{18}$$

and its first derivative, and then integrate the Zerilli equation

$$\begin{aligned} &\frac{\mathrm{d}^2 Z^{\mathrm{pol}}}{\mathrm{d}r_*^2} + [\omega^2 - V_\ell^{\mathrm{pol}}(r)]Z^{\mathrm{pol}} = 0 , \\ &V^{\mathrm{pol}}(r) = \frac{2(r-2M)}{r^4(nr+3M)^2}[n^2(n+1)r^3 + 3Mn^2r^2 + 9M^2nr + 9M^3] \end{aligned} \tag{19}$$

up to radial infinity. The asymptotic behaviour of the function Z^{pol} for large r is

$$\begin{aligned} Z^{\mathrm{ax}} \to &\left\{\alpha - \beta\frac{n+1}{\omega r} - \frac{1}{2\omega^2}\left[n(n+1)\alpha - \frac{3}{2}M\beta\omega\left(1 + \frac{2}{n}\right)\right]\frac{1}{r^2} + \dots\right\}\cos\omega r_* \\ -&\left\{\beta + \alpha\frac{n+1}{\omega r} - \frac{1}{2\omega^2}\left[n(n+1)\alpha + \frac{3}{2}M\beta\omega\left(1 + \frac{2}{n}\right)\right]\frac{1}{r^2} + \dots\right\}\sin\omega r_* , \end{aligned} \tag{20}$$

where α and β are functions of ω. In this way the complete solution is obtained.

3 The Quasi-normal Modes of a Star

The quasi-normal modes are solutions of the axial and polar equations that satisfy the following boundary conditions:

1. They behave as a pure outgoing wave at infinity.
2. All perturbed functions have a regular behaviour at $r = 0$.
3. The interior solution matches continuously with the exterior perturbation at the surface of the star.

Before describing the many interesting properties of the QNM spectrum of a star, let us see how these eigenfrequencies can be determined. We shall describe, in particular, a method which is appropriate when the imaginary part of the mode frequency is much smaller than the real part, i.e. for modes that are not very strongly damped.

3.1 A Method to Find the Frequencies of the Quasi-normal Modes

Let us consider a Schrödinger equation

$$\frac{\mathrm{d}^2 Z_c}{\mathrm{d}r_*^2} + (\omega^2 - V) Z_c = 0 \; , \tag{21}$$

where V is a short-range, central potential barrier. We want to find the complex values of the frequency, $\omega_c = \omega + \mathrm{i}\omega_i$, such that the corresponding solution of (21) is regular at $r_* = 0$ and behaves as a pure outgoing wave at radial infinity, i.e.

$$Z_c \sim \mathrm{e}^{-\mathrm{i}\omega_c r_*} \; , \text{ when } r_* \to \infty \; .$$

We shall assume that $Z_c = Z + \mathrm{i}Z_i$ and that $\omega_i \ll \omega$, i.e. the decay time of the emission of gravitational waves, $\tau = 2\pi/\omega_i$, is much longer than the pulsation period.

This condition is certainly satisfied by lower modes, which are less damped and therefore more significant from the observational point of view. By separating the real and the imaginary part in (21), we find

$$\frac{\mathrm{d}^2 Z}{\mathrm{d}r_*^2} - VZ + (\omega^2 - \omega_i^2) Z - 2\omega\omega_i Z_i = 0 \; , \tag{22}$$

$$\frac{\mathrm{d}^2 Z_i}{\mathrm{d}r_*^2} - VZ_i + (\omega^2 - \omega_i^2) Z_i + 2\omega\omega_i Z = 0 \; . \tag{23}$$

If we now put $Z_i = \omega_i Y$, and neglect terms of order $O(\omega_i^2)$ in (22) and (23), they become

$$\frac{\mathrm{d}^2 Z}{\mathrm{d}r_*^2} + (\omega^2 - V) Z = 0 \; , \tag{24}$$

$$\frac{\mathrm{d}^2 Y}{\mathrm{d}r_*^2} + (\omega^2 - V) Y + 2\omega Z = 0 \; . \tag{25}$$

From (25) it follows that

$$Y(r_*, \omega) = \frac{\partial}{\partial\omega} Z(r_*, \omega) \; , \tag{26}$$

and consequently

$$Z_c(r_*, \omega_c) = Z(r_*, \omega) + \mathrm{i}\,\omega_i \left[\frac{\partial}{\partial\omega} Z(r_*, \omega) \right] \; . \tag{27}$$

Therefore when $\omega_i \ll \omega$, we can construct the complex solution Z_c corresponding to a complex value of the frequency ω_c, by integrating exclusively (24) for the real part Z, and for real values of the frequency ω.

The Asymptotic Behaviour of Z_c. When $r_* \to \infty$, the potential V tends to zero and (24) admits two linearly independent solutions that have the following asymptotic behaviour

$$Z_1 \to \cos \omega r_* + O(r_*^{-1}) , \qquad Z_2 \to \sin \omega r_* + O(r_*^{-1}) .$$

Thus the general real solution Z is

$$Z(r_*, \omega) = \alpha(\omega) Z_1(r_*, \omega) - \beta(\omega) Z_2(r_*, \omega) , \tag{28}$$

where $\alpha(\omega)$ and $\beta(\omega)$ are functions to be determined, for example, by matching (28) with the solution (11) numerically integrated for different initially assigned values of real ω. From (27) and (28) it follows that the complete solution for Z_c, up to terms of order $O(\omega_i^2)$ is

$$\begin{aligned} Z_c &= Z + \mathrm{i}\omega_i \frac{\partial Z}{\partial \omega} \\ &= \alpha(\omega) Z_1 - \beta(\omega) Z_2 - \mathrm{i}\omega_i [\alpha'(\omega) Z_1 - \beta'(\omega) Z_2 + \alpha(\omega) Z_1' - \beta(\omega) Z_2'] , \end{aligned} \tag{29}$$

where the prime indicates differentiation with respect to ω. For sufficiently large values of r_* the behaviour of Z_c is

$$Z_c \to (\alpha + \ \omega_i \alpha' - \mathrm{i}\omega_i \beta r_*) \cos \omega r_* - (\beta + \mathrm{i}\omega_i \beta' - \mathrm{i}\omega_i \alpha r_*) \sin \omega r_* . \tag{30}$$

It is clear that the terms proportional to r_* would eventually diverge if $r_* \to \infty$. However, in the limit $\omega_i \ll \omega$, the asymptotic behaviour (28) that we use to determine α and β is established long before these terms begin to dominate.

Therefore, if the value of r_* where we start to match the numerically integrated real solution Z with the asymptotic behaviour is large enough for (28) to be applied, but not so far that the exponential growth has taken over in (30), the diverging terms can be neglected and the asymptotic form of Z_c can be written as

$$\begin{aligned} Z_c &\to \frac{1}{2}[(\alpha - \omega_i \beta') + \mathrm{i}(\beta + \omega_i \alpha')] \mathrm{e}^{\mathrm{i}\omega r_*} + \frac{1}{2}[(\alpha + \omega_i \beta') - \mathrm{i}(\beta - \omega_i \alpha')] \mathrm{e}^{-\mathrm{i}\omega r_*} \\ &= I(\omega) \mathrm{e}^{+\mathrm{i}\omega r_*} + O(\omega) \mathrm{e}^{-\mathrm{i}\omega r_*} . \end{aligned} \tag{31}$$

(That such value of r_* does indeed exist has been shown by a direct verification in [10]). We now impose the outgoing wave condition, by setting to zero the coefficient of the ingoing wave, $I(\omega)$, in (31)

$$\alpha - \omega_i \beta' = 0 , \text{ and } \beta + \omega_i \alpha' = 0 . \tag{32}$$

Eliminating ω_i we finally find

$$\alpha \alpha' + \beta \beta' = 0 . \tag{33}$$

This equation says that if there exists a value of real ω, say $\omega = \omega_0$, where the function $(\alpha^2 + \beta^2)$ has a minimum, then the solution Z_c at infinity will represent a pure outgoing wave. Therefore ω_0 is the real part of the complex characteristic frequency belonging to a quasi-normal mode.

The imaginary part can be obtained from (32) evaluated at $\omega = \omega_0$

$$\omega_i = \left.\frac{\alpha}{\beta'}\right|_{(\omega=\omega_0)} = -\left.\frac{\beta}{\alpha'}\right|_{(\omega=\omega_0)} . \tag{34}$$

Equation (34) suggests an alternative method to find ω_i. Since the function $(\alpha^2 + \beta^2)$ has a minimum when $\omega = \omega_0$, in the region $\omega \sim \omega_0$ it can be approximated by a parabola

$$\alpha^2 + \beta^2 = \text{const.} \left[(\omega - \omega_0)^2 + \omega_i^2\right] , \tag{35}$$

and ω_i can be determined by matching the values of $(\alpha^2 + \beta^2)$ obtained by numerical integration, with (35).

The application of the algorithm we have described to the axial perturbations is straightforward. We integrate the Schrödinger equation (10) for real values of ω, for sufficiently large r_* we match the integrated solution with the asymptotic behaviour (11) and determine the values of α and β. Then we find the values of $\omega = \omega_0$ where the resonance curve $(\alpha^2 + \beta^2)$ has a minimum: ω_0 will be the real part of the eigenfrequency. The imaginary part will be found from (34), or alternatively, by fitting the resonance curve with the parabola (35).

The same procedure can be applied in the case of the polar modes. The difference with respect to the axial case is that inside the star we need to integrate the system of equations (14) as described in Sect. 2.3. The purpose is to find the value of the function Z^{pol} at the boundary of the star (and its derivative), that is needed to integrate the Schrödinger equation (19) outside the star. At sufficiently large values of r_*, the integrated Z^{pol} will be matched with the asymptotic behaviour (20) and α and β will be determined. We shall then proceed as in the axial case.

3.2 Spacetime Quasi-normal Modes

In Sects. 2.1-2.2 we have shown that the equations governing the axial perturbations are not coupled to fluid pulsations and that they can be reduced to a wave equation for the gravitational variable Z^{ax}: the role of the fluid is that of determining the shape of the potential barrier, which explicitly depends on the radial profile of the energy density and of the pressure in the unperturbed star. Thus, the axial quasi-normal modes are pure gravitational modes and do not have a newtonian counterpart. In order to show how the frequency of the modes depends on the potential shape, it is interesting to consider the following illustrative example.

As in [9] we shall consider a star with uniform density and increasing compactness. Although this is an unrealistic model, it presents the advantage that the equilibrium configuration is known as an exact solution of Einstein's equations (the Schwarzschild solution). Moreover, in such models it is possible to test the effects of general relativity in a regime where they are stronger than in

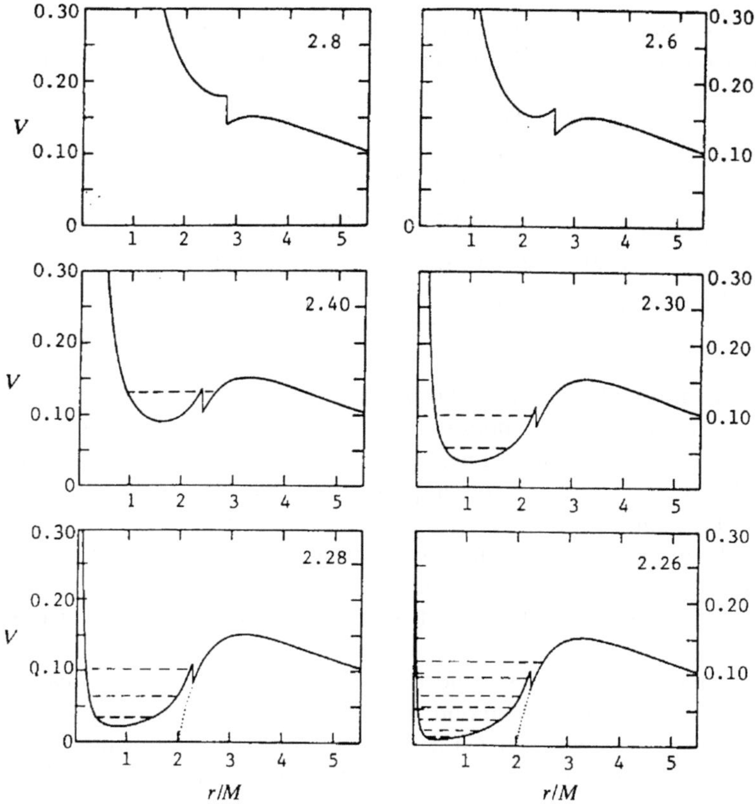

Fig. 1. The potential barrier of the axial perturbations of homogeneous stars is plotted for different values of the ratio R/M ranging from 2.8 to 2.26. The dashed lines are the values of $(\omega_0 M)^2$ corresponding to the quasi-normal modes

any other stellar model. It should be reminded that homogeneous stars can exist only if $R/M > 2.25$. The potential barrier of the axial perturbations is plotted in Fig. 1 as a function of the radial coordinate, for different values of (R/M). By integrating the axial equation, and applying the method described in Sect. 3.1, we find that if R/M is small enough ($(R/M) \lesssim 2.6$) the curve $\alpha^2 + \beta^2$ has one or more minima, i.e. there exist one or more axial QMNs and the number of modes increases as R/M approaches the limiting value.

The reason for this behaviour is that inside the star the potential is a well, which becomes deeper as the value of (R/M) decreases and the compactness increases; if the star is compact enough, the potential well becomes deep enough to allow the existence of one or more QNMs. In Table 1 we show the frequency of the lowest $\ell = 2$ quasi-normal mode for different values of (R/M). From these data we see that there is a progressive increasing of the damping time $\tau = 2\pi/\omega_i$ as the star tends to the limiting configuration $R/M = 2.25$. This means that the lowest QNMs are effectively trapped by the potential barrier and no much radiation will leak out of the star when these modes are excited.

Table 1. The frequency of the lowest $\ell = 2$ axial quasi-normal mode of homogeneous stars of different compactness. (M and ω are measured in the units $\varepsilon^{1/2}$ and $\varepsilon^{-1/2}$)

(R/M)	M	ω_0	ω_i
2.26	0.509798	0.213863874	$0.23 \cdot 10^{-8}$
2.28	0.503105	0.3689962	$0.12 \cdot 10^{-5}$
2.30	0.496557	0.473525	$0.26 \cdot 10^{-4}$
2.40	0.465848	0.7767	$0.92 \cdot 10^{-2}$

Another family of axial quasi-normal modes are the so-called **w**-modes [13]. They are strongly damped modes, i.e. the imaginary part of the frequency is comparable to the real part. These modes do exist also for the polar perturbations [14]. It is interesting to compare the eigenfrequencies of the spacetime modes of stars and black hole.

In Table 2 we give the frequencies and the damping times of the first four, $\ell = 2$, axial quasi-normal modes of homogeneous stars with mass $M = 1.35 M_\odot$ and different values of R/M and of a non-rotating black hole with the same mass.

We see that the frequencies and the damping times of the modes are higher for a star than for a black hole and increase with the order of the mode rather than decreasing. Even when the star approaches the limiting value $R/M = 2.25$, its quasi-normal mode spectrum remains different from that of a black hole with the same mass and this can be ascribed to the different boundary conditions that QNMs have to satisfy at the surface of the star and at the black hole horizon. Moreover, it may be noted that there is no smooth transition from the mathematical structure of the equations describing the perturbations of stars to those describing the perturbations of black holes.

Since the **w**-modes are strongly damped, one may argue that their physical relevance is marginal. However it has been suggested that in the first milliseconds of a gravitational collapse after the formation of a neutron star, there may be a large amount of radiation produced by the gravitational initial data the collapse itself generates.

This part of the signal should be essentially a superposition of **w**-modes, which would then be present in the emitted signal [13].

3.3 The Fluid Modes

The polar modes have a physical origin different from the axial modes, because they are essentially fluid pulsations. When the star is perturbed, each element of fluid moves under the competing action of different restoring forces, as it follows from a simple inspection of the newtonian hydrodynamical equations

$$\frac{\partial \mathbf{v}}{\partial t} = \frac{\delta \varrho}{\varrho_0^2} \nabla p_0 - \frac{1}{\varrho_0} \nabla \delta p + \delta \mathbf{f} \ , \tag{36}$$

Table 2. The characteristic frequencies and damping times of the axial quasi-normal modes of homogeneous stars with increasing compactness are compared to those of a non-rotating black hole with the same mass ($M = 1.35\ M_\odot$). We give the first four values of the frequency (in kHz) and of the damping time (in s) for $\ell = 2$. ν_0^{s} and τ^{s} refer to the *trapped* modes associated to the potential well inside the star (see text), ν_0^{w} and τ^{w} refer to the axial **w**-modes, and ν_0^{BH} and τ^{BH} to a black hole

R/M	ν_0^{s}	τ^{s}	ν_0^{w}	τ^{w}	ν_0^{BH}	τ^{BH}
2.4	8.6293	$1.52 \cdot 10^{-3}$	11.1738	$1.70 \cdot 10^{-4}$	8.9300	$7.49 \cdot 10^{-5}$
	–	–	14.2757	$8.03 \cdot 10^{-5}$	8.2848	$2.43 \cdot 10^{-5}$
	–	–	18.2232	$5.70 \cdot 10^{-5}$	7.1952	$1.39 \cdot 10^{-5}$
	–	–	22.6669	$4.88 \cdot 10^{-5}$	6.0099	$0.95 \cdot 10^{-5}$
2.28	4.4333	10.8	10.4128	$5.45 \cdot 10^{-4}$		
	6.0168	$2.50 \cdot 10^{-1}$	11.9074	$2.91 \cdot 10^{-4}$		
	7.5462	$1.44 \cdot 10^{-2}$	13.4813	$2.07 \cdot 10^{-4}$		
	8.9891	$1.83 \cdot 10^{-3}$	15.1428	$1.67 \cdot 10^{-4}$		
2.26	2.6041	$5.38 \cdot 10^{3}$	10.7852	$7.60 \cdot 10^{-4}$		
	3.5427	$1.69 \cdot 10^{2}$	11.6922	$5.34 \cdot 10^{-4}$		
	4.4802	$1.22 \cdot 10^{1}$	12.6138	$4.22 \cdot 10^{-4}$		
	5.4127	$1.37 \cdot 10^{-1}$	13.5512	$3.56 \cdot 10^{-4}$		

where

$$\delta\mathbf{f} = -\nabla\delta\phi \tag{37}$$

and $\delta\phi$ is the variation of the gravitational potential. Thus, the normal modes of oscillations are classified according to the restoring force that is prevailing: the **g**-modes, or gravity modes, if the force is due to the eulerian change in the density $\delta\varrho$, and the **p**-modes, if it is due to a change in pressure (This classification scheme was introduced by Cowling in 1942 [15]). The two classes of modes occupy well defined regions of the spectrum, and they are separated by the frequency of the **f**-mode, or 'pseudo-Kelvin' mode, as it is often referred to, since it is the generalization of the only possible mode of oscillation of an incompressible homogeneous sphere. In General Relativity this classification based on the fluid behaviour continues to hold; in addition, since the pulsation energy is dissipated, if no other dissipative mechanisms are active, through the emission of gravitational waves, the information carried by each mode is also encoded in the gravitational field; indeed the frequency of the polar QNMs modes can be found by solving (14) which are written for the perturbations of the metric tensor components only.

The frequencies of the polar QNMs depend on the internal structure of the star. For instance it is known from newtonian theory that the **f**-mode frequency scales with the mean density of the star $\nu_{\mathrm{f}} \sim \left(M/R^3\right)$ and the first p–mode frequency scales with the star's compactness M/R. This relation has been studied in General Relativity for a large number of neutron star EOSs [16–18]. For instance, the frequency and damping time of the **f**-mode are found to satisfy the following linear relations [18]

$$\frac{\nu_f}{1\ kHz} \simeq (0.78 \pm 0.01) + (1.63 \pm 0.01)\sqrt{\left(\frac{\bar{M}}{\bar{R}^3}\right)},$$
$$\frac{1\ s}{\tau_f} = \left(\frac{\bar{M}^3}{\bar{R}^4}\right)\left[(22.85 \pm 1.51) - (14.65 \pm 1.32)\left(\frac{\bar{M}}{\bar{R}}\right)\right], \qquad (38)$$

where $\bar{M} \equiv M/1.4M_\odot$ and $\bar{R} \equiv R/10$ km. Thus, if the quasi-normal modes were identified by a spectral analysis in a detected gravitational wave, these relations, and similar information on the **p**-modes, would allow to set constraints on the mass and radius of a pulsating NS.

Since the EOS in the interior of a neutron star is poorly known, it is interesting to ask whether the modes may give further information on the internal structure. It has been suggested that phase transitions from ordinary nuclear matter to quark matter may occur in the inner core of NSs [19,20]. These transitions may be associated to a density discontinuity which would affect the quasi-normal mode spectrum. In a recent study [21,22] it has been shown that, as far as the **f**-mode is concerned, the linear relation (38) between the frequency of the fundamental mode and the square root of the average density still holds for NSs with high density discontinuities. Thus, a measure of the **f**-mode frequency would be useful to constrain the neutron star average density, even when a phase transition takes place in its core. However, the presence of the discontinuity would introduce a larger error on the determination of the stellar parameters.

A discontinuity affects the QNM spectrum also in another way: due to the sudden change in density, an additional local source of buoyancy is introduced which gives rise to a discontinuity **g**-mode. If the discontinuity occurs in the inner core, these modes would have frequencies of the order of 1 kHz, somewhat lower than that of the **f**-mode, and therefore clearly distinguishable. These modes exhibit a peculiar feature: their frequencies computed for different stellar models having the same mass and different EOS, cluster in a definite region which depends exclusively on the density discontinuity $\Delta\varrho/\varrho_d$, and is independent of the other parameters of the EOS [21]. Thus, if a discontinuity **g**-mode were detected, this property could be used to infer the value of $\Delta\varrho/\varrho_d$.

Discontinuity **g**-modes are not necessarily associated to high density regions. Indeed, a density discontinuity may occur also at low density, where the characteristic chemical profile is determined by the history of the star, because shell burning, flash nuclear burning and accretion phenomena leave layers of different composition on its surface. At the interfaces between different layers, the chemical composition changes abruptly and if no significant diffusion is present, the density gradient is well approximated by a discontinuity and a low density discontinuity **g**-mode would appear. These modes would have low frequency, typically $\nu_g \lesssim 200$ Hz [23].

Finally, it should be mentioned that g-modes may have real, imaginary or zero frequency, depending on whether the considered stellar model is stable under convection; different regimes can be identified by looking at the sign of the Brunt-Väisälä frequency

$$N^2 = \frac{1}{2} e^{\nu-\mu_2} \nu' \frac{p'}{p} \left(\frac{1}{\gamma} - \frac{1}{\gamma_0} \right) , \quad \text{with} \quad \gamma_0 = \frac{\varepsilon}{p} \frac{p'}{\varepsilon'} . \tag{39}$$

A real, imaginary or zero N corresponds, respectively, to convective stability, instability or marginal instability. If the star is isentropic and/or chemically homogeneous, $\gamma = \gamma_0$ and all g-modes degenerate to zero frequency.

4 Can the Quasi-normal Modes of Stars Be Excited?

It is interesting to answer the question in the heading of this section by considering the following illustrative example.

In recent years a large number of extrasolar planetary systems have been discovered [24], which exhibit some very interesting properties. They are very close to Earth (less than 10-20 pc) and they are formed by a solar type star and one or more orbiting companions, which can be planets, super-planets or, in some cases, brown dwarfs. Many of these companions are orbiting the main star at such short distance that conflicts with the predictions of the standard theories of planetary formation and evolution. For instance, some planets [1] have orbital periods shorter than three days (for comparison, Mercury's period is 88 days).

In the light of these findings, it is interesting to ask the following question: is it possible that a planet orbits the main star so close as to excite one of its modes of oscillations? How much energy would be emitted in gravitational waves by a system in this resonant condition, compared to the energy emitted because of the orbital motion? For how long would this condition persist? These questions have been analysed in two recent papers [25] [26], whose results I shall briefly summarize.

I shall consider a simplified model of a solar type star having a polytropic EOS $p = K\varepsilon^{1+1/n}$ with $n = 3$, adiabatic exponent $\gamma = 5/3$, central density $\varepsilon_0 = 76$ g/cm^3 and $c^2\varepsilon_0/p_0 = 5.53 \cdot 10^5$. These values give a star with the same mass and radius of the Sun [26]. For this model we can compute the frequencies of the QNM, which are listed in Table 3 for $\ell = 2$. We do not include the frequencies of the p-modes because they are irrelevant for the following discussion.

If a planet moves on an orbit of radius R (we shall assume for simplicity that the orbit is circular), $\nu_{\rm k} = \sqrt{G(M_\star + M_{\rm p})/R^3}/2\pi$ is the keplerian orbital frequency, where $M_\star$ and $M_{\rm p}$ are the mass of the star and of the planet, respectively. The planet can excite a mode of frequency ν_i, only if ν_i is a specific multiple of the orbital frequency, i.e.

$$\nu_i = \ell \, \nu_{\rm k} , \tag{40}$$

[1] Here and in the following we shall indicate as "planet" also brown dwarfs and super planets.

Table 3. Values of the frequencies (in μHz) of the fundamental and of the first g-modes of oscillation of a solar type star for $\ell = 2$ (see text)

f-mode	g_1	g_2	g_3	g_4	g_5	g_6	g_7	g_8	g_9	g_{10}
285	221	168	135	113	97	85	75	68	62	57

where ℓ is the considered multipole. Introducing the dimensionless frequency

$$k_i = \frac{\nu_i}{\sqrt{GM_\star/R_\star^3}} ;$$

where $R_\star$ is the radius of the star, the resonant condition (40) can be written as

$$R_i = \left[\frac{\ell^2}{k_i^2} \cdot \left(1 + \frac{M_\mathrm{p}}{M_\star}\right)\right]^{1/3} R_\star \, . \tag{41}$$

Thus, R_i is the value of the orbital radius for a given mode to be excited; for instance, using (41) we find that in order to excite the fundamental mode of the considered star, the planet should move on an orbit of radius smaller than 2 stellar radii, therefore we first need to verify whether a planet can move on such close orbit without being disrupted by tidal interaction.

This is equivalent to establish at which distance the star starts to accrete matter from the planet (or viceversa), i.e. when the planet or the star overflow their Roche lobes. It may be noted that, following the analysis in [27], temperature effects, which may provoke the melting of the planet or the evaporation of its atmosphere, can be shown to be less stringent than the Roche-lobe limit.

Proceeding as in [26], let us assign a value of the orbital radius $R = R_i$ such that a given mode is excited according to (41) and be $R_\mathrm{RL}(R_i, M_p, M_\star)$ the Roche lobe radius; if the planet fills its Roche-lobe, its radius is equal to R_RL and its average density has a critical value, $\varrho_\mathrm{RL} = 3M_\mathrm{p}/4\pi R_\mathrm{RL}^3$. Thus, if $\varrho_\mathrm{p} > \varrho_\mathrm{RL}$, the planet does not accrete matter on the star. Using (41) and normalizing to the mass of the central star, $\varrho_\star$, the value of the critical density can be rewritten as

$$\frac{\varrho_\mathrm{RL}}{\varrho_\star} = k_i^2 \cdot \frac{M_\mathrm{p}/M_\star}{\ell^2\,(1 + M_\mathrm{p}/M_\star)\,\overline{R}_\mathrm{RL}^3} \, , \tag{42}$$

where $\overline{R}_\mathrm{RL}$ is the dimensionless quantity $\overline{R}_\mathrm{RL} = R_\mathrm{RL}/R_i$. In conclusion, a planet can excite the ith-mode of the star without overflowing its Roche lobe, only if the ratio between its mean density and that of the central star is larger than the critical ratio (42).

In [26] we have computed this ratio assuming that the central star has three different companions: two planets with the mass of the Earth and of Jupiter and a brown dwarf of 40 jovian masses, M_E, M_J and M_BD, respectively, imposing that they are on an orbit which corresponds to the resonant excitation of a given mode of the solar type star. Smaller planets produce gravitational signals that

Table 4. The critical ratio $\varrho_{\rm RL}/\varrho_\star$ is given for three planets with mass equal to that of the Earth ($M_{\rm E}$), of Jupiter ($M_{\rm J}$) and of a brown dwarf with $M_{\rm BD} = 40\ M_{\rm J}$, and for the different modes. This critical ratio corresponds to the minimum mean density that a planet should have in order to be allowed to move on an orbit which corresponds to the excitation of a **g**-mode, without being disrupted by tidal interactions

	$\varrho_{\rm RL}/\varrho_\star$									
	g_1	g_2	g_3	g_4	g_5	g_6	g_7	g_8	g_9	g_{10}
$M_{\rm E}$	12.5	7.21	4.64	3.24	2.38	1.83	1.45	1.17	0.97	0.82
$M_{\rm J}$	13.0	7.49	4.82	3.37	2.48	1.90	1.51	1.22	1.01	0.85
$M_{\rm BD}$	-	-	-	3.68	2.71	2.08	1.65	1.34	1.11	0.93

are too small to be interesting. The results of this analysis are shown in Table 4, where the value of $\varrho_{\rm RL}/\varrho_\star$ is tabulated for the three considered companions.

A planet like the Earth has a mean density such that $\varrho_{\rm E}/\varrho_\odot = 3.9$, whereas for a planet like Jupiter $\varrho_{\rm J}/\varrho_\odot = 0.9$; thus, from Table 4 we see that an Earth-like planet can orbit sufficiently close to the star to excite **g**-modes of order higher or equal to $n = 4$, whereas a Jupiter-like planet can excite only the mode g_{10} or higher.

We find that in no case the fundamental mode can be excited without the planet being disrupted by accretion. According to the brown dwarf model ("model G") by Burrows and Liebert [28] an evolved, 40 $M_{\rm J}$, brown dwarf has a radius $R_{\rm BD} = 5.9 \cdot 10^4$ km, and a corresponding mean density $\varrho_{\rm BD} = 88$ g cm^{-3}; consequently $\varrho_{\rm BD}/\varrho_\odot = 64$, and this value is high enough to allow a brown dwarf companion to excite all the **g**-modes of the central star. However, we also need to take into account the destabilizing mechanism of mass accretion *from* the central star. This imposes a further constraint and this is the reason why the slots corresponding to the excitation of the g-modes lower that g_4 in the last row of Table 4 are empty.

Having established that some **g**-modes of the central star can, in principle, be excited, we turn to the next question, i.e.: how much energy would be emitted in gravitational waves by a system in this resonant condition, compared to the energy emitted because of the orbital motion? The amplitude of the wave emitted because of the orbital motion can be computed by using the quadrupole formalism, which, in the case of circular orbit, gives (see [26] for details)

$$h_{\rm Q}(R_i) = \frac{4M_{\rm p}}{r}\sqrt{\frac{2}{15}}\,(\omega_{\rm K}R_i)^2\ . \tag{43}$$

The values of this amplitude for the three planets considered above are shown in Table 5.

However, if the stellar modes are excited the star will pulsate and emit more gravitational energy than that which can be computed by the quadrupole formula. In order to evaluate the resonant contribution we need to integrate the equations of stellar perturbations (14) by assuming that the planet is a point-

Table 5. The amplitude of the gravitational signal emitted when the three companions considered in Table 4 move on a circular orbit of radius R_i, such that the condition of resonant excitation of a **g**-mode is satisfied, is computed by the quadrupole formalism (43) for the modes allowed by the Roche-lobe analysis. The planetary systems are assumed to be at a distance of 10 pc from Earth

	g_4	g_5	g_6	g_7	g_8	g_9	g_{10}
$h_{\mathrm{Q}}^{\mathrm{E}} \cdot 10^{26}$	3.0	2.7	2.5	2.3	2.2	2.0	1.9
$h_{\mathrm{Q}}^{\mathrm{BD}} \cdot 10^{22}$	3.9	3.5	3.2	3.0	2.8	2.6	2.4
$h_{\mathrm{Q}}^{\mathrm{J}} \cdot 10^{24}$	-	-	-	-	-	-	6.1

like mass which induces a perturbation on the gravitational field and on the thermodynamical structure of the star he is orbiting.

Under these assumptions we solve the equations of stellar perturbations with a source term given by the stress-energy tensor of the planet and compute the characteristic amplitude of the gravitational wave [26]. In the following, we shall call this amplitude h_{R}, to indicate that it has been computed by solving the relativistic equations of stellar perturbations.

We have computed h_{R} for $\ell = 2$ (which is the relevant contribution), assuming that the planet moves on a circular orbit of radius $R_0 = R_i + \Delta R$, where R_i is the radius corresponding to the resonant excitation of the mode g_i, for the modes allowed by the Roche lobe analysis. We find that, as the planet approaches the resonant orbit, h_{R} grows very sharply.

In Fig. 2 we plot the ratio $h_{\mathrm{R}}(R_0)/h_{\mathrm{Q}}(R_i)$ as a function of ΔR to see how much the amplitude of the emitted wave increases, due to the excitation of a

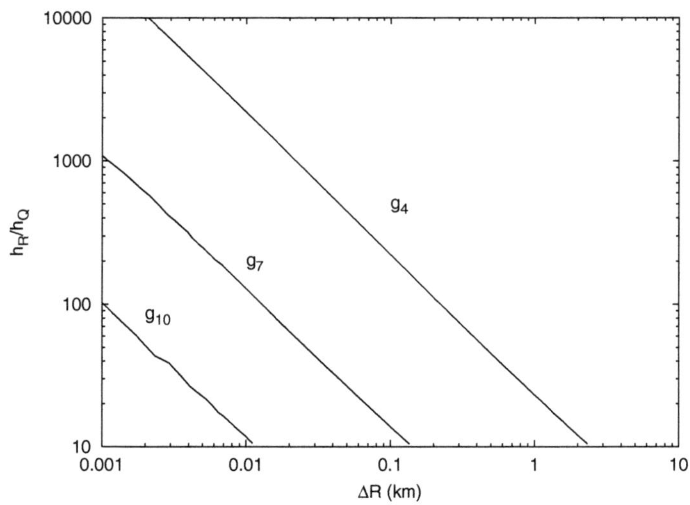

Fig. 2. The logarithm of the ratio $h_{\mathrm{R}}^{l=2}(R_i + \Delta R)/h_{\mathrm{Q}}(R_i)$ is plotted as a function of ΔR for the modes g_4, g_7, and g_{10}

g-mode, with respect to the quadrupole emission. The plot is done for the modes g_4, g_7 and g_{10} and it shows a power-law behaviour nearly independent of the order of the mode.

It should be stressed that $h_R(R_0)/h_Q(R_i)$ is also independent of the mass of the planet. From Fig. 2 we see that as the planet approaches a resonant orbit, the amplitude of the emitted wave may become significantly higher than that computed by the quadrupole formalism, h_Q. Thus the next question to answer is how long can a planet orbit near a resonance, before radiation reaction effects move it off.

Indeed, the loss of energy in gravitational waves causes a shrinking of the orbit of a planetary system, and the efficiency of this process increases as the planet approaches a resonant orbit.

On the assumption that the timescale over which the orbital radius evolves is much longer than the orbital period (adiabatic approximation), the orbital shrinking can be computed from the energy conservation law

$$M_p \left\langle \frac{dE}{dt} \right\rangle + \left\langle \frac{dE_{GW}}{dt} \right\rangle = 0 , \tag{44}$$

where E is the energy per unit mass of the planet which moves on the geodesic of radius R_0

$$E = \left(1 - \frac{2M_\star}{R_0}\right)\left(1 - \frac{3M_\star}{R_0}\right)^{-1/2} , \tag{45}$$

and $\langle dE_{GW}/dt \rangle$ is the energy emitted in gravitational waves, computed numerically by using the perturbative approach (see [26] for details).

Since $\langle dE/dt \rangle = \langle dR_0/dt \rangle \,/\, \langle dR_0/dE \rangle$, using (45) and (44) we find that the time a planet spends in the region $R_i < r < R_i + \Delta R$ is

$$\Delta T = -\frac{M_p M_\star}{2} \int_{R_i+\Delta R}^{R_i} \frac{(1 - 6M_\star/R_0)}{(1 - 3M_\star/R_0)^{3/2}} \frac{dR_0}{R_0^2 \, \langle dE_{GW}/dt \rangle} . \tag{46}$$

It should be noted that since $\langle dE_{GW}/dt \rangle$ is proportional to M_p^2, ΔT is longer for smaller planets.

The results of this calculations are summarized in Table 6. These data have to be used together with those in Table 5 as follows.

Consider for instance a planet like the Earth, orbiting its sun on an orbit resonant with the mode g_4. According to the quadrupole formalism it would emit a signal of amplitude $h_Q(R_4) = 3 \cdot 10^{-26}$, (Table 5, first row) at a frequency $\nu_{GW} = 1.13 \cdot 10^{-4}$ Hz (table 6, first row). The data of Table 6, which include the resonant contribution to the emitted radiation, indicate that before reaching that resonant orbit of radius R_4, the Earth-like planet would orbit in a region of thickness $\Delta R = 1.7$ km, slowly spiralling in, emitting waves with amplitude $h_R > 10 h_Q(R_4) = 3 \cdot 10^{-25}$, for a time interval of $5.4 \cdot 10^6$ years, and that while spanning the smaller radial region $\Delta R = 312$ m, the emitted wave would reach an amplitude $h_R > 50 h_Q(R_4) = 1.5 \cdot 10^{-24}$, for a time interval of $3.8 \cdot 10^4$ years.

Table 6. For each mode allowed by the Roche lobe analysis (see text), we give the frequency of the wave emitted when a planet moves on an orbit resonant with a **g**-mode of its sun (column 2). When the planet spans a radial region of thickness ΔR (column 3) approaching a resonant orbit, due to the stellar pulsations the amplitude of the emitted wave is amplified, with respect to the quadrupole amplitude by a factor greater than A (column 4). In the last three columns we give the time interval ΔT needed for the three companions to span the region ΔR and reach the resonance

Mode	$\nu_{\rm GW}$ (μHz)	ΔR(m)	A	$\Delta T_{\rm E}$(yrs)	$\Delta T_{\rm BD}$(yrs)	$\Delta T_{\rm J}$(yrs)
g_4	113	1700	10	$5.4 \cdot 10^6$	$4.3 \cdot 10^2$	-
		312	50	$3.8 \cdot 10^4$	3.0	-
g_5	97	616	10	$2.7 \cdot 10^6$	$2.1 \cdot 10^2$	-
		113	50	$1.9 \cdot 10^4$	1.5	-
g_6	85	240	10	$1.4 \cdot 10^6$	$1.1 \cdot 10^2$	-
		44	50	$9.6 \cdot 10^3$	$7.5 \cdot 10^{-1}$	-
g_7	75	98	10	$7.0 \cdot 10^5$	55	-
		18	50	$5.0 \cdot 10^3$	$3.9 \cdot 10^{-1}$	-
g_8	68	42	10	$3.8 \cdot 10^5$	30	-
		8	50	$2.5 \cdot 10^3$	$1.9 \cdot 10^{-1}$	-
g_9	62	18	10	$1.9 \cdot 10^5$	15	-
		3	50	$1.5 \cdot 10^3$	$1.2 \cdot 10^{-1}$	-
g_{10}	57	8	10	$1.0 \cdot 10^5$	7.8	$3.1 \cdot 10^2$
		1	50	$6.7 \cdot 10^2$	$5.3 \cdot 10^{-2}$	2.1

A jovian planet, on the other hand, could only excite modes of order $n = 10$ or higher, which would correspond to a resonant frequency of $\nu_{\rm GW} = 5.7 \cdot 10^{-5}$ Hz and a gravitational wave amplitude greater than $6 \cdot 10^{-23}$ for 310 years ($\Delta R = 8$ m), or greater than $3 \cdot 10^{-22}$ for ~ 2 years ($\Delta R = 1$m). From these data we see that the higher the order of the mode is, the more difficult is to excite it, because the region where the resonant effects become significant gets narrower and the planet transits through it for a shorter time.

Much more interesting are the data for a brown dwarf companion. In this case, for instance, the region that would correspond to the resonant excitation of the mode g_4 ($\Delta R = 1.7$ km), with a wave amplitude greater than $3.9 \cdot 10^{-21}$, would be spanned in 430 years, whereas the emitted wave would have an amplitude greater than $\sim 2 \cdot 10^{-20}$ ($\Delta R = 312$ m) over a time interval of ~ 3 years.

The results of this study show that if a brown dwarf orbits a solar type star at such distance that a **g**-mode is excited, the emitted radiation may be strong enough, and for a sufficiently long time interval, to be detectable by the space base interferometer LISA which is expecting to be operating in 2010.

5 Concluding Remarks

In this chapter I have shown how the theory of perturbations of non-rotating stars can be constructed in analogy with the theory of perturbations of non-rotating black holes. We have seen that by expanding the perturbed tensors in tensorial harmonics the equations separate and split into two decoupled sets belonging to the axial and polar parity.

As for black holes, the equations for the axial perturbations can be reduced to a wave equation with a potential barrier, the shape of which now depends on how the energy density and the pressure are distributed in the unperturbed star. Associated to this barrier there exist different classes of quasi-stationary states, i.e. of quasi-normal modes: if the star is extremely compact the potential inside the star is a well, and very slowly damped, trapped modes can exist.

In addition there always exist families of highly damped modes, the **w**-modes. The axial modes do not have a newtonian counterpart and are modes of the gravitational field.

The polar equations couple the perturbations of the fluid and of the gravitational field. Unlike the polar equations for black holes, they cannot be reduced to a single wave equation. However, they can be cast in different forms, depending on the particular problem one wants to solve. For instance, it is possible to eliminate the fluid variables and obtain a set of equations only for the gravitational variables, (see (14)), and this is useful if one is interested in computing the frequencies of the QNMs. Or, the system can be reduced to a fourth order system [29,30], or the problem can be formulated as an initial value problem for two coupled wave equations [31].

The polar QNMs have a very rich structure and their frequencies carry information on the equation of state of the star, on the mass and on the radius.

In this chapter I considered only non-rotating stars. Rotating stars are far more complex objects and their equilibrium configuration can be determined only by numerical integration. Contrary to what happens in the static, spherically symmetric case, an exact form of the metric appropriate to describe the spacetime in the exterior of the fluid configuration is not available and the Kerr metric cannot be used to this purpose; indeed, the problem of finding an interior source for the Kerr metric is still unsolved.

In the case of slow rotation the problem can be circumvented, the metric for the unperturbed spacetime was derived by J. Hartle in 1967 [32], and the perturbations can be studied in a way similar to that described in this Chap. [8,33].

It emerges that the polar and the axial perturbations, that are decoupled in the non-rotating case, couple because of the rotation, and the coupling function is the one which is responsible for the dragging of inertial frames. This is again a purely relativistic effect that does not exist in newtonian theory.

The study of slowly and fastly rotating stars is one of the main topics in relativistic astrophysics. In particular the instabilities which develop in rotating stars due to the emission of gravitational waves, firstly discovered by Chandrasekhar [34], and the viscous mechanism that may damp these instabilities, are attracting a great interest because of their obvious astrophysical relevance and because they may be associated to a large emission of gravitational waves.

A fairly complete bibliography on this subject which, however, goes beyond the purposes of this chapter, can be found in [35].

Appendix: The Explicit Expressions of the Components of the Stress-Energy Tensor

By expanding the stress-energy tensor of the fluid composing the star in tensorial harmonics, we find that the axial and polar part can be written as

$$\delta T^{\rm ax}_{\mu\nu} = \sum_{\ell m}$$

$$\begin{pmatrix} 0 & -\left(\frac{1}{2}\mathrm{i}\omega\tilde{W}_{\ell m} - \varepsilon h^{\rm ax}_{0\ell m}\right)\sin\vartheta Y_{\ell m,\vartheta} & 0 & \left(\frac{1}{2}\mathrm{i}\omega\tilde{W}_{\ell m} - \varepsilon h^{\rm ax}_{0\ell m}\right)\frac{1}{\sin\vartheta}Y_{\ell m,\varphi} \\ -\left(\frac{1}{2}\mathrm{i}\omega\tilde{W}_{\ell m} - \varepsilon h^{\rm ax}_{0\ell m}\right)\sin\vartheta Y_{\ell m,\vartheta} & 0 & -ph^{\rm ax}_{1\ell m}\sin\vartheta Y_{\ell m,\vartheta} & 0 \\ 0 & -ph^{\rm ax}_{1\ell m}\sin\vartheta Y_{\ell m,\vartheta} & 0 & ph^{\rm ax}_{1\ell m}\frac{1}{\sin\vartheta}Y_{\ell m,\varphi} \\ \left(\frac{1}{2}\mathrm{i}\omega\tilde{W}_{\ell m} - \varepsilon h^{\rm ax}_{0\ell m}\right)\frac{1}{\sin\vartheta}Y_{\ell m,\varphi} & 0 & ph^{\rm ax}_{1\ell m}\frac{1}{\sin\vartheta}Y_{\ell m,\varphi} & 0 \end{pmatrix}$$

$$\delta T^{\rm pol}_{\mu\nu} = \sum_{\ell m}$$

$$\begin{pmatrix} \mathrm{e}^{2\nu}\left(E_{\ell m} + 2\varepsilon N_{\ell m}\right)Y_{\ell m} & -\frac{1}{2}\mathrm{i}\omega W_{\ell m}Y_{\ell m,\varphi} & -\frac{1}{2}\mathrm{i}\omega U_{\ell m}Y_{\ell m} & -\frac{1}{2}\mathrm{i}\omega W_{\ell m}Y_{\ell m,\vartheta} \\ -\frac{1}{2}\mathrm{i}\omega W_{\ell m}Y_{\ell m,\varphi} & r^2\sin^2\vartheta\left(\Pi_{\ell m}Y_{\ell m} + 2pH_{11\ell m}\right) & 0 & pr^2V_{\ell m}X_{\ell m} \\ -\frac{1}{2}\mathrm{i}\omega U_{\ell m}Y_{\ell m} & 0 & \mathrm{e}^{2\mu_2}\left(\Pi_{\ell m} + 2pL_{\ell m}\right)Y_{\ell m} & 0 \\ -\frac{1}{2}\mathrm{i}\omega W_{\ell m}Y_{\ell m,\vartheta} & pr^2V_{\ell m}X_{\ell m} & 0 & r^2\left(\Pi_{\ell m}Y_{\ell m} + 2pH_{33\ell m}\right) \end{pmatrix}$$

where $H_{33} = \left[TY_{\ell m} + V\partial^2 Y_{\ell m}/\partial\vartheta^2\right]$.

The perturbations of the energy density, pressure, and lagrangian displacement are

$$\begin{aligned} \delta\varepsilon &= E_{\ell m}Y_{\ell m}\ , \qquad\qquad \delta p = \Pi_{\ell m}Y_{\ell m}\ , \\ (\varepsilon + p)\mathrm{e}^{\nu+\mu_2}\xi_r &= \frac{1}{2}U_{\ell m}Y_{\ell m}\ , \\ (\varepsilon + p)r\mathrm{e}^{\nu}\xi_\vartheta &= \frac{1}{2}\left(W_{\ell m}Y_{\ell m,\vartheta} - \tilde{W}_{\ell m}\frac{1}{\sin\vartheta}Y_{\ell m,\varphi}\right)\ , \\ (\varepsilon + p)r\sin\vartheta\mathrm{e}^{\nu}\xi_\varphi &= \frac{1}{2}\left(W_{\ell m}Y_{\ell m,\varphi} + \tilde{W}_{\ell m}\sin\vartheta Y_{\ell m,\vartheta}\right)\ . \end{aligned}$$

The hydrodynamical equations after the separation of variables give

$$\delta(T^{0\nu}{}_{;\nu}) = 0:\ E = Q\Pi + \frac{\mathrm{e}^{2\mu_2}}{2(\varepsilon + p)}(\varepsilon_{,r} - Qp_{,r})U\ ,$$

$$\delta(T^{r\nu}{}_{;\nu}) = 0: \ \Pi_{,r} + (E + \Pi)\nu_{,r} + (\varepsilon + p)N_r = -\frac{1}{2}\omega^2 \mathrm{e}^{-2\nu} U \,,$$

$$\delta(T^{\vartheta\nu}{}_{;\nu}) = 0: \ \Pi + (\varepsilon + p)N = -\frac{1}{2}\omega^2 \mathrm{e}^{-2\nu} W \,.$$

References

1. P.N. McDermott, H.M. Van Horn, C.J. Hansen: Ap. J. **325**, 725 (1988)
2. K.S. Thorne, A. Campolattaro: Ap. J. **149**, 591 (1967)
   R. Price, K.S. Thorne: Ap. J. **155**, 163 (1969)
   K.S. Thorne: Phys. Rev. Lett. **21**, 320 (1968)
   K.S. Thorne: Ap. J. **158**, 1 (1969)
   A. Campolattaro, K.S. Thorne: Ap. J. **159**, 847 (1970)
   J.R. Ipser, K.S. Thorne: Ap. J. **181**, 181 (1973)
3. K.S. Thorne: Ap. J. **158**, 997 (1969)
4. L. Lindblom, S. Detweiler: Ap. J. Suppl. **53**, 73 (1983)
5. N. Andersson, K.D. Kokkotas: Phys. Rev. Lett. **77**, 4134 (1996)
   N. Andersson, K.D. Kokkotas: MNRAS **299**, 1059 (1998)
6. S. Chandrasekhar, V. Ferrari: Proc. R. Soc. Lond. **A428**, 325 (1990)
7. S. Chandrasekhar, V. Ferrari: Proc. R. Soc. Lond. **A432**, 247 (1990)
8. S. Chandrasekhar, V. Ferrari: Proc. R. Soc. Lond. **A433**, 423 (1991)
9. S. Chandrasekhar, V. Ferrari: Proc. R. Soc. Lond. **434**, 449 (1991)
10. S. Chandrasekhar, V. Ferrari, R. Winston: Proc. R. Soc. Lond. **434**, 635 (1991)
11. S. Chandrasekhar, V. Ferrari: Proc. R. Soc. Lond. **A435**, 645 (1991)
12. S. Chandrasekhar, V. Ferrari: Proc. R. Soc. Lond. **A437**, 133 (1992)
13. K.D. Kokkotas, B.F. Schutz: Gen. Rel. Grav. **18**, 913 (1986)
    K.D. Kokkotas, B.F. Schutz: MNRAS **225**, 119 (1992)
14. K.D. Kokkotas: MNRAS **268**, 1015 (1994)
15. T.G. Cowling: MNRAS **101**, 367 (1942)
16. N. Andersson, K.D. Kokkotas: Phys. Rev. Lett. **77**, 4134 (1996)
17. N. Andersson, K.D. Kokkotas: MNRAS **299**, 1054 (1998)
18. K.D. Kokkotas, T. Apostolatos, N. Andersson: MNRAS **320**, 307 (2001)
19. H. Heiselberg, M. Hjorth-Jensen: Phys. Rep. **328**, 237 (2000)
20. M. Prakash, I. Bombaci, P.J. Ellis, J.M. Lattimer, R. Knoren: Phys. Rep. **280**, 1 (1997)
21. G. Miniutti, J. A. Pons, E. Berti, L. Gualtieri, V. Ferrari: submitted to MNRAS (2002)
22. H. Sotani, K. Tominaga, K. Maeda: Phys. Rev. **D65**, 024010 (2001)
23. L.S. Finn: MNRAS **222**, 393 (1986)
    L.S. Finn: MNRAS **227**, 265 (1987)
    P.N. McDermott: MNRAS **245**, 508 (1990)
    T.E. Strohmayer: Ap. J. **417**, 273 (1993)
24. For an updated catalogue of new extrasolar planetary systems see the web sites: http://exoplanets.org/almanacframe.html
25. V. Ferrari, M. D'Andrea, E. Berti: Int. J. Mod. Phys. **D9**, 495 (2000)
26. E. Berti, V. Ferrari: Phys. Rev. **D63**, 064031 (2001)
27. T. Guillot, A. Burrows, W.B. Hubbard, J.I. Lunine, D. Saumon: Ap. J **459**, L35 (1996)
28. A. Burrows, J. Liebert: Rev. Mod. Phys. **65**, 301 (1993)

29. R.H. Price, J.R. Ipser: Phys. Rev. **D43**, 1768 (1991)
30. L.Lindblom,S.Detweiler: Ap. J. **292**, 12 (1985)
31. G. Allen, N. Andersson, K.D. Kokkotas, B. Schutz: Phys. Rev. **D58**, 124012 (1998)
32. J.B. Hartle: Ap. J. **150**, 1005 (1967)
33. Y. Kojima: Phys. Rev. **D46**, 4289 (1992)
    Y. Kojima: Ap. J. **414**, 247 (1993)
    Y. Kojima: Prog. Theor. Phys. **90**, 977 (1993)
34. S. Chandrasekhar: Phys. Rev. Lett. **24**, 611 (1970)
35. N. Andersson, K.D. Kokkotas: IJMP **10 n.4**, 381 (2001)
36. N. Andersson, G.L. Comer, 2001: Phys. Rev. Lett. **24**, 241101 (2001)

Numerical Relativistic Hydrodynamics

José María Ibáñez

Departamento de Astronomía y Astrofísica, Universidad de Valencia,
Edificio de Investigación, C/ Dr. Moliner, s/n 46100 Burjassot, Spain.
E-mail: jose.m.ibanez@uv.es

Abstract. *High Resolution Shock Capturing* (HRSC) techniques achieve highly accurate numerical approximations (formally second order or better) in smooth regions of the flow, and capture the motion of unresolved steep gradients without creating spurious oscillations. I will show how these techniques have been extended to relativistic hydrodynamics, making it possible to explore some challenging astrophysical scenarios. I will review recent literature concerning the main properties of different *special relativistic Riemann solvers*, and discuss several test problems which are commonly used to evaluate the performance of numerical methods in relativistic hydrodynamics. In the second part, I will illustrate the use of HRSC methods in various astrophysical applications, where special and general relativistic hydrodynamical processes play a crucial role, such as: relativistic extragalactic jets, accretion onto compact objects, models of formation of Gamma-Ray Bursts, and stellar core collapse.

1 Introduction

Astrophysical scenarios involving relativistic flows have drawn the attention and efforts of many researchers since the pioneering studies of May and White [1] and Wilson [2]. Relativistic jets, accretion onto compact objects (in X-ray binaries or in the inner regions of active galactic nuclei), stellar core collapse, coalescing compact binaries (neutron star and/or black holes) and recent models of formation of gamma-ray bursts (GRBs) are examples of systems in which the evolution of matter is described within the framework of the theory of relativity (special or general).

Since 1991 [3] the use of numerical schemes based upon *Riemann solvers*, i.e., algorithms designed to solve Riemann problems (see definition, below) in computational relativistic hydrodynamics has proved successful in handling complex flows, with high Lorentz factors and strong shocks, superseding traditional methods which failed to describe ultrarelativistic flows [4]. By exploiting the hyperbolic and conservative character of the relativistic hydrodynamic equations, and following the approach developed in Newtonian hydrodynamics, we extended high-resolution shock-capturing (HRSC) methods to the relativistic case, first in one-dimensional calculations [3], and, later on, in multidimensional special relativistic [5], [6] and multidimensional general relativistic hydrodynamics [7], [8], [9].

Our approach made use of a linearized Riemann solver based on the knowledge of the *spectral decomposition* of the Jacobian matrices of the system. Unlike the case of classical fluid dynamics the use of HRSC techniques in relativistic fluid dynamics is very recent and has yet to cover the full set of possible applications. In the second part of this chapter I will summarize some of the most recent applications in modelling relativistic astrophysical systems. The task of developing robust, stable and accurate (special or general) relativistic hydrocodes is a challenge in the field of Relativistic Astrophysics. A general relativistic hydrocode is a useful research tool for studying flows which evolve in a background spacetime. Furthermore, when appropriately coupled with Einstein equations, such a general relativistic hydrocode is crucial to model the evolution of matter in a dynamical spacetime. The coupling between geometry and matter arises through the sources of the corresponding system of equations. Such a marriage between numerical relativity and numerical relativistic hydrodynamics is useful, for example, to analyze the dynamics (and the physics) of coalescing compact binaries. These are one of the most promising sources of gravitational radiation to be detected by the near future Earth-based laser-interferometer observatories of gravitational waves. Readers interested in deepening into the contents of this chapter are addressed to the following reviews: [10], [11], [12] and [13].

2 The Equations of General Relativistic Hydrodynamics as a Hyperbolic System of Conservation Laws

For the sake of consistency, in this section I am going to summarize the basics (definitions and properties) of hyperbolic systems of conservation laws (Sect. 2.1), in connection with HRSC techniques (Sect. 2.2), and I will apply them to the particular system of equations of relativistic hydrodynamics (Sects. 2.3, 2.4 and 2.5). Further mathematical details can be found in the following textbooks: [14], [15], [16] and [17].

2.1 Hyperbolic Systems of Conservation Laws: Basics

Let us start by considering the system of p equations of conservation laws

$$\frac{\partial \mathbf{u}}{\partial t} + \sum_{j=1}^{d} \frac{\partial \mathbf{f}_j(\mathbf{u})}{\partial x_j} = 0 \;\; (= \mathbf{s}(\mathbf{u})) \tag{1}$$

where $\mathbf{u} = (u_1, u_2, \ldots, u_p)^T$ is the vector of unknowns, function of $\mathbf{x}$ and t, with $\mathbf{x} = (x_1, x_2, \ldots, x_d) \in \mathbb{R}^d$ and $\mathbf{f}_j(\mathbf{u}) = (f_{1j}, f_{2j}, \ldots, f_{pj})^T$ is the vector of fluxes. Formally, (1) expresses the conservation of the vector $\mathbf{u}$. Let D be an arbitrary domain of $\mathbb{R}^d$ and let $\mathbf{n} = (n_1, \ldots, n_d)$ be the outward unit normal to the boundary ∂D of D. Then, from (1), it follows that

$$\frac{\mathrm{d}}{\mathrm{d}t} \int_D \mathbf{u} \, \mathrm{d}\mathbf{x} + \sum_{j=1}^{d} \int_{\partial D} \mathbf{f}_j(\mathbf{u}) n_j \, \mathrm{d}S = 0 \, . \tag{2}$$

This balance equation establishes that the time variation of $\int_D \mathbf{u} \, \mathrm{d}\mathbf{x}$ is equal to the losses through the boundary ∂D. Now, for all $j = 1, \ldots, d$ let

$$\mathbf{A}_j(\mathbf{u}) = \frac{\partial \mathbf{f}_j(\mathbf{u})}{\partial \mathbf{u}} \tag{3}$$

be the Jacobian matrix of $\mathbf{f}_j(\mathbf{u})$. The system (1) is called *hyperbolic* if, for any $\omega = (\omega_1, \ldots, \omega_d) \in \mathbb{R}^d$, and for any $\mathbf{u}$, the matrix

$$\mathbf{A}(\mathbf{u}, \omega) = \sum_{j=1}^{d} \omega_j \mathbf{A}_j(\mathbf{u}) \tag{4}$$

has p real eigenvalues $\lambda_1(\mathbf{u}, \omega) \leq \lambda_2(\mathbf{u}, \omega) \leq \cdots \leq \lambda_p(\mathbf{u}, \omega)$ and p linearly independent (right) eigenvectors $\mathbf{r}_1(\mathbf{u}, \omega)$, $\mathbf{r}_2(\mathbf{u}, \omega)$, ..., $\mathbf{r}_p(\mathbf{u}, \omega)$. If, in addition, the eigenvalues $\lambda_k(\mathbf{u}, \omega)$ are all different, the system (1) is called *strictly hyperbolic.*

In most cases one is concerned with the so-called *initial value problem* (IVP), i.e., the solution of (1) with the initial condition

$$\mathbf{u}(\mathbf{x}, t = 0) = \mathbf{u}_0(\mathbf{x}) \, . \tag{5}$$

A key property of hyperbolic systems is that features in the solution propagate at *characteristic speeds* given by the eigenvalues of the Jacobian matrices. The characteristic curves associated to (1) are the integral curves of the differential equations

$$\frac{\mathrm{d}x}{\mathrm{d}t} = \lambda_k(\mathbf{u}(x,t)), \qquad k = 1, \ldots, p \, , \tag{6}$$

($d = 1$). It can be easily proven that, along these curves the so-called *characteristic variables* (a combination of the components of $\mathbf{u}$) are constant. Essentially, characteristic curves give information about the propagation of the initial data, which formally allows one to reconstruct the solution for the initial value problem (1) with (5) at $t > 0$.

Continuous and differentiable solutions that satisfy (1) and (5) pointwise are called *classical solutions.* However, for nonlinear systems, classical solutions do not exist in general even when the initial condition $\mathbf{u}_0$ is a smooth function, and discontinuities develop after a finite time. Then one seeks generalized solutions that satisfy the integral form of the conservation system (2) which are classical solutions where they are continuous and have a finite number of discontinuities (*weak solutions*). The following theorem characterizes these solutions.

Let $\mathbf{u}$ be a piecewise smooth function. Then, $\mathbf{u}$ is a solution of the integral form of the conservation system if and only if the two following conditions are satisfied:

1. $\mathbf{u}$ is a classical solution in the domains where it is continuous.
2. Across a given surface of discontinuity, Σ, it satisfies the jump conditions (Rankine-Hugoniot conditions)

$$(\mathbf{u}_R - \mathbf{u}_L) n_t + \sum_{j=1}^{d} \left[\mathbf{f}_j(\mathbf{u}_R) - \mathbf{f}_j(\mathbf{u}_L)\right] n_{xj} = 0 \, , \tag{7}$$

where $\mathbf{u}_L$ and $\mathbf{u}_R$ stand, respectively, for the values of $\mathbf{u}$ on the left and right hand sides of Σ, and $\mathbf{n} = (n_t, n_{x1}, n_{x2}, \ldots, n_{xd})$ denotes a vector normal to Σ.

For 1D systems, the *Rankine-Hugoniot jump condition* (7) reduces to

$$s(\mathbf{u}_R - \mathbf{u}_L) = \mathbf{f}(\mathbf{u}_R) - \mathbf{f}(\mathbf{u}_L) \tag{8}$$

where s is the *propagation velocity of the discontinuity.*

Rankine-Hugoniot conditions follow from the conservation of fluxes across the surfaces of discontinuity. They can be used in combination with standard finite-difference methods for the smooth regions and special procedures for tracking the location of discontinuities to solve the equations in the presence of shocks (*shock-tracking* approach). In 1D this is often a viable approach. However, in multidimensional applications the discontinuities lie along curves (in 2D) or surfaces (in 3D) and in realistic problems there may be many such discontinuities interacting in complicated ways, making the use of shock-tracking methods much more difficult.

The class of all weak solutions is too wide in the sense that there is no uniqueness for the initial value problem. Therefore, an effort should be made to develop numerical methods which converge to the physically admissible solution. Mathematically, this solution is characterized by the so-called *entropy condition* (in the language of fluids, the condition that the entropy of any fluid element should increase when running into a discontinuity). The characterization of the *entropy-satisfying solutions* for scalar equations follows [18], whereas for systems of conservation laws it has been developed by Lax [15]. Most HRSC methods are based on exact or approximate solutions of Riemann problems between contiguous numerical cells. Consider the hyperbolic system of conservation laws in 1D

$$\frac{\partial \mathbf{u}}{\partial t} + \frac{\partial \mathbf{f}(\mathbf{u})}{\partial x} = 0 \tag{9}$$

with initial data $\mathbf{u}(x,0) = \mathbf{u}_0(x)$. A Riemann problem for (9) is an initial value problem with discontinuous data, i.e.,

$$\mathbf{u}_0 = \begin{cases} \mathbf{u}_L \text{ if } x < 0 \\ \mathbf{u}_R \text{ if } x > 0 \end{cases} \tag{10}$$

The Riemann problem is invariant under similarity transformations $(x,t) \rightarrow (ax, at)$, $a > 0$, so that the solution is constant along the straight lines $x/t =$ constant and, hence, self-similar. It consists of constant states separated by *rarefaction waves* (continuous self-similar solutions of the differential system), *shocks* and *contact discontinuities* [15]. In the following I will denote the *Riemann solution* for the left and right states $\mathbf{u}_L$ and $\mathbf{u}_R$, respectively, as $\mathbf{u}(x/t; \mathbf{u}_L, \mathbf{u}_R)$.

2.2 High-Resolution Shock-Capturing Schemes

Let us start by considering an IVP for (9). Finite-difference methods are based on a discretization of the $x-t$ plane defined by the discrete mesh points (x_j, t^n)

$$x_j = (j - 1/2)\Delta x, \quad j = 1, 2, \ldots \tag{11}$$

$$t^n = n\Delta t, \quad n = 0, 1, 2, \dots \;, \tag{12}$$

where Δx and Δt are, respectively the cell width and the time step. A finite-difference scheme is a time-marching procedure allowing one to obtain approximations to the solution in the mesh points, $\mathbf{u}_j^{n+1}$, from the approximations in previous time steps $\mathbf{u}_j^n$. Quantity $\mathbf{u}_j^n$ is an approximation to $\mathbf{u}(x_j, t^n)$ but, in the case of a conservation law, it is often preferable to view it as an approximation to the average of $\mathbf{u}(x,t)$ within the numerical cell $[x_{j-1/2}, x_{j+1/2}]$ $(x_{j\pm 1/2} = (x_j + x_{j\pm 1})/2)$

$$\mathbf{u}_j^n \approx \frac{1}{\Delta x} \int_{x_{j-1/2}}^{x_{j+1/2}} \mathbf{u}(x, t^n) \mathrm{d}x \;, \tag{13}$$

consistent with the integral form of the conservation law.

For hyperbolic systems of conservation laws, methods in *conservation form* are preferred as they guarantee that the convergence (if it exists) is to one of the weak solutions of the original system of equations (*Lax-Wendroff theorem* [19]). Conservation form means that the algorithm is written as

$$\begin{aligned} \mathbf{u}_j^{n+1} = \mathbf{u}_j^n - \frac{\Delta t}{\Delta x} \Big(& \hat{\mathbf{f}}(\mathbf{u}_{j-r}^n, \mathbf{u}_{j-r+1}^n, \dots, \mathbf{u}_{j+q}^n) \\ & - \hat{\mathbf{f}}(\mathbf{u}_{j-r-1}^n, \mathbf{u}_{j-r}^n, \dots, \mathbf{u}_{j+q-1}^n) \Big) \end{aligned} \tag{14}$$

where $\hat{\mathbf{f}}$ is a consistent (i.e., $\hat{\mathbf{f}}(\mathbf{u}, \mathbf{u}, \dots, \mathbf{u}) = \mathbf{f}(\mathbf{u})$) *numerical flux function.* The Lax-Wendroff theorem does not state whether the method converges. To guarantee convergence, some form of stability is required, as for linear problems (*Lax equivalence theorem*, [20]). In this direction, the notion of *total-variation stability* has proven very successful although powerful results have only been obtained for scalar conservation laws. The total variation of a solution at $t = t^n$, $\mathrm{TV}(\mathbf{u}^n)$, is defined as

$$\mathrm{TV}(\mathbf{u}^n) = \sum_{j=0}^{\infty} |\mathbf{u}_{j+1}^n - \mathbf{u}_j^n| \tag{15}$$

and a numerical scheme is said to be TV-stable if $\mathrm{TV}(\mathbf{u}^n)$ is bounded for all Δt at any time for each initial data. For a non-linear scalar conservation law, TV-stability is a sufficient condition for convergence of numerical schemes in conservation form with consistent numerical flux functions [16].

In recent years a very interesting line of research has focused on developing high-order, accurate methods in conservation form satisfying the condition of TV-stability. The conservation form is ensured by starting with the integral version of the partial differential equation in conservation form. Integrating the PDE in a spacetime computational cell $[x_{j-1/2}, x_{j+1/2}] \times [t^n, t^{n+1}]$ and comparing with (14), the numerical flux function $\hat{\mathbf{f}}_{j+1/2}$ is seen to be an approximation to the time-averaged flux across the interface, i.e.,

$$\hat{\mathbf{f}}_{j+1/2} \approx \frac{1}{\Delta t} \int_{t^n}^{t^{n+1}} \mathbf{f}(\mathbf{u}(x_{j+1/2}, t)) \, \mathrm{d}t \;. \tag{16}$$

In the above expression, the flux integral depends on the solution at the numerical interfaces, $\mathbf{u}(x_{j+1/2}, t)$, during the time step. Hence, a possible procedure is to calculate $\mathbf{u}(x_{j+1/2}, t)$ by solving Riemann problems at every numerical interface to obtain

$$\mathbf{u}(x_{j+1/2}, t) = \mathbf{u}(0; \mathbf{u}_j^n, \mathbf{u}_{j+1}^n) . \quad (17)$$

This is the approach followed by an important subset of shock-capturing methods, the so-called *Godunov-type methods* [21], [22].

These methods are written in conservation form and use different procedures (Riemann solvers) to compute approximations to $\mathbf{u}(0; \mathbf{u}_j^n, \mathbf{u}_{j+1}^n)$. High-order of accuracy is usually achieved by using conservative polynomial functions to interpolate the approximate solutions within the numerical cells. The idea is to produce more accurate left and right states for the Riemann problems by substituting the mean values $\mathbf{u}_j^n$ (that give only first-order accuracy) for better approximations of the true flux near the interfaces, $\mathbf{u}_{j+1/2}^{\mathrm{L}}$, $\mathbf{u}_{j+1/2}^{\mathrm{R}}$ (the *flux-corrected-transport* algorithm [23] constitutes an alternative procedure where higher accuracy is obtained by adding an anti-diffusive flux term to the first-order numerical flux). The interpolation algorithms have to preserve the TV-stability of the scheme and this is usually achieved by using monotonic functions which lead to a decrease in the total variation (*total-variation-diminishing schemes*, TVD; [24]).

If R is an interpolant function for the approximate solution $\mathbf{u}^n$ and $\tilde{\mathbf{u}}(x, t^n)$ is the interpolated function within the cells, i.e., $\tilde{\mathbf{u}}(x, t^n) = \mathrm{R}(\mathbf{u}^n; x)$, satisfying $\mathrm{TV}(\tilde{\mathbf{u}}(\cdot, t^n)) \leq \mathrm{TV}(\mathbf{u}^n)$ then it can be proven that the whole scheme verifies $\mathrm{TV}(\mathbf{u}^{n+1}) \leq \mathrm{TV}(\mathbf{u}^n)$. High-order TVD schemes were first constructed by van Leer [25] who obtained second-order accuracy by using monotonic *piecewise linear* slopes for cell reconstruction. The *piecewise parabolic method* (PPM) of Colella and Woodward [26] provides higher accuracy. The TVD property implies TV-stability but can be too restrictive. In fact, TVD methods degenerate to first order accuracy at extreme points [27]. Hence, other reconstruction alternatives have been developed in which some growth of the total variation is allowed. This is the case of the *total-variation-bounded* schemes [28], *essentially nonoscillatory* (ENO) schemes [29] and the *piecewise-hyperbolic method* [30].

2.3 The Equations of General Relativistic Hydrodynamics

The evolution of a relativistic fluid is governed by a system of equations which summarize *local conservation laws*: the local conservation of baryon number, $\nabla \cdot \mathbf{J} = 0$, and the local conservation of energy-momentum, $\nabla \cdot \mathbf{T} = 0$ ($\nabla \cdot$ stands for the covariant divergence).

If $\{\partial_{\mathbf{t}}, \partial_{\mathbf{i}}\}$ define the coordinate basis of 4-vectors which are tangent to the corresponding coordinate curves, then, the *current of rest-mass*, $\mathbf{J}$, and the *energy-momentum tensor*, $\mathbf{T}$, for a perfect fluid, have the components: $J^\mu = \varrho u^\mu$, and $T^{\mu\nu} = \varrho h u^\mu u^\nu + p g^{\mu\nu}$, respectively, ϱ being the rest-mass density, p the pressure and h the specific enthalpy, defined by $h = 1 + \varepsilon + p/\varrho$, where ε is the specific internal energy. u^μ is the four-velocity of the fluid and $g_{\mu\nu}$ defines the metric of the spacetime $\mathcal{M}$ where the fluid evolves. As usual, Greek (Latin) indices run from 0 to 3 (1 to 3) – or, alternatively, they stand for the general coordinates $\{t, x, y, z\}$ ($\{x, y, z\}$) – and the system of units is the so-called geometrized ($c = G = 1$). An equation of state $p = p(\varrho, \varepsilon)$ closes the system. Accordingly, the local sound velocity c_s satisfies: $hc_s^2 = \chi + (p/\varrho^2)\kappa$, with $\chi = \partial p/\partial\varrho|_\epsilon$ and $\kappa = \partial p/\partial\epsilon|_\varrho$. Following Banyuls [7], let $\mathcal{M}$ be a general spacetime described by the four dimensional metric tensor $g_{\mu\nu}$. According to the $\{3+1\}$ formalism of General Relativity (see, e.g., [31]), the metric is split into the objects α (*lapse*), β^i (*shift*) and γ_{ij}, keeping the line element in the form:

$$\mathrm{d}s^2 = -(\alpha^2 - \beta_i\beta^i)\,\mathrm{d}t^2 + 2\beta_i\,\mathrm{d}x^i\mathrm{d}t + \gamma_{ij}\,\mathrm{d}x^i\mathrm{d}x^j \tag{18}$$

If $\mathbf{n}$ is a unit timelike vector field normal to the spacelike hypersurfaces Σ_t (t = const.), then, by definition of α and β^i: $\partial_t = \alpha\mathbf{n} + \beta^i\partial_i$, with $\mathbf{n}\cdot\partial_i = 0$, $\forall i$. Observers, $\mathcal{O}_E$, at rest in the slice Σ_t, i.e., those having $\mathbf{n}$ as four-velocity (*Eulerian observers*), measure the following velocity of the fluid

$$v^i = \frac{u^i}{\alpha u^t} + \frac{\beta^i}{\alpha} \tag{19}$$

where $W \equiv -(\mathbf{u}\cdot\mathbf{n}) = \alpha u^t$ is the Lorentz factor which satisfies $W = (1-\mathrm{v}^2)^{-1/2}$ with $\mathrm{v}^2 = v_i v^i$ ($v_i = \gamma_{ij}v^j$).

Let us define a basis adapted to the observer $\mathcal{O}_E$, and the following five four-vector fields $\{\mathbf{J},\, \mathbf{T}\cdot\mathbf{n},\, \mathbf{T}\cdot\partial_1,\, \mathbf{T}\cdot\partial_2,\, \mathbf{T}\cdot\partial_3\}$.

Hence, the system of equations of general relativistic hydrodynamics can be written as a first-order, flux-conservative system

$$\nabla\cdot\mathbf{A} = s\,, \tag{20}$$

where $\mathbf{A}$ denotes any of the above 5 vector fields, and s is the corresponding source term. The set of *conserved variables* gathers those quantities which are directly measured by $\mathcal{O}_E$, i.e., the rest-mass density (D), the momentum density in the j-direction (S_j), and the total energy density (E). In terms of the *primitive variables* $\mathbf{w} = (\varrho, v_i, \varepsilon)$ ($v_i = \gamma_{ij}v^j$) they are

$$D = \varrho W\,, \qquad S_j = \varrho h W^2 v_j\,, \qquad E = \varrho h W^2 - p \tag{21}$$

Taking all the above relations together, (20) reads

$$\frac{1}{\sqrt{-g}}\left(\frac{\partial\sqrt{\gamma}\,\mathbf{F}^0(\mathbf{w})}{\partial x^0} + \frac{\partial\sqrt{-g}\,\mathbf{F}^i(\mathbf{w})}{\partial x^i}\right) = \mathbf{s}(\mathbf{w}) \tag{22}$$

where the quantities $\mathbf{F}^\alpha(\mathbf{w})$ are

$$\mathbf{F}^0(\mathbf{w}) = (D, S_j, \tau) \tag{23}$$

$$\mathbf{F}^i(\mathbf{w}) = \left(D\left(v^i - \frac{\beta^i}{\alpha}\right),\; S_j\left(v^i - \frac{\beta^i}{\alpha}\right) + p\delta^i_j,\; \tau\left(v^i - \frac{\beta^i}{\alpha}\right) + pv^i\right) \tag{24}$$

and the corresponding sources $\mathbf{s}(\mathbf{w})$ are

$$\mathbf{s}(\mathbf{w}) = \left(0, T^{\mu\nu}\left(\frac{\partial g_{\nu j}}{\partial x^{\mu}} - \Gamma^{\delta}_{\nu\mu} g_{\delta j}\right), \; \alpha\left(T^{\mu 0}\frac{\partial \ln\alpha}{\partial x^{\mu}} - T^{\mu\nu}\Gamma^{0}_{\nu\mu}\right)\right) \tag{25}$$

with $\tau \equiv E - D$, and $\Gamma^{\delta}_{\nu\mu}$ being the 4-dimensional Christoffel symbols associated to the 4-metric $g_{\mu\nu}$. The determinant (g) of the 4-metric ($g_{\mu\nu}$) verifies $\sqrt{-g} = \alpha\sqrt{\gamma}$.

2.4 The Characteristic Fields

Modern HRSC schemes use the characteristic structure of the hyperbolic system of conservation laws. In many Godunov-type schemes, the characteristic structure is used to compute either an exact or an approximate solution of a family of local Riemann problems at each cell interface. In characteristic based methods the characteristic structure is used to compute the local characteristic fields, which define the directions along which the characteristic variables propagate. In both approaches, the characteristic decomposition of the Jacobian matrices of the nonlinear system of conservation laws is important, not only to compute the numerical fluxes at the interfaces, but because experience has shown that it facilitates a robust upgrading of the order of a numerical scheme.

The three 5×5-Jacobian matrices $\mathcal{B}^i$ associated to (22) are

$$\mathcal{B}^i = \alpha \frac{\partial \mathbf{F}^i}{\partial \mathbf{F}^0} \; . \tag{26}$$

The full spectral decomposition of the above three 5×5–Jacobian matrices $\mathcal{B}^i$ can be found in [10]. For the sake of completeness let me include the expressions of the eigenvalues corresponding to the matrix $\mathcal{B}^x$:

$$\lambda_0 = \alpha v^x - \beta^x \quad \text{(triple)} \tag{27}$$

which defines the *material waves*, and

$$\lambda_{\pm} = \frac{\alpha\left\{v^x(1-c_s^2) \pm c_s\sqrt{(1-v^2)[\gamma^{xx}(1-v^2c_s^2) - v^x v^x(1-c_s^2)]}\right\}}{1-v^2c_s^2} - \beta^x \tag{28}$$

which are associated with the *acoustic waves*. Similar expressions can be derived for matrices $\mathcal{B}^y$ and $\mathcal{B}^z$. As the reader can easily notice from (27) and (28), the speed of the material waves not only depend on the fluid velocity components in the wave propagation direction (v^x in this case), but also on the normal velocity components (v^y and v^z). This coupling has its origin in the presence of the Lorentz factor in the equations themselves. Two crucial implications arise from that coupling

1. The system of equations of relativistic fluid dynamics is more nonlinear than its Newtonian counterpart. It possesses new numerical difficulties which are specific to relativistic fluid dynamics.

2. In the Newtonian case, the material waves in the x-direction, propagate with the speed $\lambda_\pm = v^x \pm c_s$, independently of the values of the other components (v^y, v^z), i.e., the transversal flow velocity. On the contrary, in the relativistic case, the influence of the transversal components of the flow velocity in the value of the speed propagation of the material waves will lead to a totally different dynamical behaviour. This has been emphasized by Pons, Martí and Müller [32] in their derivation of the exact solution of the relativistic Riemann problem in the multidimensional case. The exact *Special Relativistic Riemann Solver* derived by these authors will be a keystone not only in the way of deepening into the theoretical knowledge about the dynamics of relativistic flows, but also in the way of building up robust multidimensional relativistic hydro-codes.

Let me end this section pointing out that *covariant formulations* of the general relativistic hydrodynamic equations, alternative to the one described in Sect. 2.3 are available in the literature [33], [34]. These formulations are also suited for the used of advanced HRSC schemes. The corresponding characteristic structure can be found in the above references.

2.5 Riemann Solvers in Relativistic Hydrodynamics

The scientific literature devoted to special relativistic Riemann Solvers (SRRS) has witnessed a rapid progress since the second half of 1990s. Although some of the SRRS proposed are a straightforward extension of the corresponding Riemann solvers in classical fluid dynamics, most of them have been specifically designed to handle the Riemann problem of the equations of (special) relativistic hydrodynamics (for perfect fluids). An up-to-dated list of the SRRS can be found in [11]:

1. Roe-type [35], SRRS [3].
2. HLLE (from the following authors: Harten, Lax and van Leer [21] and Einfeldt [22]) SRRS (Schneider et al. [36]).
3. The exact SRRS (Martí and Müller [37]).
4. Two-Shock Approximation (Balsara [38]).
5. ENO (Essentially Non-Oscillatory, Shu and Osher [39]) SRRS (Dolezal and Wong [40]).
6. General relativistic extension of Roe RS (Eulderink and Mellema [33]).
7. Upwind SRRS (Falle and Komissarov [41]).
8. Relativistic extension of PPM (Piecewise Parabolic Method, Martí, Müller [42]).
9. Glimm SRRS (Wen, Panaitescu and Laguna [43]).
10. Iterative SRRS (Dai and Woodward [44]).
11. Marquina's flux formula (Donat et al. [6]).
12. The exact SRRS for non-zero transversal velocities (Pons, Martí and Müller [32]).

To end this first part, let me briefly mention the paper by Pons *et al.* [45] in which we show how to extend any SRRS, well established in the framework of the Special Relativity theory, to the field of general-relativistic hydrodynamics.

3 Astrophysical Applications

In this second part I am going to summarize the main results obtained by our group in the study of astrophysical systems where flows evolve reaching velocities near the speed of light and/or in the presence of strong gravitational fields (background or dynamical).

3.1 Relativistic Jets

In terms of the distance to the central object (a supermassive black hole) powering the nuclear activity in radio loud active galactic nuclei we can distinguish, in their associated relativistic jets, three main regions: *Subparsec scale*, *Parsec scale* and *Kiloparsec scale*. At kiloparsec scales, the implications of relativistic flow speeds and/or relativistic internal energies in the morphology and dynamics of jets have been the subject of a detailed investigation: [46], [47], [48], [49], [50]. Beams with large internal energies (*hot jets*) show little internal structure and relatively smooth cocoons allowing the terminal shock (the hot spot in the radio maps) to remain well-defined during the evolution. Their morphologies resemble those observed in naked quasar jets like 3C273 [51]. Highly supersonic models, in which kinematic relativistic effects dominate due to high beam Lorentz factors (*cold jets*), display extended overpressured cocoons. As noted by [49], these overpressured cocoons can help to confine the jets during the early stages of evolution and even cause their deflection when propagating through non-homogeneous environments. The cocoon overpressure causes the formation of a series of oblique shocks within the beam in which the synchrotron emission is enhanced. In long term simulations [52] the evolution is dominated by a strong deceleration phase during with large lobes of jet material, like the ones observed in many FRIIs (e.g., Cyg A, see [53]), start to inflate around the jet's head. The numerical simulations reproduce some properties observed in powerful extragalactic radio jets (lobe inflation, hot spot advance speeds and pressures, deceleration of the beam flow along the jet) and can help to constrain the values of basic parameters (such as the particle density and the flow speed) in the jets of real sources.

The development of multidimensional relativistic hydrodynamic codes has allowed the simulation of parsec scale jets and superluminal radio components for the first time. The presence of emitting flows at almost the speed of light enhance the importance of relativistic effects in the appearance of these sources (relativistic Doppler boosting, light aberration, time delays). Hence, models should use a combination of hydrodynamics and synchrotron radiation transfer to compare them with observations. In these models, moving radio components are obtained from perturbations in steady relativistic jets. These jets propagate through pressure decreasing atmospheres causing them to expand and accounting for the observed jet opening angles. Where pressure mismatches exist between the jet and the surrounding atmosphere, reconfinement shocks are produced. The energy density enhancement produced downstream from these shocks can give rise to stationary radio knots like those observed in many VLBI sources. Superluminal components are produced by triggering small perturbations in these steady jets which propagate at almost the jet flow speed. The first radio emission simulations from high-resolution three-dimensional relativistic hydrodynamic jets have been presented in [54], [55]. They have been generated running GENESIS [56], an optimized and parallelized 3D special relativistic hydro-code, which is suited for massively parallel computers with distributed memory.

3.2 Jets from Collapsars

Various different types of catastrophic collapse events have been proposed to explain the energy released in a GRB (see, e.g. [57] and [58] for a review on GRBs) including mergers of compact binaries [59], [60], [61], [62], collapsars [63] and hypernovae [64]. According to the current view these models require a common engine, namely a stellar mass black hole (BH) which accretes up to several solar masses of matter on a dynamical timescale of a few seconds. A fraction of the gravitational binding energy released by accretion is thought to power a pair fireball. If the baryon load of the fireball is not too large, the baryons are accelerated together with the e^+e^- pairs to Lorentz factors $> 10^2$ [65]. Such relativistic flows are supported by radio observations of GRB 980425 [66]. MacFadyen and Woosley [67] have explored the evolution of rotating helium stars ($M_\alpha \approx 10\,M_\odot$) whose iron core collapse does not produce a successful outgoing shock, but instead forms a BH surrounded by a compact accretion torus. Assuming that the efficiency of energy deposition is higher in the polar regions, [67] obtain relativistic jets along the rotation axis, which remain highly focused and seem capable of penetrating the star. However, as their simulations were performed with a Newtonian code, they obtained jet speeds which are appreciably superluminal.

Using a collapsar progenitor, provided by MacFadyen and Woosley, we have simulated [68] the propagation of an axisymmetric jet through a collapsing rotating massive star with the multidimensional relativistic hydrodynamic code GENESIS. The jet forms as a consequence of an assumed (neutrino) energy deposition in the range $10^{50} - 10^{51}$ ergs s^{-1} within a 30^0 cone around the rotation axis. We have considered a background spacetime corresponding to a Schwarzschild BH.

Effects due to the self-gravity of the star on the dynamics are neglected. The equation of state includes the contributions of non-relativistic nucleons treated as a mixture of Boltzmann gases, radiation, and an approximate correction due to e^+e^-–pairs.

Complete ionization is assumed, and the effects due to degeneracy are neglected. We advect nine non-reacting nuclear species which are present in the initial model: C^{12}, O^{16}, Ne^{20}, Mg^{24}, Si^{28}, Ni^{56}, He^4, neutrons and protons. The jet flow is strongly beamed, spatially inhomogeneous and time dependent. The jet reaches the surface of the stellar progenitor intact. At breakout, the maximum Lorentz factor of the jet flow is 33. After breakout, the jet accelerates into the circumstellar medium, whose density is assumed to first decrease exponentially and then become constant ($\approx 10^{-5}$ g cm^{-3}). Outside the star, the flow begins to expand laterally but the beam remains very well collimated. When the simulation ends, the Lorentz factor has increased to 44.

3.3 Relativistic Bondi-Hoyle Accretion

Recent discoveries made in the field of X-ray Astronomy have greatly increased interest in the physics of accretion flows around compact objects (neutron stars and black holes). Analysis of quasi-periodic oscillations (QPOs), in the kHz range, observed in neutron star X-ray binaries [69] may lead to measurements of the precession of the accretion disk, due to the Lense-Thirring effect [70]. Same line of reasoning applied to QPOs observed in the black hole candidate GRS 1915+105 [71] suggests that GRO J1655-40 and GRS 1915+105 are spinning at a rate close to the maximum theoretical limit [72]. The iron Kα emission line in the active galaxy MCG-6-30-15 [73] furnishes further evidence that (rotating) black holes are at the center of active galactic nuclei.

Historically, the canonical astrophysical scenario in which matter is accreted in a non-spherical way by a compact object was suggested by Hoyle and Lyttleton [74] and Bondi and Hoyle [75]. In the following this will be referred to as the *Bondi-Hoyle-Lyttleton* accretion onto a moving object. Using Newtonian gravity these authors studied the accretion onto a gravitating point mass moving with constant velocity through a nonrelativistic gas which is at rest and has a uniform density at infinity. Since then, this pioneering analytic work has been numerically investigated, for a finite size accretor, by a great number of authors over the years (see, e.g. [76] for an up-to-date reference list).

In a series of papers [76], [77], [78], [79] the authors made a comprehensive numerical study of the relativistic extension of the Bondi-Hoyle-Lyttleton scenario. In particular, in [78], [79] a detailed analysis is made of the morphology and dynamics of the flow evolving in the equatorial plane of a Kerr black hole. The analysis made is novel in its use of *the Kerr-Schild (KS) coordinate system*, which is the simplest within the family of *horizon adapted coordinate systems*, introduced in [80] where all fields, i.e., metric, fluid and electromagnetic fields, are free of coordinate singularities at the event horizon. This procedure allows to perform accurate numerical studies of relativistic accretion flows around black holes since it is possible to extend the computational grid inside the black hole horizon. A HRSC scheme (which makes use of a linearized Riemann solver) has been used to solve (22) in Boyer-Lindquist (BL) and also KS coordinates.

In BL coordinates, for a near-extreme Kerr black hole, the shock wraps many times before reaching the horizon due to coordinate effects. This is a pathology of the BL system associated to the collapse of the lapse function at the horizon. The wrapping in the shock wave has an important and immediate consequence: its computation in BL coordinates, although possible in principle, would be much more challenging than in KS coordinates, and the numerical difficulties would increase the closer to the horizon one would impose the boundary conditions in the BL framework.

Finally, in [81] we made some studies of the spherical and axisymmetric accretion onto a dynamic black hole, the fully dynamical evolution of imploding shells of dust with a black hole, the evolution of matter in rotating spacetimes, the gravitational radiation induced by the presence of the matter fields and the behaviour of apparent horizons through the evolution.

3.4 General Relativistic Stellar Core Collapse

In the case of spherically symmetric spacetimes the general relativistic equations can be written in a simple way which resembles Newtonian hydrodynamics. To this aim the choice of coordinates is crucial. The use of Schwarzschild-type coordinates [82] leads to a simple general relativistic extension of Eulerian Newtonian hydrodynamics. In terms of slicing of space-time, Schwarzschild-type coordinates are the realization of a *polar time slicing* and *radial gauge* [83]. We have studied [84] the general-relativistic spherically symmetric stellar core collapse, paying particular attention to the numerical treatment of the formation and propagation of strong shocks (in the framework of the so-called prompt mechanism of type II Supernova) extending HRSC techniques to the general-relativistic hydrodynamic equations. Details on the particular equations to be solved can be found in the above reference. A very simple way of modelling the essential features of the stellar core collapse of a massive star is to incorporate a simple equation of state like that of an ideal gas, but taking an adiabatic exponent, $\varGamma$, which depends on the density according to:

$$\varGamma = \varGamma_{min} + \eta(\log \varrho - \log \varrho_b) \quad , \tag{29}$$

with $\eta = 0$ if $\varrho < \varrho_b$ and $\eta > 0$ otherwise [85]. In [84] we have considered the collapse of a white dwarf-like configuration, with a gravitational mass of 1.3862 $M_\odot$, and two sets of values for the parameters $\varGamma_{min}$, η and ϱ_b: {1.33, 1, 2.5×10^{14} gcm^{-3} } (*model A*) and {1.33, 5, 2.5×10^{15} gcm^{-3} } (*model B*). The following main results obtained by [84] merit to be pointed out:

1. The formation and evolution of a strong shock is sharply solved in one or two zones and is free of spurious oscillations.
2. As it should be, the conservative features of the code preserve, during the evolution, the values of total baryonic mass and energy (or gravitational mass).

Let me draw reader's attention towards a recent paper by Dimmelmeier, Font and Müller [86] where, for first time, the authors employ HRSC schemes to study multidimensional relativistic core collapse in a dynamical, axisymmetric space-time. In this reference, gravity is described assuming the so-called *conformal flatness condition* for the spatial part of the spacetime metric. The authors make an exhaustive analysis of the gravitational waveforms generated during axisymmetric relativistic rotational core collapse. Figure 1, taken from [87], displays three of the many cases analyzed by the authors (the interested reader is addressed to [88]). The different curves show the time evolution of the central density (left panel) and the quadrupole gravitational wave amplitude (right

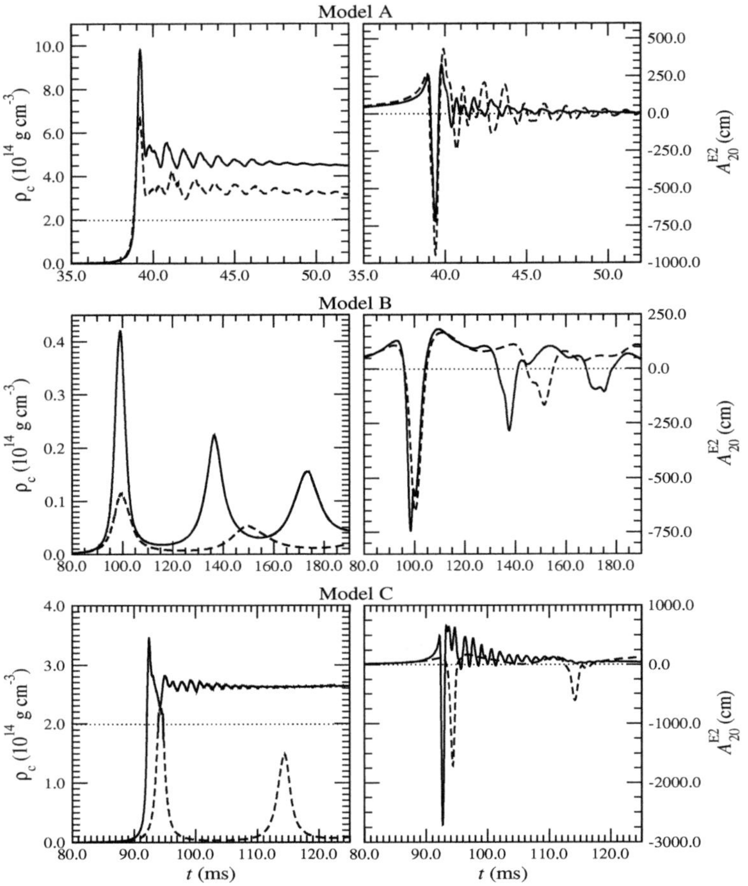

Fig. 1. Central density versus time (left panel) and quadrupole gravitational wave amplitude (right panel) for three models of rotating configurations. Solid (dotted) lines correspond to the relativistic (Newtonian) calculations

panel), comparing –in each case– the Newtonian (dotted lines) and the Einsteinian (continuous line) theories of gravity. The general trend found in these curves is that, although relativistic gravity leads to higher central densities than in Newtonian gravity, the ϱr^4-term in the quadrupole formula damps, in some way, that behaviour and enhances the contribution of the mass shells in the exterior envelope.

The study made by Dimmelmeier, Font and Müller [86] is not only consistent with what was already found in the early one-dimensional calculations of stellar core collapse, but goes beyond the purely one-dimensional case, and is, without any doubt, an outstanding step towards further studies of fully multidimensional general-relativistic stellar core collapse.

Finally, astrophysical applications using the *characteristic formulation* of general relativity and hydrodynamics [34] in investigations of collapse of supermassive stars and gravitational waves from accreting black holes can be found in [89] and [90], respectively.

Acknowledgments

The author acknowledges the organizers of the Spanish Relativity Meeting their extreme kindness and their success in creating a friendly environment during all the days of the meeting. This contribution relies on extensive analytical and numerical work done in collaboration with M.A. Aloy, R. Donat, J.A. Font, J.L. Gómez, A. Marquina, A.P. Marscher, J.M. Martí, J.A. Miralles, E. Müller, P. Papadopoulos, J.A. Pons and J.V. Romero. Present work has been partially supported by the Spanish MCyT (grant AYA2001-3490-C02-01) and by the EU Programme 'Improving the Human Research Potential and the Socio-Economic Knowledge Base' (Research Training Network Contract HPRN-CT-2000-00137).

References

1. M.M. May, R.H. White: Math. Comp. Phys. **7**, 219 (1967)
2. J.R. Wilson: Astrophys. Journ. **173**, 431 (1972)
3. J.M. Martí, J.M. Ibáñez, J.A. Miralles: Phys. Rev. **D43**, 3794 (1991)
4. M.L. Norman, K.-H.A. Winkler: In: *Astrophysical Radiation Hydrodynamics.* ed. by M.L. Norman, K.-H.A. Winkler (Reidel, 1986)
5. J.A. Font, J.M. Ibáñez, A. Marquina, J.M. Martí: Astron. Astrophys. **282**, 304 (1994)
6. R. Donat, J.A. Font, J. M. Ibáñez, A. Marquina: J. Comp. Phys. **146**, 58 (1998)
7. F. Banyuls, J.A. Font, J.M. Ibáñez, J.M. Martí, J.A. Miralles: Astrophys. Journ. **476**, 221 (1997)
8. J.A. Font, M. Miller, W.M. Suen, M. Tobias: Phys. Rev. **D61**, 044011 (2000)
9. J.M. Ibáñez, M.A. Aloy, J.A. Font, J.M. Martí, J.A. Miralles, J.A. Pons: In: *Godunov Methods: Theory and Applications.* ed. by E.F. Toro (Kluwer Academic/Plenum Publishers, New York, 2001) pp. 485–496
10. J.M. Ibáñez, J.M. Martí: J. Comp. Appl. Math. **109**, 173 (1999)
11. J.M. Martí, E. Müller: In: *Living Reviews in Relativity* (1999) http://www.livingreviews.org/Articles/Volume2
12. J.A. Font: In: *Living Reviews in Relativity* (2000) http://www.livingreviews.org/Articles/Volume3
13. M.A. Aloy, J.M. Martí: In: *Relativistic Flows in Astrophysics* (Lecture Notes in Physics Series, Springer-Verlag, Berlin, 2001)
14. A.M. Anile: *Relativistic Fluids and Magnetofluids* (Cambridge University Press, Cambridge, 1989)
15. P. Lax: *Regional Conference Series Lectures in Applied Math.* **11** (SIAM, Philadelphia, 1972)
16. R.J. LeVeque: *Numerical methods for conservation laws* (Birkhäuser, Basel, 1992)
17. E.F. Toro: *Riemann Solvers and Numerical Methods for Fluid Dynamics* (Springer-Verlag, Berlin, 1997)
18. O.A. Oleinik: Am. Math. Soc. Transl. Ser. **2**, 26 (1963)
19. P. Lax, B. Wendroff: Comm. Pure Appl. Math. **13**, 217 (1960)

20. R.D. Richtmyer, K.W. Morton: *Difference methods for initial-value problems* (Wiley-Interscience, New York, 1967)
21. A. Harten, P.D. Lax, B. van Leer: SIAM Review **25**, 35 (1983)
22. E. Einfeldt: SIAM J. Num. Anal. **25**, 294 (1988)
23. J.P. Boris, D.L. Book: J. Comp. Phys. **11**, 38 (1973)
24. A. Harten: SIAM J. Num. Anal. **21**, 1 (1984)
25. B. van Leer: J. Comp. Phys. **32**, 101 (1979)
26. P. Colella, P.R. Woodward: J. Comp. Phys. **54**, 174 (1984)
27. S. Osher, S. Chakravarthy: SIAM J. Num. Anal. **21**, 995 (1984)
28. C.W. Shu: Math. Comput. **49**, 105 (1987)
29. A. Harten, B. Engquist, S. Osher, S. Chakravarthy: J. Comp. Phys. **71**, 231 (1987)
30. A. Marquina: SIAM J. Sci. Comput. **15**, 892 (1994)
31. R. Arnowitt, S. Deser, C.W. Misner: 'The Dynamics of General Relativity'. In: *Gravitation: An Introduction to Current Research*. ed. by L. Witten (John Wiley, New York, 1962) pp. 227–265
32. J.A. Pons, J.M. Martí, E. Müller: J. Fluid Mech. **422**, 125 (2000)
33. F. Eulderink, G. Mellema: Astron. Astrophys. Suppl. **110**, 587 (1995)
34. P. Papadopoulos, J.A. Font: Phys. Rev. **D61**, 024015 (2000)
35. P. Roe: J. Comp. Phys. **43**, 357 (1981)
36. V. Schneider, U. Katscher, D.H. Rischke, B. Waldhauser, J.A. Maruhn, C.-D. Munz: J. Comp. Phys. **105**, 92 (1993)
37. J.M. Martí, E. Müller: J. Fluid Mech. **258**, 317 (1994)
38. D.S. Balsara: J. Comp. Phys. **114**, 284 (1994)
39. C.W. Shu, S.J. Osher: J. Comp. Phys. **83**, 32 (1989)
40. A. Dolezal, S.S.M. Wong: J. Comp. Phys. **120**, 266 (1995)
41. S.A.E.G. Falle, S.S. Komissarov: Mon. Not. Roy. Astron. Soc. **278**, 586 (1996)
42. J.M. Martí, E. Müller: J. Comp. Phys. **123**, 1 (1996)
43. L. Wen, A. Panaitescu, P. Laguna: Astrophys. Journ. **486**, 919 (1997)
44. W. Dai, P.R. Woodward: SIAM J. Sci. Comput. **18**, 982 (1997)
45. J.A. Pons, J.A. Font, J.M. Martí, J.M. Ibáñez, J.A. Miralles: Astron. Astrophys. **339**, 638 (1998)
46. M.H.P.M. van Putten: Astrophys. Journ. **408**, L21 (1993)
47. J.M. Martí, E. Müller, J.M. Ibáñez: Astron. Astrophys. **281**, L9 (1994)
48. G.C. Duncan, P.A. Hughes: Astrophys. Journ. **436**, L119 (1994)
49. J.M. Martí, E. Müller, J.A. Font, J.M. Ibáñez: Astrophys. Journ. **448**, L105 (1995)
50. J.M. Martí, E. Müller, J.A. Font, J.M. Ibáñez, A. Marquina: Astrophys. Journ. **479**, 151 (1997)
51. R.J. Davis, T.W.B. Muxlow, R.G. Conway: Nature **318**, 343 (1985)
52. L. Scheck, M.A. Aloy, J.M. Martí, J.L. Gómez, E. Müller: Mon. Not. Roy. Astron. Soc. (submitted) (2001)
53. C.L. Carilli, R.A. Perley, N. Bartel, B. Sorathia: In: *Energy Transport in Radio Galaxies and Quasars*. ed. by P.E. Hardee, A.H. Bridle, J.A. Zensus (P.A.S.P., 1996)
54. M.A. Aloy, J.M. Ibáñez, J.M. Martí, J.L. Gómez, E. Müller: Astrophys. Journ. **523**, L125 (1999),
55. M.A. Aloy, J.L. Gómez, J.M. Ibáñez, J.M. Martí, E. Müller: Astrophys. Journ. **528**, L85 (2000)
56. M.A. Aloy, J.M. Ibáñez, E. Müller: Astrophys. Journ. Suppl. **122**, 151 (1999)
57. T. Piran: Phys. Rep. **314**, 575 (1999)
58. M.H.P.M. van Putten: Phys. Rep. **345**, 1 (2001)

59. B. Pacyński: Astrophys. Journ. **308**, L43 (1986)
60. J. Goodman: Astrophys. Journ. **308**, L47 (1986)
61. D. Eichler, M. Livio, T. Piran, D.N. Schramm: Nature **340**, 126 (1989)
62. R. Mochkovitch, M. Hernanz, J. Isern, X. Martin: Nature **361**, 236 (1993)
63. S.E. Woosley: Astrophys. Journ. **405**, 273 (1993)
64. B. Pacyński: Astrophys. Journ. **494**, L45 (1998)
65. G. Cavallo, M.J. Rees: Mon. Not. Roy. Astron. Soc. **183**, 359 (1978)
66. S.R. Kulkarni *et al*: Nature **395**, 663 (1998)
67. A.I. MacFadyen, S.E. Woosley: Astrophys. Journ. **524**, 262 (1999)
68. M.A. Aloy, E. Müller, J.M. Ibáñez, J.M. Martí, A. MacFadyen: Astrophys. Journ. **531**, L119 (2000)
69. M. van der Klis: In: *Proceedings NATO ASI. The many faces of neutron stars* (1997) astro-ph/9710016
70. L. Stella, M. Vietri: Astrophys. Journ. **492**, L59 (1998)
71. E.H. Morgan, R.A. Remillard, J. Greiner: Astrophys. Journ. **482**, 993 (1997)
72. W. Cui, S.N. Zhang, W. Chen: Astrophys. Journ. **492**, L53 (1998)
73. B.C. Bromley, W.A. Miller, V.I. Pariev: Nature **391**, 54 (1998)
74. F. Hoyle, R.A. Lyttleton: Proc. Cambridge Philos. Soc. **35**, 405 (1939)
75. H. Bondi, F. Hoyle: Mon. Not. Roy. Astron. Soc. **104**, 273 (1944)
76. J.A. Font, J.M. Ibáñez: Astrophys. Journ. **494**, 297 (1998)
77. J.A. Font, J.M. Ibáñez: Mon. Not. Roy. Astron. Soc. **298**, 835 (1998)
78. J.A. Font, J.M. Ibáñez, P. Papadopoulos: Astrophys. Journ. **507**, L67 (1998)
79. J.A. Font, J.M. Ibáñez, P. Papadopoulos: Mon. Not. Roy. Astron. Soc. **305**, 920 (1999)
80. P. Papadopoulos, J.A. Font: Phys. Rev. **D58**, 024005 (1999)
81. S. Brandt, J.A. Font, J.M. Ibáñez, J. Massó, E. Seidel: Comp. Phys. Comm. **124**, 169 (2000)
82. H. Bondi: Proc. Roy. Soc. London **A281**, 39 (1964)
83. E. Gourgoulhon: Astron. Astrophys. **252**, 651 (1991)
84. J.V. Romero, J.M. Ibáñez, J. M. Martí, J. A. Miralles: Astrophys. Journ. **462**, 839 (1996)
85. K.A. van Riper: Astrophys. Journ. **232**, 558 (1979)
86. H. Dimmelmeier, J.A. Font, E. Müller: Astrophys. Journ. **560**, L163 (2001)
87. H. Dimmelmeier, J.A. Font, E. Müller: Class. Quantum Grav. **19**, 1291 (2002)
88. H. Dimmelmeier: General Relativistic Collapse of Rotating Stellar Cores in Axisymmetry. Ph.D. Thesis, Technische Universität, München (2001)
89. F. Linke, J.A. Font, H.-T. Janka, E. Müller, P. Papadopoulos: Astron. Astrophys. **376**, 568 (2001)
90. P. Papadopoulos, J.A. Font: Phys. Rev. **D63**, 044016 (2001)

Flux Limiter Methods in 3D Numerical Relativity

Carles Bona and Carlos Palenzuela

Universitat de les Illes Balears, Grup de Relativitat, cta. de Bunyola km 7.5, 07071 Palma de Mallorca, Baleares, Spain

Abstract. New numerical methods have been applied in relativity to obtain a numerical evolution of Einstein equations much more robust and stable. Starting from 3+1 formalism and with the evolution equations written as a FOFCH (first-order flux conservative hyperbolic) system, advanced numerical methods from CFD (Computational Fluid Dynamics) have been successfully applied. A flux limiter mechanism has been implemented in order to deal with steep gradients like the ones usually associated with black hole spacetimes. As a test bed, the method has been applied to 3D metrics describing propagation of nonlinear gauge waves. Results are compared with the ones obtained with standard methods, showing a great increase in both robustness and stability of the numerical algorithm.

1 Introduction

From the very beginning, 3D numerical relativity has not been an easy domain. Difficulties arise either from the computational side (the large amount of variables to evolve, the large number of operations to perform, the stability of the evolution code) or from the physical side, like the complexity of the Einstein equations themselves, boundary conditions, singularity avoiding gauge choices, and so on. Sometimes there is a connection between both sides. For instance, the use of singularity avoiding slicings generates large gradients in the vicinity of black holes. Numerical instabilities can be produced by these steep gradients. The reason for this is that the standard evolution algorithms are unable to deal with sharp profiles. The instability shows up in the form of spurious oscillations which usually grow and break down the code.

Numerical advanced methods from CFD (Computational Fluid Dynamics) can be used to avoid this. Stable codes are obtained which evolve in a more robust way, without too much dissipation, so that the shape of the profiles of the evolved quantities is not lost. These advanced methods are then specially suited for the problem of shock propagation, but they apply only to strongly hyperbolic systems, where one is able to get a full set of eigenfields which generates all the physical quantities to be evolved. In the 1D case, these methods usually fulfill the TVD (Total Variation Diminishing) condition when applied to transport equations. This ensures that no new local extreme appears in the profiles of the eigenfields, so that spurious oscillations are ruled out ab initio (monotonicity preserving condition). Unluckily, there is no general method with this property in the 3D case, mainly because the eigenfield basis depends on the direction of propagation. We will show how this can be achieved at least in some cases.

The specific methods we will use are known as flux limiter methods. We will consider plane waves in 3D as a first generalization of the 1D case, because the propagation direction is constant. This specific direction leads then to an specific eigenfield basis, so that the 1D numerical method can be easily generalized to the 3D case.

The algorithm will be checked with a "Minkowski waves" metric. It can be obtained by a coordinate transformation from Minkowski space-time. All the metric components are transported while preserving their initial profiles. The line element has the following form:

$$\mathrm{d}s^2 = -H(x-t)\,\mathrm{d}t^2 + H(x-t)\,\mathrm{d}x^2 + \mathrm{d}y^2 + \mathrm{d}z^2 \tag{1}$$

where $H(x-t)$ is any positive function. We can choose a periodic profile with sharp peaks so both the space and the time derivatives of $H(x-t)$ will have discontinuous step-like profiles. If we can solve well this case (the most extreme), we can hope that the algorithm will work as well in more realistic cases where discontinuities do not appear.

Minkowski waves are a nice test bed because the instabilities can arise only from the gauge (these are pure gauge after all!). This is a first step to deal with evolution instabilities in the Einstein equations by the use of flux limiter methods. This will allow us to keep all our gauge freedom available to deal with more physical problems, like going to a co-rotating frame or adapting to some special geometry. Advanced numerical methods take care of numerical problems so that 'physical' gauge choices can be used to take care of physics requirements.

2 The System of Equations

We will use the well known 3+1 description of spacetime [1–3] which starts by decomposing the line element as follows:

$$\mathrm{d}s^2 = -\alpha^2\,\mathrm{d}t^2 + \gamma_{ij}\,(\mathrm{d}x^i + \beta^i\,\mathrm{d}t)\,(\mathrm{d}x^j + \beta^j\,\mathrm{d}t) \quad i,j = 1,2,3 \tag{2}$$

where γ_{ij} is the metric induced on the three-dimensional slices and α, β^i are the lapse and the shift, respectively. For simplicity the case $\beta^i = 0$ (normal coordinates) will be considered here. The intrinsic curvature of the slices is then given by the three-dimensional Ricci tensor ${}^{(3)}R_{ij}$, whereas their extrinsic curvature K_{ij} is given by:

$$\partial_t \gamma_{ij} = -2\,\alpha\,K_{ij}\;. \tag{3}$$

In what follows, all the geometrical operations (index raising, covariant derivations, etc) will be performed in the framework of the intrinsic three-dimensional geometry of every constant time slice. With the help of the quantities defined in (2,3), the ten fourdimensional field equations can be expressed as a set of six evolution equations:

$$\partial_t K_{ij} = -\nabla_i \alpha_j + \alpha\,\left[{}^{(3)}R_{ij} - 2K_{ij}^2 + \mathrm{tr}\,K\,K_{ij} - 8\,\pi\,\left(T_{ij} - \frac{T}{2}\gamma_{ij}\right)\right] \tag{4}$$

plus four constraint equations

$$ {}^{(3)}R - \mathrm{tr}\,(K^2) + (\mathrm{tr}\,K)^2 = 16\pi\alpha^2 T^{00} \tag{5}$$

$$ \nabla_k K^k{}_i - \partial_i(\mathrm{tr}\,K) = 8\pi\alpha T^0_i \,. \tag{6}$$

The evolution system (4) has been used by numerical relativists since the very beginning of the field (see for instance the seminal work of Eardley and Smarr [4]), both in spherically symmetric (1D) and axially symmetric (2D) spacetimes. By the turn of the century, the second order system (4) has been rewritten as a first-order flux conservative hyperbolic (FOFCH) system [5–7] in order to deal with the generic (3D) case, where no symmetries are present. But the second order system (4) is still being used in 3D numerical calculations [8], mainly when combined with the conformal decomposition of K_{ij} as introduced by Shibata and Nakamura [9,10]. In the system (3,4) there is a degree of freedom to be fixed because the evolution equation for the lapse function α is not given. In the study of Black Holes, the slicing is usually chosen in order to avoid the singularity [11]:

$$ \partial_t \ln\alpha = -\alpha Q \tag{7}$$

where:

$$ Q = f(\alpha)\mathrm{tr}\,K \,. \tag{8}$$

Three basic steps are needed to obtain a FOFCH system from the ADM system. First, one must introduce some new auxiliary variables to express the second order derivatives in space as first order. These new quantities correspond to the space derivatives:

$$ A_k = \partial_k \ln\alpha \,, \quad D_{kij} = 1/2\; \partial_k \gamma_{ij} \,. \tag{9}$$

The evolution equations for these variables are:

$$ \partial_t A_k + \partial_k(\alpha f \mathrm{tr}\,K) = 0 \tag{10}$$

$$ \partial_t D_{kij} + \partial_k(\alpha K_{ij}) = 0 \,. \tag{11}$$

At the second step the system is expressed in a first order balance law form

$$ \partial_t \boldsymbol{u} + \partial_k F^k(\boldsymbol{u}) = S(\boldsymbol{u}) \,, \tag{12}$$

where the array $\boldsymbol{u}$ displays the set of independent variables to evolve and both "fluxes" F^k and "sources" S are vector valued functions. At the third step another additional independent variable is introduced to obtain a strongly hyperbolic system [11]:

$$ V_i = D_{ir}{}^r - D^r{}_{ri} \tag{13}$$

and its evolution equation is obtained using the definition of K_{ij} from (3) and switching space and time derivatives in the momentum constraint (6). The result is an independent evolution equation for V_i while the previous definition (13) in terms of space derivatives can instead be considered as a first integral of the extended system. The extended array $\boldsymbol{u}$ will then contain the following 37 functions $\boldsymbol{u} = (\alpha,\, \gamma_{ij},\, K_{ij},\, A_i,\, D_{kij},\, V_i)$.

3 The Numerical Algorithm

Due to the structure of the equations, the evolution (represented by the operator $E(\Delta t)$) described by (12) can be decomposed into two separate processes; the first one is a transport process and the second one is the contribution of the sources.

The sources step (represented by the operator $S(\Delta t)$) does not involve space derivatives of the fields, so that it consists in a system of coupled non-linear ODE (Ordinary Differential Equations):

$$\partial_t \boldsymbol{u} = S(\boldsymbol{u}) \,. \tag{14}$$

The transport step (represented by the operator $T(\Delta t)$) contains the principal part and it is given by a set of quasi-linear transport equations:

$$\partial_t \boldsymbol{u} + \partial_k F^k(\boldsymbol{u}) = 0 \,. \tag{15}$$

The numerical implementation of these separated processes is quite easy. Second order accuracy in Δt can be obtained by using the well known Strang splitting.

$$E(\Delta t) = S(\Delta t/2)\, T(\Delta t)\, S(\Delta t/2) \,. \tag{16}$$

According to (3,7) the lapse and the metric have no flux terms. It means that a reduced set of 30 quantities $\boldsymbol{u} = (K_{ij}, A_i, D_{kij}, V_i)$ are transported in the second step over an inhomogeneous static background composed by (α, γ_{ij}). The equations for the transport step (15) are given by:

$$\partial_t K_{ij} + \partial_k(\alpha \lambda^k{}_{ij}) = 0 \tag{17}$$

$$\partial_t A_k + \partial_k(\alpha f(\alpha) \mathrm{tr}\, K) = 0 \tag{18}$$

$$\partial_t D_{kij} + \partial_k(\alpha K_{ij}) = 0 \tag{19}$$

$$\partial_t V_k = 0 \tag{20}$$

where:

$$\lambda^k{}_{ij} = D^k_{ij} - \frac{m}{2} V^k \gamma_{ij} + \frac{1}{2}\delta^k_i (A_j + 2V_j - D_{jr}{}^r) + \frac{1}{2}\delta^k_j (A_i + 2V_i - D_{ir}{}^r) \tag{21}$$

and m is an arbitrary parameter.

To evolve the transport step, we will consider flux-conservative numeric algorithms [12], obtained by applying the balance to a single computational cell. In the 1D case the cell goes from n to n+1 in time ($t = n \cdot \Delta t$) and from $j - 1/2$ to $j + 1/2$ in space ($x_j = j \cdot \Delta x$), so that we have:

$$U_j^{n+1} = U_j^n - \frac{\Delta t}{\Delta x}\left[F_{j+1/2}^{n+1/2} - F_{j-1/2}^{n+1/2}\right] \,. \tag{22}$$

Interface fluxes can be calculated in many different ways, leading to different numerical methods. We will use here the well known MacCormack method. This

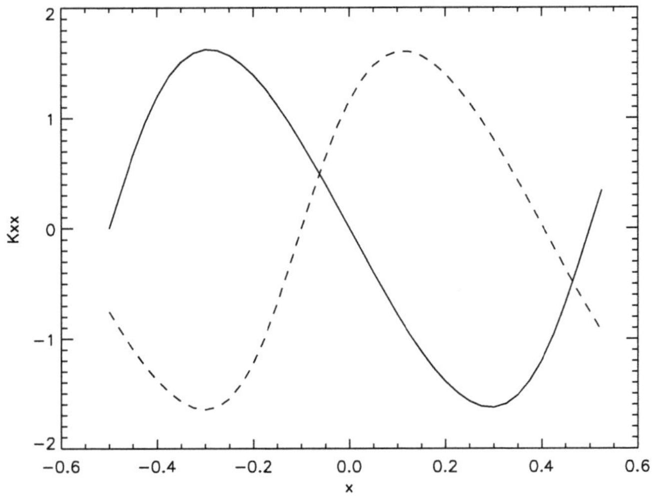

Fig. 1. Plot of K_{xx} for the initial metric given by (1) with $H(x-t) = 1 + A\ \cos[\omega\ (x-t))]$ with periodic boundaries. Continuous line is the initial condition. Dashed line is after 40 iterations

flux-conservative standard algorithm works well for smooth profiles, as it can be appreciated in Fig. 1.

But this standard algorithm is not appropriate for step-like profiles because it produces spurious oscillations near the steep regions, as it can be appreciated in Fig. 2.

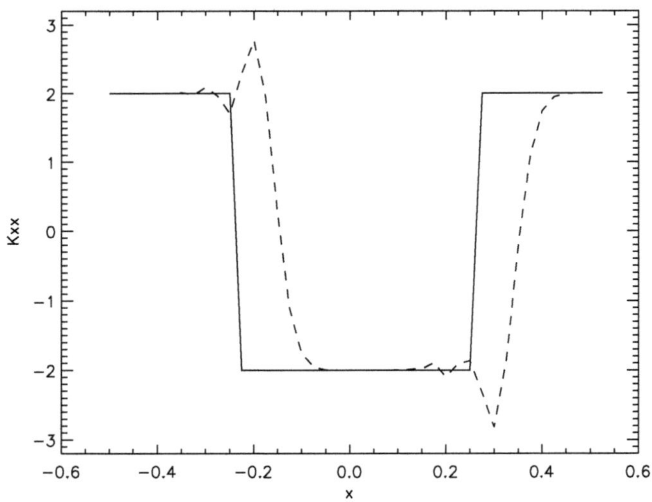

Fig. 2. Same as in Fig. 1 with the step-like initial data for K_{xx}. Continuous line is the initial condition. Dashed line is after 10 iterations. Note the spurious oscillations around the corners

More advanced numerical methods must be used to eliminate (or at least to reduce) these oscillations. These advanced methods use information about the eigenfields and the propagation direction, so the flux characteristic matrix along the propagation direction must be diagonalized.

4 Eigenfields

We will use a convenient method to compute the eigenfields. Let us study the propagation of a step-like discontinuity in the transported variables $\boldsymbol{u}$ which will move along a specific direction n with a given velocity v. Information about the corresponding eigenfields can be extracted from the well known Rankine-Hugoniot shock conditions:

$$v[u] = n_k[F^k(u)] \tag{23}$$

where [ ] represents the jump in the discontinuity. In our case

$$\begin{aligned} v[K_{ij}] &= n_r[\alpha\lambda^r{}_{ij}] && (24)\\ v[A_k] &= n_k[\alpha f(\alpha)\mathrm{tr}\, K] && (25)\\ v[D_{kij}] &= n_k[\alpha K_{ij}] && (26)\\ v[V_k] &= 0 && (27) \end{aligned}$$

where we must note that both background metric coefficients and propagation direction are assumed to be continuous, so they are transparent to the [ ] symbol.

If we develop these expressions we arrive at the following conclusions, where $S_n = n^r S_r$ is the projection of the quantity S over n and $S_\perp = S_k - S_n\, n_k$ are the transverse components:

1. $[V_k], [A_\perp], [D_{\perp ij}]$ and $[A_n - f\ \mathrm{tr}\, D_n]$ propagate along n with speed $v = 0$. There are eighteen such eigenfields. For the line element given by (1) n^k is along the x axis and all these fields are actually zero.
2. $[\lambda^n{}_{ij} - \mathrm{tr}\,\lambda^n n_i n_j]$ and $[K_{ij} - \mathrm{tr}\, K n_i n_j]$ do generate eigenfields propagating along n with speed $v = \pm\alpha$ (light cones). There are only ten such eigenfields because all of them are traceless. For Minkowski waves, where there is only gauge, all these combinations are zero. This indicates that the correct way to get the traceless part of a given tensor S_{ij} in this context is just to take $S_{ij} - \mathrm{tr}\, S n_i n_j$, so that the contribution of gauge modes will disappear.
3. $[A_n]$ and $[\mathrm{tr}\, K]$ do generate eigenfields propagating along n with speed $v = \pm\sqrt{f}\alpha$ (gauge cones). There are two such eigenfields corresponding to the gauge sector. For Minkowski waves, there are the only non-zero components. We are left with:

$$\begin{aligned} v[\mathrm{tr}\, K] &= \alpha[A_n] && (28)\\ v[A_k] &= \alpha f(\alpha)[\mathrm{tr}\, K] n_k && (29) \end{aligned}$$

so that $[A_k]$ is proportional to n_k. Now we can get the gauge eigenfields:

$$\sqrt{f} n_k F^k(\mathrm{tr}\, K) \pm F(A_n)\ . \tag{30}$$

These eigenfields propagate along n according simple advection equations, a familiar situation in the 1D case. Although this decomposition and diagonalization is trivial in 1D, it is very useful in the multidimensional case for a generic direction n.

5 Flux Limiter Methods

The flux limiter methods [12] we will use can be decomposed into some basic steps. First of all the interface fluxes have to be calculated with any standard second order accurate method (MacCormack in our case). Then, the propagation direction $\boldsymbol{n}$ and the corresponding eigenfields can be properly identified at every cell interface using the straight relation between n^k and A^k (which is strongly related with $F^k(\mathrm{tr}\, K)$) and the general decomposition from the previous section. Two advection equations (one for every sense of propagation) are now available for the gauge eigenfluxes (30).

At the 1D case the procedure is quite well known. Let us choose for instance the eigenflux which propagates to the right (an equivalent process will be valid for the other eigenflux propagating to the left). This interface eigenflux $F^{n+1/2}_{j+1/2}$ can be understood as the grid point flux F^n_j plus some increment $\Delta_{j+1/2} = F^{n+1/2}_{j+1/2} - F^n_j$. In general, the purpose of the limiter is to use a mixture of the increments $\Delta_{j+1/2}$ and $\Delta_{j-1/2}$ to ensure monotonicity. In our case we are using a robust mixture which goes by applying the well known minmod rule to $\Delta_{j+1/2}$ and $2\Delta_{j-1/2}$. In that way, the limiter acts only in steep regions, where the ratio between neighbouring increments exceeds a factor of two.

We can apply this method to the step-like initial data propagating along the x axis. We can see in Fig. 3 that the result is much better than before. It can be (hardly) observed a small deviation from the TVD condition, which is produced by the artificial separation produced by the Strang splitting into transport and non-linear source steps.

This method can be applied, with the general decomposition described in section 4, to discontinuities which propagate along any constant direction, and not only to the trivial case, aligned with the x axis, that we have considered until now.

To prove it, we have rotated the metric of Minkowski waves in the $x-z$ plane to obtain a diagonal propagation of the profile. The line element in this case has the following form:

$$
\begin{aligned}
\mathrm{d}s^2 = &- H\left(\frac{x+z}{\sqrt{2}} - t\right)\, \mathrm{d}t^2 + \frac{1}{2}\left[1 + H\left(\frac{x+z}{\sqrt{2}} - t\right)\right]\, (\mathrm{d}x^2 + \mathrm{d}z^2) + \mathrm{d}y^2 \\
&+ \frac{1}{2}\left[-1 + H\left(\frac{x+z}{\sqrt{2}} - t\right)\right]\, (\mathrm{d}x\, \mathrm{d}z + \mathrm{d}z\, \mathrm{d}x)
\end{aligned}
\tag{31}
$$

We show the results in Fig. 4.

We can also see in Fig. 5 a $z =$constant section of the same results to allow a more detailed comparison with the initial data. We can also compare with the 1D case in Fig. 3 and check the closer evolved profiles.

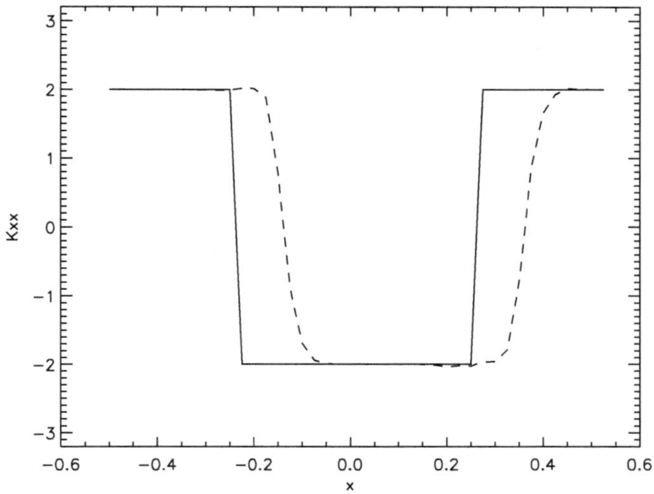

Fig. 3. Same as Fig.2 where the methods presented in this paper are applied. Continuous line is the initial condition. Dashed line is after 10 iterations

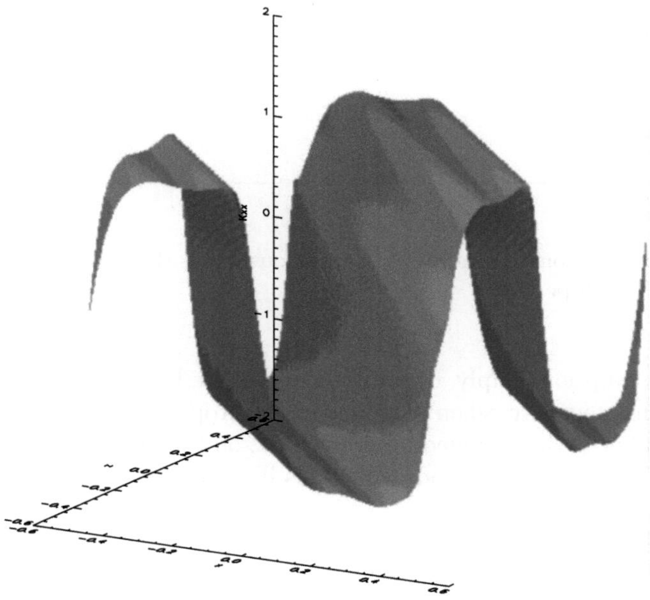

Fig. 4. $3D$ plot of K_{xx}. The step-like profile has been propagated with periodic boundary conditions until one full period (about 80 iterations in this case) has elapsed and it has returned to the initial position

Now we are in position to compare the simple Flux-limiter method we are using here with the advanced 3D methods used in CFD. From the theoretical point of view, we know that none of these methods can be shown to be really TVD, because the eigenfield decomposition is direction-dependent and the characteristic matrices for different directions do not commute. From the numerical point of view, our methods are based on two simplifying assumptions, namely:

1. One is able to identify one specific direction in which gauge propagation takes place. In our case, this is taken to be $n_k = [A_k]$.
2. The propagation equation along that direction $\boldsymbol{n}$ can be approximated by an advection equation. In our case, this implies that $\boldsymbol{n}$ can be locally considered as a constant vector.

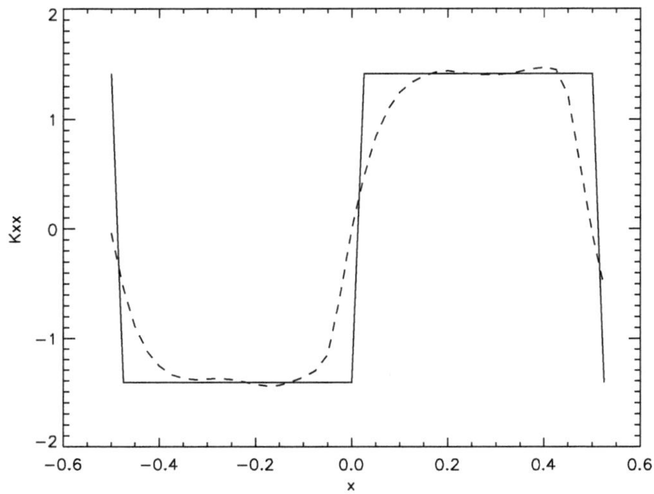

Fig. 5. Section $z =$const. from Fig. 4. Continuous line is the initial condition. Dashed line is after one full period

These assumptions imply, of course, that we are loosing generality. This can be a problem, specially when studying wave propagation in the near region, where $\boldsymbol{n}$ can hardly be assumed constant. But, on the other hand, they allow us the use of a very simple method, which has its own advantages:

1. The Flux corrections can be selectively applied to the gauge sector only, without affecting other degrees of freedom.
2. Our limiter choice amounts to define a numerical discontinuity where the ratio between neighbouring slopes exceeds a factor of two, so that the limiter is applied only at these regions, having most of the grid points unaffected.

Allowing for that, we can conclude that our results are a first try to incorporate 3D CFD methods to Numerical Relativity. Our results can then be considered as a preliminary view of what can be expected from a more systematic use of such powerful methods in the near future.

Acknowledgements

This work has been supported by the EU Programme 'Improving the Human Research Potential and the Socio-Economic Knowledge Base' (Research Training Network Contract (HPRN-CT-2000-00137) and by a grant from the Conselleria d'Innovacio i Energia of the Govern de les Illes Balears.

References

1. Y. Choquet-Bruhat: C. R. Acad. Sc. Paris **226**, 1071 (1948). J. Rat. Mec. Analysis **5**, 951 (1956)
2. Y. Choquet-Bruhat: 'Cauchy Problem', In *Gravitation: an introduction to Current Research.* ed. by L. Witten (Wiley, New York, 1962) pp. 130–168
3. R. Arnowitt, S. Deser, C. W. Misner: 'The dynamic of general relativity'. In *Gravitation: an introduction to Current Research.* ed. by L. Witten (Wiley, New York, 1962) pp. 227–265
4. D. M. Eardley, L. Smarr: Phys. Rev. **D19**, 2239 (1979)
5. C. Bona, J. Massó: Phys. Rev. Lett. **68**, 1097 (1992)
6. Y. Choquet-Bruhat, J.W.York: C. R. Acad. Sc. Paris **321** 1089 (1995)
7. S. Frittelli, O. A. Reula: Phys. Rev. Lett. **76**, 4667 (1996)
8. M. Alcubierre, B. Bruegmann, D. Pollney, E. Seidel, R. Takahashi: Phys. Rev. **D64**, 061501 (2001)
9. M. Shibata, T. Nakamura: Phys. Rev. **D52**, 5428 (1995)
10. T. Baumgarte and S. Shapiro, Phys. Rev. **D59**, 024007 (1999)
11. C. Bona, J. Massó, E. Seidel, J. Stela: Phys. Rev. **D56**, 3405 (1997)
12. R. J. LeVeque: *Numerical methods for conservation laws*, (Birkhauser, Basel, 1992)

Gauge Conditions for Long-Term Numerical Black Hole Evolution With or Without Excision

Miguel Alcubierre[1], Bernd Brügmann[1], Denis Pollney[1], Edward Seidel[1,2], and Ryoji Takahashi[1]

[1] Max-Planck-Institut für Gravitationsphysik, Albert-Einstein-Institut, Am Mühlenberg 1, 14476 Golm, Germany
[2] National Center for Supercomputing Applications, Beckman Institute, 405 N. Mathews Ave., Urbana, IL 61801

Abstract. We extend previous work on 3D black hole excision to the case of distorted black holes, with a variety of dynamic gauge conditions that are able to respond naturally to the spacetime dynamics. We show that the combination of excision and gauge conditions we use is able to drive highly distorted, rotating black holes to an almost static state at late times, with well behaved metric functions, without the need for any special initial conditions or analytically prescribed gauge functions. Further, we show for the first time that one can extract accurate waveforms from these simulations, with the full machinery of excision or *no excision* and dynamic gauge conditions. The evolutions can be carried out for long times, far exceeding the longevity and accuracy of even better resolved 2D codes. While traditional 2D codes show errors in quantities such as apparent horizon mass of over 100% by $t \approx 100M$, and crash by $t \approx 150M$, with our new techniques the same systems can be evolved for more than hundreds of M's in full 3D with errors of only a few percent.

1 Introduction

The long term numerical evolution of black hole systems is one of the most challenging and important problems in numerical relativity. For black holes, the difficulties of accuracy and stability in solving Einstein's equations numerically are exacerbated by the special problems posed by spacetimes containing singularities. At a singularity, geometric quantities become infinite and cannot be handled easily by a computer.

Traditionally, in the 3+1 approach the freedom in choosing the slicing is used to slow down the approach of the time slices towards the singularity ("singularity avoidance"), while allowing them to proceed outside the black hole. Singularity avoiding slicings are able to provide accurate evolutions, allowing one to study black hole collisions and extract waveforms [1], but only for limited cases and evolution times. Combining short full numerical evolutions with perturbation methods, one can even study the plunge from the last stable orbit of two black holes [2]. But a dramatic breakthrough is required to push numerical simulations far enough to study orbiting black holes, requiring accurate evolutions exceeding time scales of $t \approx 100M$. In 3D, traditional approaches have not been able to reach such time scales, even in the case of Schwarzschild black holes.

A more promising approach involves cutting away the singularity from the calculation ("singularity excision"), assuming it is hidden inside an apparent horizon (AH) [3,4]. Although this work has been progressing, from early spherical proof of principle in [4] to recent 3D developments [5–8], beyond a few spherical test cases [9,10] it has yet to be used in conjunction with appropriate live gauge conditions designed to respond to both the dynamics of the black hole and the coordinate motion through the spacetime.

In this paper we extend recent excision work [7] to the case of distorted, dynamic black holes in 3D, using a new class of gauge conditions. These gauge conditions, which not only respond naturally to the true spacetime dynamics, but also *drive the system towards an almost static state at late times*, allow us to handle black holes without considering special initial coordinate systems, such as the Kerr-Schild type, which may be difficult or impossible to find during a generic black hole evolution. We show that not only are the evolutions accurate as indicated by the mass associated with the apparent horizon, but also that very accurate waveforms can be extracted with excision or *without* excision, even when the waves carry only a tiny fraction of the energy of the spacetime. We also show that the 3D evolutions of dynamic black holes we are now able to perform, are superior, in terms of accuracy, stability, and longevity, to previous dynamic 3+1 black hole simulations, whether they were carried out in full 3D or even when restricted to 2D. These results indicate that black hole evolution with new gauge conditions can be made to work under rather general circumstances, and can dramatically improve both the length of the evolutions, and the accuracy of the waveforms extracted, which will be crucial for gravitational wave astronomy.

2 Initial Data

For this paper we consider a series of single distorted black hole spacetimes [11,12] that have been used to model the late stages of black hole coalescence [13,14]. Following [11,12], the initial three-metric γ_{ab} is chosen to be

$$\mathrm{d}s^2 = \psi^4 \left[\mathrm{e}^{2q} \left(\mathrm{d}\eta^2 + \mathrm{d}\theta^2\right) + \sin^2\theta \, \mathrm{d}\phi^2\right] \ , \tag{1}$$

where the "Brill wave" function q is a general function of the spatial coordinates, subject to certain regularity and fall off restrictions, that can be tailored to produce very distorted 3D black holes interacting with nonlinear waves. The radial coordinate η is logarithmic in the cartesian radius r. There are two classes of data sets used here corresponding to even- and odd-parity distortions. The even-parity data have vanishing extrinsic curvature, while the cases containing an odd-parity component have nontrivial extrinsic curvature K_{ij}. As shown in [15,16], these distorted black hole data sets can include rotation as well, corresponding to spinning, distorted black holes that mimic the early merger of two orbiting black holes. Hence they make an ideal test case for the development of our techniques. We leave the details of the construction of these black hole initial data sets to [15,16].

An important point that we wish to emphasize is that such data are *not* of the Kerr-Schild form with ingoing coordinates at the horizon. That particular form of black hole initial data sets has been recently advocated as providing a more natural treatment for black hole excision since the coordinate system is adapted to inward propagation of quantities at the horizon [17]. However, it is not obvious that the physically desired initial data can always be written in the Kerr-Schild form (or, for that matter, in any other particular form). Furthermore, during an evolution, even if similar such coordinates are somehow actively enforced, it is probably not possible to have such a coordinate system in place at all times, when a new black hole forms. Hence, we prefer to be able to handle black hole data in any coordinate system, and apply coordinate conditions that naturally drive the system into a static state as the black hole system settles down to Kerr, from any starting point.

3 Evolution and Excision Procedures

Our simulations have been performed using what we refer to as the "BSSN" version of the 3+1 evolution equations [18–21], which we have found to have superior stability properties when compared to standard formulations.

The standard variables in the 3+1 formulation of ADM (Arnowitt-Deser-Misner, see [22]) are the 3-metric γ_{ij} and its extrinsic curvature K_{ij}. The gauge is determined by the lapse function α and the shift vector β^i. We will only consider the vacuum case. The evolution equations are

$$(\partial_t - \mathcal{L}_\beta)\,\gamma_{ij} = -2\alpha K_{ij}\ , \tag{2}$$

$$(\partial_t - \mathcal{L}_\beta)\,K_{ij} = -D_iD_j\alpha + \alpha(R_{ij} + KK_{ij} - 2K_{ik}K^k{}_j)\ , \tag{3}$$

and the constraints are

$$\mathcal{H} \equiv R + K^2 - K_{ij}K^{ij} = 0\ , \tag{4}$$

$$\mathcal{D}^i \equiv D_j(K^{ij} - \gamma^{ij}K) = 0\ . \tag{5}$$

Here $\mathcal{L}_\beta$ is the Lie derivative with respect to the shift vector β^i, D_i is the covariant derivative associated with the 3-metric γ_{ij}, R_{ij} is the three-dimensional Ricci Tensor, R the Ricci scalar, and K is the trace of K_{ij}.

We will use the BSSN form of these equations (Baumgarte, Shapiro [19], and Shibata, Nakamura [18]). One introduces new variables based on a trace decomposition of the extrinsic curvature and a conformal rescaling of both the metric and the extrinsic curvature. The trace-free part A_{ij} of the extrinsic curvature is defined by

$$A_{ij} = K_{ij} - \frac{1}{3}\gamma_{ij}K\ . \tag{6}$$

Assuming that the metric γ_{ij} is obtained from a conformal metric $\tilde{\gamma}_{ij}$ by a conformal transformation,

$$\gamma_{ij} = \psi^4\tilde{\gamma}_{ij}\ , \tag{7}$$

we can choose a conformal factor ψ such that the determinant of $\tilde{\gamma}_{ij}$ is 1:

$$\psi = \gamma^{1/12} , \quad (8)$$
$$\tilde{\gamma}_{ij} = \psi^{-4}\gamma_{ij} = \gamma^{-1/3}\gamma_{ij} , \quad (9)$$
$$\tilde{\gamma} = 1 , \quad (10)$$

where γ is the determinant of γ_{ij} and $\tilde{\gamma}$ is the determinant of $\tilde{\gamma}_{ij}$. Instead of γ_{ij} and K_{ij} we can therefore use the variables

$$\phi = \ln\psi = \frac{1}{12}\ln\gamma , \quad (11)$$
$$K = \gamma_{ij}K^{ij} , \quad (12)$$
$$\tilde{\gamma}_{ij} = \mathrm{e}^{-4\phi}\gamma_{ij} , \quad (13)$$
$$\tilde{A}_{ij} = \mathrm{e}^{-4\phi}A_{ij} , \quad (14)$$

where $\tilde{\gamma}_{ij}$ has determinant 1 and $\tilde{A}_{ij}$ has vanishing trace. Furthermore, we introduce the conformal connection functions

$$\tilde{\Gamma}^i = \tilde{\gamma}^{jk}\tilde{\Gamma}^i{}_{jk} = -\partial_j\tilde{\gamma}^{ij} , \quad (15)$$

where $\tilde{\Gamma}^i{}_{jk}$ is the Christoffel symbol of the conformal metric. The second equality holds if the determinant of the conformal 3-metric $\tilde{\gamma}$ is actually unity (which is true analytically but may not be numerically). We call ϕ, K, $\tilde{\gamma}_{ij}$, $\tilde{A}_{ij}$, and $\tilde{\Gamma}^i$ the BSSN variables.

In terms of the BSSN variables the evolution equation (2) becomes

$$(\partial_t - \mathcal{L}_\beta)\,\tilde{\gamma}_{ij} = -2\alpha\tilde{A}_{ij} , \quad (16)$$
$$(\partial_t - \mathcal{L}_\beta)\,\phi = -\frac{1}{6}\alpha K , \quad (17)$$

while (3) leads to

$$(\partial_t - \mathcal{L}_\beta)\,\tilde{A}_{ij} = \mathrm{e}^{-4\phi}[-D_iD_j\alpha + \alpha R_{ij}]^{TF} + \alpha(K\tilde{A}_{ij} - 2\tilde{A}_{ik}\tilde{A}^k{}_j) , \quad (18)$$
$$(\partial_t - \mathcal{L}_\beta)\,K = -D^iD_j\alpha + \alpha(\tilde{A}_{ij}\tilde{A}^{ij} + \frac{1}{3}K^2) , \quad (19)$$

where TF denotes the trace-free part of the expression in brackets. On the right-hand side of (19) we have used the Hamiltonian constraint (4) to eliminate the Ricci scalar,

$$R = K_{ij}K^{ij} - K^2 = \tilde{A}_{ij}\tilde{A}^{ij} - \frac{2}{3}K^2 . \quad (20)$$

An evolution equation for $\tilde{\Gamma}^i$ can be obtained from (15) and (16),

$$\partial_t\tilde{\Gamma}^i = -2(\alpha\partial_j\tilde{A}^{ij} + \tilde{A}^{ij}\partial_j\alpha) - \partial_j\mathcal{L}_\beta\tilde{\gamma}^{ij} . \quad (21)$$

In this equation we use the momentum constraint (5) to substitute for the divergence of $\tilde{A}^{ij}$,

$$\partial_j\tilde{A}^{ij} = -\tilde{\Gamma}^i{}_{jk}\tilde{A}^{jk} - 6\tilde{A}^{ij}\partial_j\phi + \frac{2}{3}\tilde{\gamma}^{ij}\partial_jK . \quad (22)$$

One subtlety in obtaining numerically stable evolutions with the BSSN variables is precisely the question of how the constraints are used in the evolution equations. Several choices are possible and have been studied, see [23].

Note that in the preceding equations we are computing Lie derivatives of tensor densities. If the weight of a tensor density T is w, that is if T is a tensor times $\gamma^{w/2}$, then

$$\mathcal{L}_\beta T = [\mathcal{L}_\beta T]_\partial^{w=0} + wT\partial_k\beta^k \ , \tag{23}$$

where the first term denotes the tensor formula for Lie derivatives with the derivative operator ∂ and the second is the additional contribution due to the density factor. The density weight of $\psi = \mathrm{e}^\phi$ is $1/6$, so the weight of $\tilde\gamma_{ij}$ and $\tilde A_{ij}$ is $-2/3$ and the weight of $\tilde\gamma^{ij}$ is $2/3$. To be explicit,

$$\begin{aligned}
\mathcal{L}_\beta\phi &= \beta^k\partial_k\phi + \frac{1}{6}\partial_k\beta^k \ , & (24)\\
\mathcal{L}_\beta\tilde\gamma_{ij} &= \beta^k\partial_k\tilde\gamma_{ij} + \tilde\gamma_{ik}\partial_j\beta^k + \tilde\gamma_{jk}\partial_i\beta^k - \frac{2}{3}\tilde\gamma_{ij}\partial_k\beta^k \ , & (25)\\
\mathcal{L}_\beta\tilde\gamma^{ij} &= \beta^k\partial_k\tilde\gamma^{ij} - \tilde\gamma^{ik}\partial_k\beta^j - \tilde\gamma^{jk}\partial_k\beta^i + \frac{2}{3}\tilde\gamma^{ij}\partial_k\beta^k \ . & (26)
\end{aligned}$$

The evolution equation (21) for $\tilde\Gamma^i$ therefore becomes

$$\begin{aligned}
\partial_t\tilde\Gamma^i = {}& \tilde\gamma^{jk}\partial_j\partial_k\beta^i + \frac{1}{3}\tilde\gamma^{ij}\partial_j\partial_k\beta^k + \beta^j\partial_j\tilde\Gamma^i - \tilde\Gamma^j\partial_j\beta^i + \frac{2}{3}\tilde\Gamma^i\partial_j\beta^j \\
& -2\tilde A^{ij}\partial_j\alpha + 2\alpha(\tilde\Gamma^i{}_{jk}\tilde A^{jk} + 6\tilde A^{ij}\partial_j\phi - \frac{2}{3}\tilde\gamma^{ij}\partial_j K) \ . \qquad (27)
\end{aligned}$$

In the second line we see the formula for a vector density of weight $2/3$, but since $\tilde\Gamma^i$ is derived from the Christoffel symbols we obtain extra terms involving second derivatives of the shift (the first line in the equation above).

On the right-hand sides of the evolution equations for $\tilde A_{ij}$ and K, (18) and (19), there occur covariant derivatives of the lapse function, and the Ricci tensor of the non-conformal metric. Since

$$\Gamma^k{}_{ij} = \tilde\Gamma^k{}_{ij} + 2(\delta^k_i\partial_j\phi + \delta^k_j\partial_i\phi - \tilde\gamma_{ij}\tilde\gamma^{kl}\partial_l\phi) \ , \tag{28}$$

where $\tilde\Gamma^k{}_{ij}$ is the Christoffel symbol of the conformal metric, we have for example

$$D^iD_i\alpha = \mathrm{e}^{-4\phi}(\tilde\gamma^{ij}\partial_i\partial_j\alpha - \tilde\Gamma^k\partial_k\alpha + 2\tilde\gamma^{ij}\partial_i\phi\partial_j\alpha) \ . \tag{29}$$

The Ricci tensor can be separated in two parts,

$$R_{ij} = \tilde R_{ij} + R^\phi_{ij} \ , \tag{30}$$

where $\tilde R_{ij}$ is the Ricci tensor of the conformal metric and R^ϕ_{ij} denotes additional terms depending on ϕ:

$$\begin{aligned}
R^\phi_{ij} = {}& -2\tilde D_i\tilde D_j\phi - 2\tilde\gamma_{ij}\tilde D^k\tilde D_k\phi \\
& +4\tilde D_i\phi\,\tilde D_j\phi - 4\tilde\gamma_{ij}\tilde D^k\phi\,\tilde D_k\phi \ , \qquad (31)
\end{aligned}$$

where $\tilde{D}_i$ is the covariant derivative associated with the conformal metric. The conformal Ricci tensor can be written in terms of the conformal connection functions as

$$\begin{aligned}\tilde{R}_{ij} = &-\frac{1}{2}\tilde{\gamma}^{lm}\partial_l\partial_m\tilde{\gamma}_{ij} + \tilde{\gamma}_{k(i}\partial_{j)}\tilde{\Gamma}^k + \tilde{\Gamma}^k\tilde{\Gamma}_{(ij)k} \\ &+\tilde{\gamma}^{lm}\left(2\tilde{\Gamma}^k{}_{l(i}\tilde{\Gamma}_{j)km} + \tilde{\Gamma}^k{}_{im}\tilde{\Gamma}_{klj}\right) . \end{aligned} \qquad (32)$$

A key observation here is that if one introduces the $\tilde{\Gamma}^i$ as independent variables, then the principal part of the right-hand side of (18) contains the wave operator $\tilde{\gamma}^{lm}\partial_l\partial_m\tilde{\gamma}_{ij}$ but no other second derivatives of the conformal metric. This brings the evolution system one step closer to being hyperbolic.

One of the reasons why we have written out the BSSN system in such detail is to point out a subtlety that arises in the actual implementation if one wants to achieve numerical stability. In the computer code we do not use the numerically evolved $\tilde{\Gamma}^i$ in all places, but follow this rule:

- Partial derivatives $\partial_j\tilde{\Gamma}^i$ are computed as finite differences of the independent variables $\tilde{\Gamma}^i$ that are evolved using (27).
- In expressions that require $\tilde{\Gamma}^i$, not its derivative, we substitute $\tilde{\gamma}^{jk}\tilde{\Gamma}^i{}_{jk}(\tilde{\gamma})$, that is we do not use the independently evolved variable $\tilde{\Gamma}^i$ but recompute $\tilde{\Gamma}^i$ according to its definition (15) from the current values of $\tilde{\gamma}_{ij}$.

In practice we have found that the evolutions are far less stable if either $\tilde{\Gamma}^i$ is treated as an independent variable everywhere, or if $\tilde{\Gamma}^i$ is recomputed from $\tilde{\gamma}_{ij}$ before each time step. The rule just stated helps to maintain the constraint $\tilde{\Gamma}^i = -\partial_j\tilde{\gamma}^{ij}$ well behaved without removing the advantage of reformulating the principal part of the Ricci tensor.

In summary, we evolve the BSSN variables $\tilde{\gamma}_{ij}$, ϕ, $\tilde{A}_{ij}$, K, and $\tilde{\Gamma}^i$ according to (16), (17), (18), (19), and (27), respectively. The Ricci tensor is separated as shown in (30) with each part computed according to (31) and (32) respectively. The Hamiltonian and momentum constraints have been used to write the equations in a particular way. The evolved variables $\tilde{\Gamma}^i$ are only used when their partial derivatives are needed (the one term in the conformal Ricci tensor (32) and the advection term $\beta^k\partial_k\tilde{\Gamma}^i$ in the evolution equation for the $\tilde{\Gamma}^i$ themselves, (27)).

We use the simple excision approach described in [7]. Our excision algorithm is based on the following ideas: (a) Excise a *cube* contained inside the AH that is well adapted to cartesian coordinates; (b) Use a simple but stable boundary condition at the sides of the excised cube: copying of time derivatives from their values one grid point out along the normal directions; (c) Use standard centered (non-causal) differences in all terms except for advection terms on the shift (those that look like $\beta^i\partial_i$). For these terms we use second order upwind along the shift direction. These simplifications in excision reduce the complexity in the algorithm, avoid delicate interpolation issues near the excision boundary, and have allowed us to make rapid progress.

4 Numerics

The numerical time integration in our code uses an iterative Crank-Nicholson scheme with 3 iterations, see e.g. [23]. Derivatives are represented by second order finite differences on a Cartesian grid. We use standard centered difference stencils for all terms, except in the advection terms involving the shift vector (terms that look like $\beta^i \partial_i$). For these terms we use second order upwind along the shift direction. We have found the use of an upwind scheme in such advection-type terms crucial for the stability of our code. Notice that this is the only place in our implementation where any information about causality is used (i.e. the direction of the tilt in the light cones).

At the outer boundary we use a radiation (Sommerfeld) boundary condition. We start from the assumption that near the boundary all fields behave as spherical waves, namely we impose the condition

$$f = f_0 + \frac{u(r - vt)}{r} \; . \tag{33}$$

Where f_0 is the asymptotic value of a given dynamical variable (typically 1 for the lapse and diagonal metric components, and zero for everything else), and v is some wave speed. If our boundary is sufficiently far away one can safely assume that the speed of light is 1, so $v = 1$ for most fields. However, the gauge variables can easily propagate with a different speed implying a different value of v (see below where we discuss the gauge conditions).

In practice, we do not use the boundary condition (33) as it stands, but rather we use it in differential form:

$$\partial_t f + v \partial_r f - v \, \frac{(f - f_0)}{r} = 0 \; . \tag{34}$$

Since our code is written in Cartesian coordinates, we transform the last condition to

$$\frac{x_i}{r} \, \partial_t f + v \partial_i f + \frac{v x_i}{r^2} \, (f - f_0) = 0 \; . \tag{35}$$

We finite difference this condition consistently to second order in both space and time and apply it to all dynamic variables (with possible different values of f_0 and v) at all boundaries.

There is a final subtlety in our boundary treatment. Wave propagation is not the only reason why fields evolve near a boundary. Simple infall of the coordinate observers will cause some small evolution as well, and such evolution is poorly modeled by a propagating wave. This is particularly important at early times, when the above boundary condition introduces a bad transient effect. In order to minimize the error at our boundaries introduced by such non-wavelike evolution, we allow for boundary behavior of the form:

$$f = f_0 + \frac{u(r - vt)}{r} + \frac{h(t)}{r^n} \; , \tag{36}$$

with h a function of t alone and n some unknown power. This leads to the differential equation

$$\partial_t f + v\partial_r f - \frac{v}{r}\,(f - f_0) = \frac{vh(t)}{r^{n+1}}\,(1 - nv) + \frac{h'(t)}{r^n} \simeq \frac{h'(t)}{r^n} \quad \text{for large } r\;, \tag{37}$$

or in Cartesian coordinates

$$\frac{x_i}{r}\,\partial_t f + v\partial_i f + \frac{vx_i}{r^2}\,(f - f_0) \simeq \frac{x_i h'(t)}{r^{n+1}}\;. \tag{38}$$

This expression still contains the unknown function $h'(t)$. Having chosen a value of n, one can evaluate the above expression one point away from the boundary to solve for $h'(t)$, and then use this value at the boundary itself. Empirically, we have found that taking $n = 3$ almost completely eliminates the bad transient caused by the radiative boundary condition on its own.

5 Gauge Conditions

We will consider families of gauge conditions that can be used in principle with any 3+1 form of the Einstein's equations that allows a general gauge. However, the specific family we test in this paper is best motivated by considering the BSSN system introduced above. For the present purposes, of special importance are the following two properties of this formulation:

- The trace of the extrinsic curvature K is treated as an independent variable. For a long time it has been known that the evolution of K is directly related to the choice of a lapse function α. Thus, having K as an independent field allows one to impose slicing conditions in a much cleaner way.
- The appearance of the "conformal connection functions" $\tilde{\Gamma}^i$ as independent quantities. As already noted by Baumgarte and Shapiro [19], the evolution equation for these quantities can be turned into an elliptic condition on the shift which is related to the minimal distortion condition. More generally, one can relate the shift choice to the evolution of these quantities, again allowing for a clean treatment of the shift condition.

Our aim is to look for gauge conditions that at late times, once the physical system under consideration has settled to a final stationary state, will be able to drive the coordinate system to a frame where this stationarity is evident. In effect, we are looking for "symmetry seeking" coordinates of the type discussed by Gundlach and Garfinkle [24] that will be able to find the approximate Killing field that the system has at late times. In order to achieve this we believe that the natural approach is to relate the gauge choice to the evolution of certain combinations of dynamic quantities in such a way that the gauge will either freeze completely the evolution of those quantities (typically by solving some elliptic equations), or will attempt to do so with some time delay (by solving instead parabolic or hyperbolic equations).

We will consider the lapse and shift conditions in turn. Special cases of the gauge conditions that we will introduce here were recently used together with black hole excision with remarkable results in [7], but as we will show below, the gauge conditions are so powerful that in the cases tested, they work even without excision.

5.1 Slicing Conditions

The starting point for our slicing conditions is the "K-freezing" condition $\partial_t K = 0$, which in the particular case when K=0 reduces to the well known "maximal slicing" condition. The K-freezing condition leads to the following elliptic equation for the lapse

$$\nabla^2 \alpha = \beta^i \partial_i K + \alpha K_{ij} K^{ij} \ , \tag{39}$$

with ∇^2 the Laplacian operator for the spatial metric γ_{ij}. In the BSSN formulation, once we have solved the elliptic equation for the lapse, the K-freezing condition can be imposed at the analytic level by simply not evolving K.

One can construct parabolic or hyperbolic slicing conditions by making either $\partial_t \alpha$ or $\partial_t^2 \alpha$ proportional to $\partial_t K$. We call such conditions "K-driver" conditions (see [25]). The hyperbolic K-driver condition has the form [7]

$$\partial_t^2 \alpha = -\alpha^2 f(\alpha)\, \partial_t K = \alpha^2 f(\alpha) \left[\nabla^2 \alpha - \beta^i \partial_i K - \alpha K_{ij} K^{ij} \right] \ , \tag{40}$$

where $f(\alpha)$ is an arbitrary positive function of α. From the above equation it is clear the lapse obeys a wave equation with a source term. The corresponding wave speed can be easily seen to be $v_\alpha = \alpha \sqrt{f(\alpha)}$, which explains the need for $f(\alpha)$ to be positive. Notice that, depending on the value of $f(\alpha)$, this wave speed can be larger or smaller than the physical speed of light. This represents no problem, as it only indicates the speed of propagation of the coordinate system, i.e. it is only a "gauge speed". The hyperbolic K-driver condition is closely related to the Bona-Massó family of slicing conditions [26]: $\partial_t \alpha = \alpha^2 f(\alpha) K$. Our new condition has the advantage of allowing for static solutions for which K itself is non-zero.

In our evolutions, we normally take $f = 2/\alpha$, since empirically we have found that such a choice has excellent singularity avoiding properties. Notice that inside a black hole, where the lapse typically collapses to very small values, this choice of f implies that the gauge speed v_α will be very large, much larger than the physical speed of light.

5.2 Shift Conditions

In the BSSN formulation, an elliptic shift condition is easily obtained by imposing the "Gamma-freezing" condition $\partial_t \tilde{\Gamma}^k$=0, or

$$\begin{aligned} &\tilde{\gamma}^{jk} \partial_j \partial_k \beta^i + \frac{1}{3} \tilde{\gamma}^{ij} \partial_j \partial_k \beta^k - \tilde{\Gamma}^j \partial_j \beta^i + \frac{2}{3} \tilde{\Gamma}^i \partial_j \beta^j + \beta^j \partial_j \tilde{\Gamma}^i \\ &- 2 \tilde{A}^{ij} \partial_j \alpha - 2\alpha \left(\frac{2}{3} \tilde{\gamma}^{ij} \partial_j \mathrm{tr} K - 6 \tilde{A}^{ij} \partial_j \phi - \tilde{\Gamma}^i_{jk} \tilde{A}^{jk} \right) = 0 \ . \end{aligned} \tag{41}$$

Notice that, just as with the K-freezing condition for the lapse, once we have solved the previous elliptic equations for the shift, the Gamma-freezing condition can be enforced at an analytic level by simply not evolving the $\tilde{\Gamma}^k$.

The Gamma-freezing condition is closely related to the well known minimal distortion shift condition [27]. In order to see exactly how these two shift conditions are related, we write here the minimal distortion condition

$$\nabla_j \Sigma^{ij} = 0 \,, \tag{42}$$

where Σ_{ij} is the so-called "distortion tensor" defined as

$$\Sigma_{ij} := \frac{1}{2} \gamma^{1/3} \partial_t \tilde{\gamma}_{ij} \,, \tag{43}$$

with $\tilde{\gamma}_{ij}$ the same as before. A little algebra shows that the evolution equation for the conformal connection functions (27) can be written in terms of Σ_{ij} as

$$\partial_t \tilde{\Gamma}^i = 2 \partial_j \left(\gamma^{1/3} \Sigma^{ij} \right) \,. \tag{44}$$

More explicitly, we have

$$\partial_t \tilde{\Gamma}^i = 2 e^{4\phi} \left[\nabla_j \Sigma^{ij} - \tilde{\Gamma}^i_{jk} \Sigma^{jk} - 6 \Sigma^{ij} \partial_j \phi \right] \,. \tag{45}$$

We then see that the minimal distortion condition $\nabla^j \Sigma_{ij} = 0$, and the Gamma-freezing condition $\partial_t \tilde{\Gamma}^i = 0$ are equivalent up to terms involving first spatial derivatives of the spatial metric multiplied with the distortion tensor itself. In particular, all terms involving second derivatives of the shift are identical in both cases (but not so terms with first derivatives of the shift).

Just as it was the case with the lapse, we obtain parabolic and hyperbolic shift prescriptions by making either $\partial_t \beta^i$ or $\partial_t^2 \beta^i$ proportional to $\partial_t \tilde{\Gamma}^i$. We call such conditions "Gamma-driver" conditions. The parabolic Gamma driver condition has the form

$$\partial_t \beta^i = k_p \, \partial_t \tilde{\Gamma}^i \,, \qquad (k_p > 0) \,, \tag{46}$$

and the hyperbolic one

$$\partial_t^2 \beta^i = k_h \, \partial_t \tilde{\Gamma}^i - \eta \, \partial_t \beta^i \,, \qquad (k_h, \eta > 0) \,, \tag{47}$$

where k_p, k_h and η are positive functions of space and time. In the case of the hyperbolic Gamma-driver we have found it useful to add a dissipation term with coefficient η. Experience has shown that by tuning the value of this dissipation coefficient we can manage to almost freeze the evolution of the system at late times.

An important point that needs to be considered when using the hyperbolic Gamma-driver condition is that of the gauge speeds. Just as it happened with the lapse, the use of a hyperbolic equation for the shift introduces new "gauge speeds" associated with the propagation of the shift. In order to get an idea of how these gauge speeds behave, we will consider for a moment the shift condition (47) for small perturbations of flat space (and taking η=0). From the form of $\partial_t \tilde{\Gamma}^i$ given by (27) we see that in such a limit the principal part of the evolution equation for the shift reduces to

$$\partial_t^2 \beta^i = k_h \left(\delta^{jk} \partial_j \partial_k \beta^i + \frac{1}{3} \delta^{ij} \partial_j \partial_k \beta^k \right) . \tag{48}$$

Consider now only derivatives in a given direction, say x. We find

$$\partial_t^2 \beta^i = k_h \left(\partial_x^2 \beta^i + \frac{1}{3} \delta^{ix} \partial_x \partial_x \beta^x \right) , \tag{49}$$

which implies

$$\partial_t^2 \beta^x = \frac{4}{3} k_h \partial_x^2 \beta^x , \tag{50}$$

$$\partial_t^2 \beta^q = k_h \partial_x^2 \beta^q \qquad q \neq x . \tag{51}$$

We can then see that in regions where spacetime is almost flat, the longitudinal part of the shift propagates with speed $v_{\mathrm{long}} = 2\sqrt{k_h/3}$ while the transverse part propagates with speed $v_{\mathrm{trans}} = \sqrt{k_h}$. In all the simulations presented below, we have chosen:

$$k_h = \frac{3}{4} \frac{\alpha^{n_1}}{\psi^{n_2}} , \tag{52}$$

with ψ the conformal factor coming from the initial data. The division by ψ^{n_2} (in this paper, all simulations are done by $n_1 = 0$ and $n_2 = 4$) helps to keep the shift small near the vicinity of the horizon. Since far from the black hole both α and ψ are close to 1, our choice implies that the longitudinal part of the shift will propagate with a speed of 1 (the speed of light), and the transverse part will propagate with a speed equal to $\sqrt{3}/2$. At the boundaries, we simply use the speed of light for all shift components. This will introduce an error for the transverse components, but in all our simulations those components are typically very small close to the boundaries.

6 Results

The first example we show is Schwarzschild, written in the standard isotropic coordinates used in many black hole evolutions. Note that with this initial data and our starting gauge conditions, the black hole should evolve rapidly. If α and β^i were held fixed at their initial values, the slice would hit the singularity at $t = \pi M$ and crash. Instead, α and β^i work together with the excision to rapidly drive the system towards a static state, without any special choice of initial conditions.

In Fig. 1 we show the radial metric function g_{rr}/ψ^4 vs. time. The grid covers an octant with 128^3 points ($\Delta x = 0.1M$, $M = 2$). Notice that the metric begins

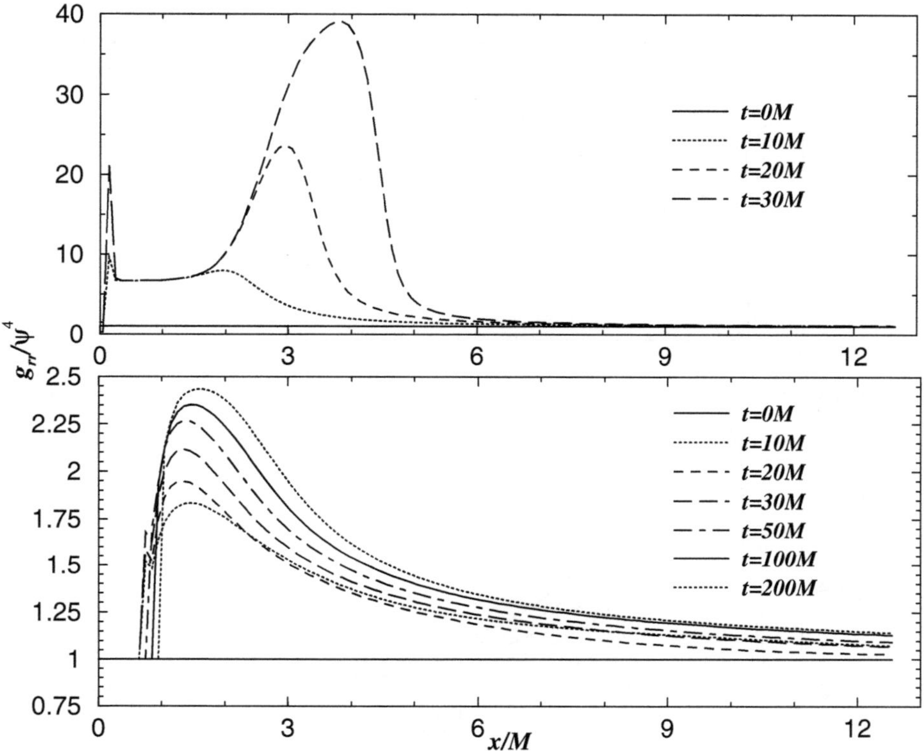

Fig. 1. We show the radial metric function g_{rr}/ψ^4 for a Schwarzschild black hole along the $x-$axis, constructed from the cartesian metric components, as it evolves with time. The upper panel shows the grid-stretching in the metric for singularity avoiding slicing with vanishing shift and no excision, while the lower panel shows the metric for the new gauge conditions with an excision box inside a sphere of radius $1M$. Note the difference in the vertical scales. Without shift and excision the metric grows out of control, while with shift and excision a peak begins to form initially as grid stretching starts, but later freezes in as the shift drives the black hole into a static configuration (note the time labels)

to grow, as it does without a shift, but as the shift builds up the growth slows down significantly. At this stage, the system is effectively static, even though we started in the highly dynamic isotropic coordinates. We also show the time development of α and β^r in Fig. 2, which evolve rapidly at first but then effectively freeze, bringing the system towards an almost static configuration by $t = 10M$. The evolution then proceeds only very slowly until the simulation is stopped well after $t = 200M$.

In Fig. 3 we show the AH mass $M_{\rm AH}$, determined with a 3D AH finder [28]. For comparison, we also show the value of $M_{\rm AH}$ obtained from a highly resolved 2D simulation with zero shift and no excision, and for the 3D run without shift. While the 3D simulation with shift and excision continues well beyond $t = 200M$, the 2D result becomes very inaccurate and the code crashes due to axis

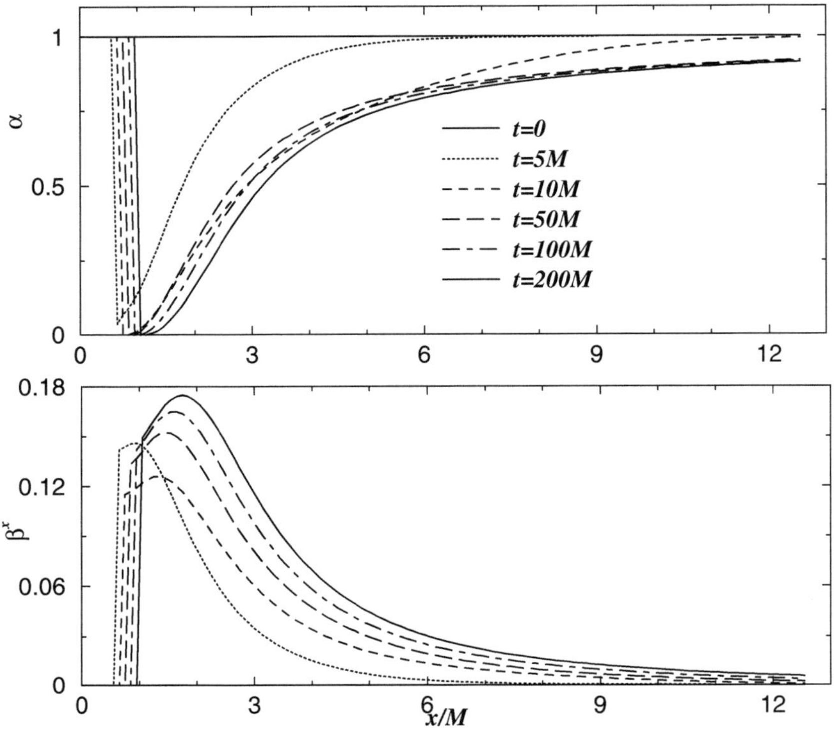

Fig. 2. We show the lapse and shift for the excision evolution of a Schwarzschild black hole. After around 10M, the lapse and shift freeze in as the system is driven to a static configuration. The size of the excision box was allowed to grow with the change in the coordinate location of the AH

instabilities by $t = 150M$, and the 3D run without shift crashes already by $t = 50M$. Notice that in the 2D case, after around $t = 35M$, $M_{\rm AH}$ grows rapidly due to numerical errors associated with grid stretching, and the AH finder ultimately fails as the code crashes. With excision and our new gauge conditions, the 3D run has less than a few percent error by $t = 200M$, while the 2D case has more than 100% error before it crashes at $t \approx 150M$. For the excision run, notice also that while there is some initial evolution in the metric and the coordinate size of the AH (see Figs. 1 and 2) the AH mass changes only very little. With new gauge conditions, we also find out that the 3D run without excision produces same results as with excision!

Next, we turn to a truly dynamic, even-parity distorted black hole. This system contains a strong gravitational wave that distorts the black hole, causing it to evolve, first nonlinearly, and then oscillating at its quasi-normal frequency, finally settling down to a static Schwarzschild black hole. This provides a test case for our techniques with dynamic, evolving black hole spacetimes, and allows us to test our ability to extract gravitational waves with excision for the first time. In this case, in the language of [15], we choose the Brill wave parameters

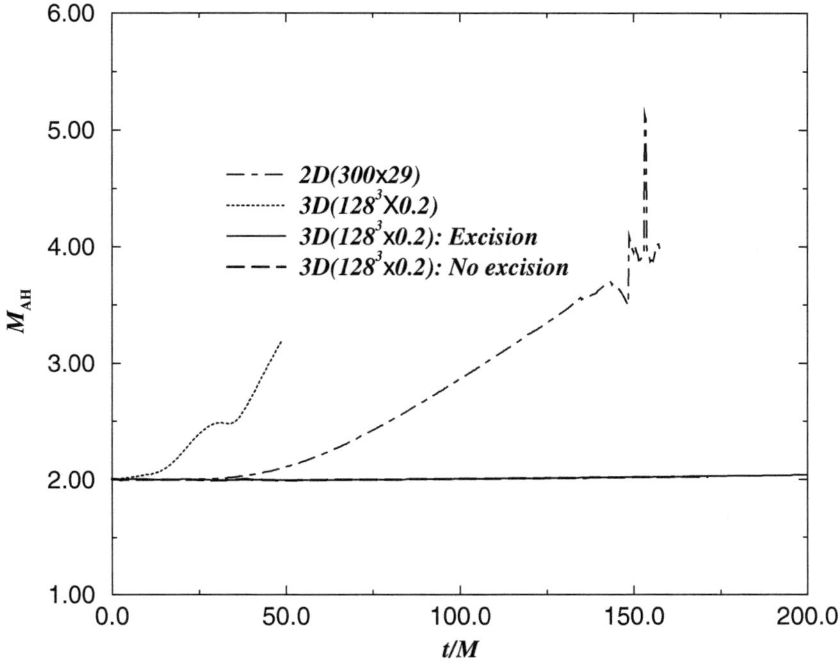

Fig. 3. The solid and long dashed lines show the development of the AH mass $M_{\rm AH}$, determined through a 3D AH finder, for the excision and no *no* excision simulation of a Schwarzschild black hole shown above, while the dot-dashed and dot lines show the AH mass obtained using 2D and 3D codes with zero shift and no excision. The 2D code crashes at around $t = 150M$, the 3D run without shift crashes around $t = 50M$, while the 3D runs with shift and excision (or) *no* excision reach an effectively static state and the error remains less than a few percent even after $t = 200M$

to be $Q_0 = 0.5$, $\eta_0 = 0$, $\sigma = 1$, corresponding to a highly distorted black hole with $M = 1.83$.

In Fig. 4 we show the AH mass $M_{\rm AH}$ as a function of time for the distorted black hole simulations carried out in both 2D and 3D. $M_{\rm AH}$ grows initially as a nonlinear burst of gravitational waves is absorbed by the black hole, distorting it strongly, but then levels off as the black hole goes into a ring-down phase towards Schwarzschild.

In the 3D cases, the dynamic gauge conditions and excision or *no* excision quickly drive the evolution towards an almost static configuration, as the system itself evolves towards a static Schwarzschild black hole. The evolution is continued until terminated at around $t = 300M$. Even in this highly dynamic system, no specialized form of initial data or lapse and shift are needed; our gauge choices naturally drive the system to a static state as desired. To our knowledge, distorted black holes of this type have never been evolved for so long, nor with such accuracy, in either 2D or 3D. By comparison, in the more highly resolved 2D case with zero shift and no excision, the familiar grid stretching effects allowed by the

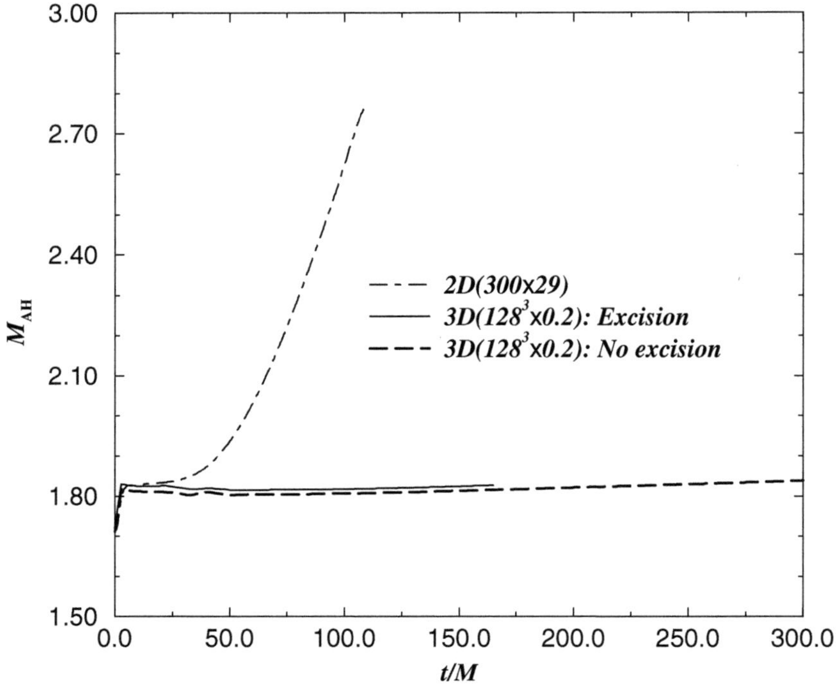

Fig. 4. We show the AH masses M_{AH} for a black hole with even-parity distortion for the 2D (no excision, no shift) and 3D (excision and *no* excision, shift) cases. The 3D result continues well past $300M$, while the 2D result becomes very inaccurate and crashes by $t = 100M$

gauge choice lead to highly inaccurate evolutions after some time with the error in M_{AH} again approaching 100% when the code finally crashes at $t \approx 100M$.

In Fig. 5, we show the results of extracting waves from the evolution of this highly distorted black hole. Using the standard gauge-invariant waveform extraction technique, the Zerilli function is shown for both the 2D and 3D simulations discussed above. There is a slight but physically irrelevant phase difference in the two results due to differences in the slicing; otherwise the results are remarkably similar.

This shows conclusively that the excision or *no* excision and live gauge conditions do not adversely affect the waveforms, even if they carry a small amount of energy (around $10^{-3}M_{ADM}$ in this case).

We now turn to a rather different type of distorted black hole, including rotation and general even- and odd-parity distortions. In the language of [15,16], the parameters for this simulation are $Q_0 = 0.5$, $\eta_0 = 0$, $\sigma = 1$, $J = 35$, corresponding to a rotating distorted black hole with $M = 7.54$ and an effective rotation parameter $J/M^2 = 0.62$. Previously, such data sets could be evolved only to about $40M$ [14]. Again, for the purposes of this paper we have chosen an axisymmetric case so that we can compare the results to those obtained with a 2D code. Since this example is much more demanding, we have found it im-

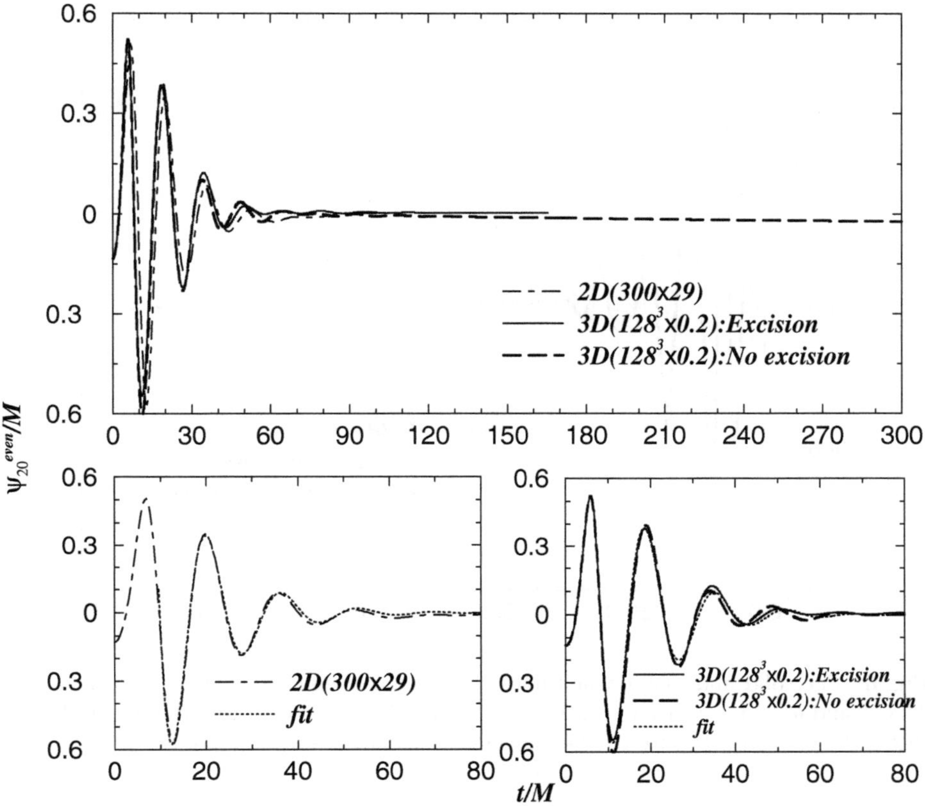

Fig. 5. The solid and long dashed lines show the result of the $\ell = 2, m = 0$ waveform extraction at a radius $5.45M$ for the even-parity distorted black hole described on the text, while the dashed line shows the result of the same simulation carried out in the 2D code. We also show a fit to the two lowest QNM's of the black hole for 2D and 3D separately, using numerical data from $t = 9M$ to $t = 80M$

portant in order to increase the accuracy of our runs to perform a single initial maximal solve to reduce the initial gauge dynamics. The gauge conditions used work well even in the presence of rotation: the shift drives the system towards a static *Kerr* black hole spacetime after the true dynamics settle down. The metric functions (not shown) evolve in a similar way to those shown before, essentially freezing at late times.

In Fig. 6, we show the extracted waveforms, now computed using the imaginary part of the Newman-Penrose quantity Ψ_4 (e.g. [2]), which includes contributions from all $\ell-$modes at the same time. The results from the 2D and 3D codes agree very closely, except for a slight phase shift due to slicing differences, until the 2D code becomes inaccurate and later crashes. The 3D simulation continues well beyond this point, and is terminated at $t = 140M$.

Figure 7 shows the snapshots of the apparent horizon with shift vectors for rotating 3D distorted. T shows the coordinate time so that the last picture

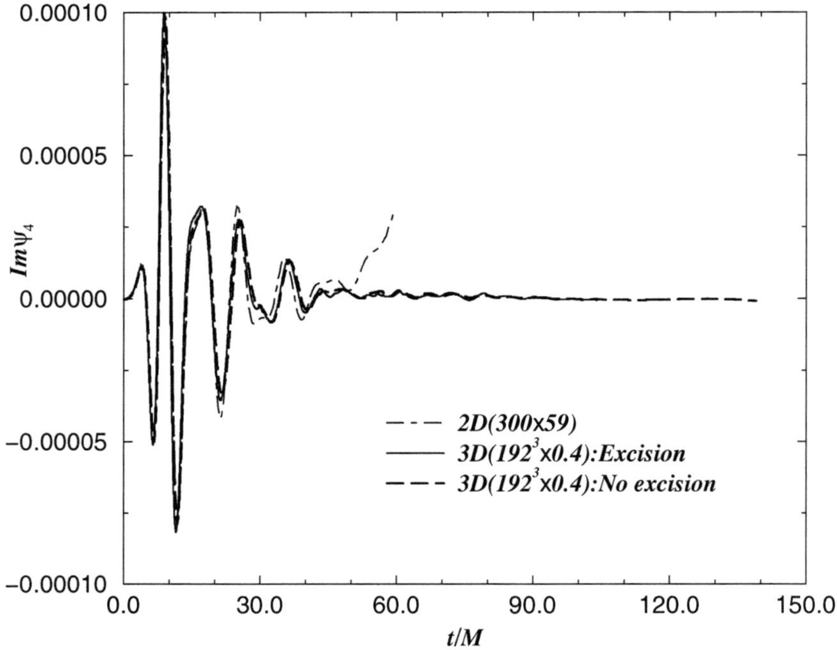

Fig. 6. The solid and long dashed lines show the a imaginary part of ψ_4 at a radius $3.94M$ and $\theta = \phi = \pi/4$ for rotating 3D distorted black hole, while the dot-dash line shows the same initial date by the 2D code which crashes around $60M$

shows them at around $t = 40M$. As the shift drives the system towards a static Kerr black hole spacetime, horizon grows until certain time and then oscillates towards a static Kerr black hole.

7 Conclusions

We have extended recently developed 3D black hole excision techniques, using a new class of live gauge conditions that *dynamically drive* the black hole system towards an essentially static state at late times, when the system itself settles to a stationary Kerr black hole. Our techniques have been tested on highly distorted, rotating black holes, are shown to be very robust, and require no special coordinate systems or special forms of initial data. For the first time, excision is tested with wave extraction, and waveforms are presented and verified. The results are shown to be more accurate, and much longer lived, than previous 3D simulations and even better resolved 2D simulations of the same initial data. Such improvements in black hole excision are badly needed for more astrophysically realistic black hole collision simulations, which are in progress and will be reported elsewhere.

Furthermore, we have found that the new gauge conditions can bring the evolution to an almost static state even *without* excision. Although we could

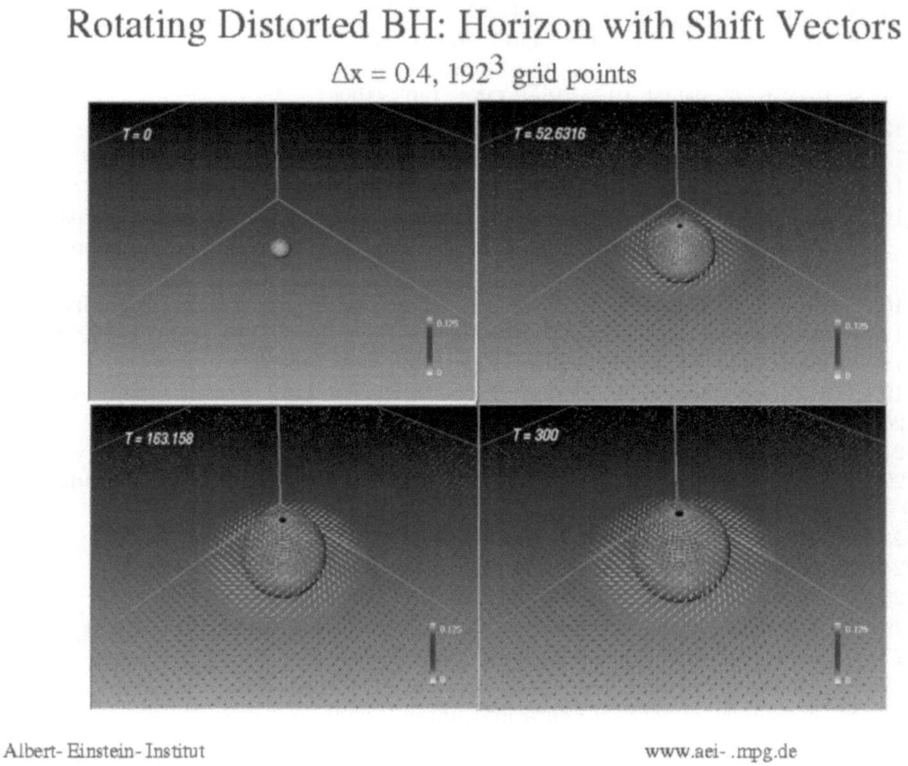

Fig. 7. We show snapshots of the apparent horizon with shift vectors for rotating 3D distorted black hole. T shows the coordinate time so that the last picture shows them at around $t = 40M$

show some primitive results, we are currently investigating properties of new gauge conditions [29]. We will also report these results in further publications.

Acknowledgements

Calculations were performed using the Cactus code at AEI, NCSA, PSC, and RZG.

References

1. M. Alcubierre *et al.*: Phys. Rev. Lett. **87**, 271103 (2001)
2. J. Baker *et al.*: Phys. Rev. Lett. **87**, 121103 (2001)
3. J. Thornburg: Class. Quantum Grav. **4**, 1119 (1987)
4. E. Seidel, W.-M. Suen: Phys. Rev. Lett. **69**, 1845 (1992)
5. G. B. Cook *et al.*: Phys. Rev. Lett **80**, 2512 (1998)
6. R. Gomez *et al.*: Phys. Rev. Lett. **80**, 3915 (1998)
7. M. Alcubierre, B. Brügmann: Phys. Rev. **D63**, 104006 (2001)

8. S. Brandt *et al.*: Phys. Rev. Lett. **85**, 5496 (2000)
9. P. Anninos *et al.*: Phys. Rev. **D52**, 2059 (1995)
10. G. E. Daues: Ph. D. Thesis, Washington University, St. Louis, Missouri (1996)
11. S. Brandt, E. Seidel: Phys. Rev. **D54**, 1403 (1996)
12. S. Brandt, E. Seidel: Phys. Rev. **D52**, 856 (1995)
13. K. Camarda, E. Seidel, Phys. Rev. **D59**, 064019 (1999)
14. J. Baker *et al.*: Phys. Rev. **D62**, 127701 (2000)
15. S. Brandt, K. Camarda, E. Seidel: In *Proceedings of the 8th Marcel Grossmann Meeting on General Relativity*, ed. by T. Piran (World Scientific, Singapore, 1999) pp. 741–743
16. S. Brandt, K. Camarda, E. Seidel, R. Takahashi: in preparation (unpublished)
17. R. A. Matzner, M. F. Huq, D. Shoemaker: Phys. Rev. **D59**, 024015 (1999)
18. M. Shibata, T. Nakamura: Phys. Rev. **D52**, 5428 (1995)
19. T. W. Baumgarte, S. L. Shapiro: Physical Review **D59**, 024007 (1999)
20. M. Alcubierre *et al.*: Phys. Rev. **D61**, 041501 (R) (2000)
21. M. Alcubierre *et al.*: Phys. Rev. **D62**, 124011 (2000)
22. J. York: In *Sources of Gravitational Radiation*, ed. by L. Smarr (Cambridge University Press, Cambridge, England, 1979)
23. M. Alcubierre *et al.*: Phys. Rev. **D62**, 044034 (2000)
24. D. Garfinkle, C. Gundlach: Class. Quantum Grav. **16**, 4111 (1999)
25. J. Balakrishna *et al.*: Class. Quantum Grav. **13**, L135 (1996)
26. A. Arbona, C. Bona, J. Massó, J. Stela: Phys. Rev. **D60**, 104014 (1999)
27. L. Smarr, J. York: Phys. Rev. **D17**, 2529 (1978)
28. M. Alcubierre *et al.*: Class. Quantum Grav. **17**, 2159 (2000)
29. M. Alcubierre *et al.*: in preparation (unpublished)

Numerical Relativity with the Conformal Field Equations

Sascha Husa

Max-Planck-Institut für Gravitationsphysik, 14476 Golm, Germany

Abstract. I discuss the conformal approach to the numerical simulation of radiating isolated systems in general relativity. The method is based on conformal compactification and a reformulation of the Einstein equations in terms of rescaled variables, the so-called "conformal field equations" developed by Friedrich. These equations allow to include "infinity" on a finite grid, solving regular equations, whose solutions give rise to solutions of the Einstein equations of (vacuum) general relativity. The conformal approach promises certain advantages, in particular with respect to the treatment of radiation extraction and boundary conditions. I will discuss the essential features of the analytical approach to the problem, previous work on the problem – in particular a code for simulations in 3+1 dimensions, some new results, open problems and strategies for future work.

1 Introduction

In order to understand the physical content of the theory of general relativity, it is desirable to both *mathematically* understand its solutions and *observationally* understand the physical phenomena for which the theory is relevant. The latter effort typically requires predictions from the theory, both qualitative and quantitative – such as gravitational wave templates or binary pulsar deceleration parameters. The lack of genericity in available exact solutions then naturally leads to the use of approximation methods such as post-Newtonian approximations, perturbation theory or numerical analysis, which allows very general non-perturbative approximations. Concrete solutions do however also play an important role in the quest for a mathematical understanding of the solution space. The experience gained from such solutions can suggest theorems, test conjectures, or lead to the discovery of previously unknown phenomena. For some particularly interesting examples see [1], [2] or [3]. The construction and study of solutions, be it with approximate or exact methods, obviously profits from a sound mathematical basis in the form of well-posed equations, analytic estimates and the likes. Eventually – hopefully – it will also profit from observational evidence!

In the following I will discuss a particular approach to the numerical solution of the Einstein field equations, which addresses the problems associated with the treatment of asymptotic regions by conformal compactification. The interest in asymptotic regions is rooted in the problem of describing isolated systems. Physical intuition suggests that many astrophysical processes (whether they are of actual astrophysical relevance or rather hypothetical) should essentially be independent of the large-scale structure of the universe, or, say, the local galaxy. The idealization of an isolated system, where the geometry approaches a Minkowski geometry at large distances, thus forms the basis for the general-relativistic analysis of processes which are essentially of non-cosmological nature. The mathematical formalization of the physical idea of isolated systems is the concept of *asymptotically flat spacetimes.* This formalization is already nontrivial, due to the lack of a preferred background geometry or coordinate system – with respect to which one could define "distance" and the appropriate limits. Conformal compactification, however, renders possible a discussion of asymptotically flat spacetimes in terms of local differential geometry. In this approach, pioneered by Penrose [4], an unphysical Lorentzian metric g_{ab} is introduced on an unphysical manifold $\mathcal{M}$ which gives rise to the physical metric $\tilde{g}_{ab}$ by the rescaling $\tilde{g}_{ab} = \Omega^{-2} g_{ab}$. The physical manifold $\tilde{\mathcal{M}}$ is then given by $\tilde{\mathcal{M}} = \{p \in \mathcal{M} \mid \Omega(p) > 0\}$. In this picture physical "infinity" corresponds to a three-dimensional boundary of a four-dimensional region in $\mathcal{M}$, defined by $\Omega = 0$. Limiting procedures and approximations can thus be replaced by local differential geometry on the boundary.

In gravitational theory, quantities such as the total mass, (angular) momentum or emitted gravitational radiation can only consistently be defined at "infinity". In the conformal approach the unambiguous extraction of gravitational waves from a numerical spacetime is straightforward. In the "traditional" approach to dealing with asymptotic falloff in numerical relativity, where one introduces an arbitrary spatial cutoff, matters are much more complicated and ambiguities are introduced which one would have to get rid off by complicated limiting procedures. Without at least being able to define a clean concept of radiation leaving or entering a system, it is furthermore very hard to define physically realistic and consistent boundary conditions at finite distance. The traditional approach is thus not completely satisfactory both from a mathematical but also from a practical point of view. Here we discuss the principal ideas of the envisioned "conformal cure", the technical and conceptual problems associated with it and the current status of this approach.

It is easy to see that the conformal cure cannot be straightforward, by writing Einstein's vacuum equations in terms of Ω and g_{ab}:

$$\tilde{G}_{ab}[\Omega^{-2} g_{ab}] = G_{ab}[g_{ab}] + \frac{2}{\Omega}\left(\nabla_a \nabla_b \Omega + g_{ab} \nabla_c \nabla^c \Omega\right) + \frac{3}{\Omega^2} g_{ab} \left(\nabla_c \Omega\right) \nabla^c \Omega \,. \quad (1)$$

This expression is singular for $\Omega = 0$, multiplication by Ω^2 also does not help here because then the principal part of the partial differential equations encoded in G_{ab} would degenerate at $\Omega = 0$. The conformal compactification approach thus cannot be carried to the level of the field equations in a straightforward way. This step however has been achieved by Friedrich, who has developed a judicious reformulation of the equations [5–9].

These *conformal field equations* are *regular* equations for g_{ab} and certain additional independent variables.

In analytical work, such global methods have proven to provide essential simplifications leading to new results and insights. Already by providing a different point of view on some of the essential problems in numerical relativity, the conformal picture is quite helpful and can stimulate new ideas. Certainly, we desire more – to make this approach also a practical tool. There is significant hope, that global methods will eventually show advantages for practical numerical work and, despite the small number of researchers involved so far (may there be more!), some significant progress in this direction has been made.

In the present chapter I will try to sketch the present status of the quest for the conformal cure and discuss some important open questions. We will start with a brief introduction of the concepts of asymptotic flatness in terms of conformal compactification in Sect. 2, highlighting some important features of "future null infinity" and then discuss the conformal field equations. In Sect. 3 I will discuss some explicit examples of compactifying Minkowski spacetime, both to paint a more concrete picture of our scenario and to set the arena for some numerical code tests. Sect. 4 contains a brief overview of the history of numerical work on the conformal field equations, leading to a description of a 3D code written by Hübner [10–13]. New results from 3D calculations performed with this code will be presented in Sect. 5 and a discussion will be given in Sect. 6, concluding with a roadmap for future work.

2 Compactification and the Mathematical Description of Isolated Systems

The material in this section is intended to present some essential ideas in a condensed form. The reader should be aware that I am not doing justice here to subtleties and long history of the mathematical description of isolated systems in general relativity – rather this section intends to motivate to look into more complete reviews such as [14–16].

2.1 Asymptotic Flatness and Compactification

As noted above the formulation of the concept of asymptotic flatness is far from straightforward in GR, due to the absence of a background metric or preferred coordinate system, in terms of which falloff rates can be specified. A resolution of this problem is provided by a definition of asymptotic flatness, where, after a suitable conformal rescaling of the metric, "points at infinity" are added to the manifold. One thus works on a compactified auxiliary manifold and local differential geometry can be used to study the asymptotic properties of the gravitational field. We will give a simple definition of asymptotic flatness here, which for our purposes catches all essential features. For alternative definitions and more detailed explanations compare for example [4,14,15,17].

Definition 1 (asymptotic simplicity)
A smooth spacetime $(\tilde{\mathcal{M}}, \tilde{g}_{ab})$ *is called asymptotically simple, if there exists another smooth spacetime* $(\mathcal{M}, g_{ab})$ *and a scalar function* Ω *such that:*

1. *$\tilde{\mathcal{M}}$ is an open submanifold of $\mathcal{M}$ with smooth boundary $\partial\tilde{\mathcal{M}} = \mathscr{I}$ (Scri).*

2. *$g_{ab} = \Omega^2 \tilde{g}_{ab}$ on $\tilde{\mathcal{M}}$, with $\Omega > 0$ on $\tilde{\mathcal{M}}$, $\Omega = 0$ on $\mathscr{I}$ and $\nabla_a \Omega \neq 0$ on $\mathscr{I}$.*

3. *Every null geodesic in $\tilde{\mathcal{M}}$ acquires two end points on $\mathscr{I}$.*

Definition 2 (asymptotic flatness)
Asymptotically simple spacetimes are called asymptotically flat if their Ricci tensor $\tilde{R}_{ab}$ vanishes in a neighbourhood of $\mathscr{I}$.

Examples of asymptotically simple spacetimes, which are not asymptotically flat are the de Sitter and anti-de Sitter solutions. Correspondingly to asymptotically flat spacetimes one can consider asymptotically de Sitter and anti-de Sitter spacetimes. Note that the completeness condition 3 in Def. 1, which ensures that the entire boundary is included, excludes black-hole spacetimes. For modifications to weaken condition 3, thus allowing black holes, see the definitions of [18] or [15]. For example, the definition of *weak asymptotic simplicity* [18] requires condition 3 to hold only in a neighbourhood of $\mathscr{I}$. See e.g. [15] for a discussion of asymptotic flatness at spacelike infinity (i.e. the part of infinity which is reached along spacelike geodesics) versus null infinity (i.e. the part of infinity which is reached along null curves). The notion of asymptotic flatness at timelike infinity does not make much sense in a general situation, because then all energy would have to be radiated away, leaving only flat space behind – excluding black holes or "stars". For weak data however, in vacuum say, where all radiation eventually disperses, once expects asymptotic flatness to hold also at timelike infinity, this issue will be discussed below in application to concrete spacetimes.

The notion of asymptotic flatness of isolated systems turns out to be intimately related to the possibility of defining the total energy-momentum for such systems in general relativity – remember that no well-defined local energy density of the gravitational field is known (compare e.g. Sect. 11.2 of the textbook of Wald [15]). However, total energy-momentum quantities, which transform as a 4-vector under asymptotic Lorentz transformations, can be assigned to null and spatial infinity of asymptotically flat spacetimes. If a manifold has more than one asymptotically flat end, e.g. in the presence of wormholes of the Einstein-Rosen-bridge type, then different energy-momenta can be associated with each of these asymptotic regions.

The expression for the energy-momentum four-vector at spatial infinity has been given first by Arnowitt, Deser and Misner in 1962 [19] in the context of the Hamiltonian formalism and is usually called the ADM momentum, the time component being called ADM mass. The ADM energy corresponds to the energy of some Cauchy surface, i.e. a snapshot of the spacetime at some fixed time. It is a constant of motion and can therefore be expressed in terms of the initial data on an asymptotically flat Cauchy hypersurface.

The expression for the energy-momentum at null infinity, usually referred to as the Bondi energy-momentum, can be associated with a fixed retarded time, i.e. some asymptotically null surface. The decrease of this quantity measures the energy-momentum carried away by gravitational radiation. For a brief introduction and references to original work on different definitions of the Bondi mass see e.g. the textbook of Wald [15]. The formulation most appropriate for usage in numerical codes based on the conformal field equations was given by Penrose [4], and defines the Bondi mass in terms of the behaviour of certain projections of the Weyl tensor at $\mathscr{I}^+$ and the shear of the outgoing congruence of null geodesics orthogonal to $\mathscr{I}$ in the gauge defined below by (2). It was already shown in 1962 by Bondi, van der Burg and Metzner [20] that the Bondi mass M_{B} can only *decrease* with time: gravitational radiation always carries positive energy away from a radiating system. Note that this means in particular, that while compactification at spatial infinity would lead to a "piling up" of waves, at $\mathscr{I}^+$ this effect does not appear. In the compactified picture the waves leave the physical spacetime through the boundary $\mathscr{I}^+$.

A fundamental issue of general relativity is the positivity of the ADM and Bondi energies. Although it is trivial to write down a metric with negative mass if no conditions on the energy-momentum tensor are imposed, for *reasonable* matter fields with nonnegative energy density (thus satisfying the dominant energy condition), non-negativity of the ADM and Bondi energies is expected on physical grounds: if the energy of an isolated system could be negative, it would most likely be unstable and decay to lower and lower energies. Indeed, a proof of the positive definiteness of the ADM energy has been given in 1979 by Schoen and Yau [21] (several simplified proofs have been given later) and was extended to the Bondi mass in 1982 by Horowitz and Perry [22].

2.2 What is $\mathscr{I}$?

We will now have a closer look at $\mathscr{I}$ and discuss some of its features, which will allow us to understand the basic ideas of radiation extraction and help us to understand some issues related with choosing boundary conditions for numerical solutions of the conformal field equations.

Looking at (1) and multiplying by Ω^2, one can see that for a vacuum spacetime, $\tilde{G}_{ab} = 0$, $(\nabla_c\Omega)\nabla^c\Omega = 0$ at $\mathscr{I}$, which thus must consist of null surfaces. In fact, one can then prove (see e.g. [17]), that

1. $\mathscr{I}$ has two connected components, each with topology $S^2 \times \mathbb{R}$.

2. The connected components of $\mathscr{I}$ are smooth null hypersurfaces in $\mathcal{M}$ and as such are generated by null geodesics.

3. The congruence of null geodesic generators of $\mathscr{I}$ is shear free.

The two connected components are called future null infinity ($\mathscr{I}^+$) and past null infinity ($\mathscr{I}^-$) and provide the future and past endpoints for null geodesics in $\tilde{\mathcal{M}}$. In a naive picture they could be viewed as emanating from a point i^0

which represents spatial infinity[1]. These features will become more graphic when dealing with explicit examples below.

Note that there is gauge freedom in the choice of the conformal factor: one is free to rescale the conformal factor Ω by some $\omega > 0$ such that $\hat{\Omega} = \omega\Omega$, $\hat{g}_{ab} = \omega^2 g_{ab} = \hat{\Omega}^2\tilde{g}_{ab}$. It is an interesting exercise (see Sect. 11.1 of [15]) to prove that outside any neighbourhood of i^0 – on $\mathscr{I}^+$ say – one can always use this conformal gauge freedom to achieve

$$\hat{\nabla}_a \hat{\nabla}_b \hat{\Omega} = 0 \qquad \text{on } \mathscr{I}^+ , \tag{2}$$

where $\hat{\nabla}_a$ is the derivative operator compatible with the metric $\hat{g}_{ab}$. This conformal gauge implies, that the null tangent $n^a = \hat{g}^{ab}\nabla_b\hat{\Omega}$ to the null geodesic generators of $\mathscr{I}$ satisfies the affinely parameterized geodesic equation,

$$n^a \hat{\nabla}_a \hat{n}^b = 0 . \tag{3}$$

Consequently, expansion of the generators of $\mathscr{I}$ vanishes in addition to the shear and twist (n_a is a gradient). Using the remaining gauge freedom of ω, we can choose coordinates such that the metric on $\mathscr{I}$ takes the form

$$\mathrm{d}\hat{s}^2|_{\mathscr{I}^+} = 2\mathrm{d}\Omega\,\mathrm{d}u + \mathrm{d}\theta^2 + \sin^2\theta\mathrm{d}\phi^2 , \tag{4}$$

where u is the affine parameter of the null geodesic generators, scaled such that $n^a\hat{\nabla}_a u = 1$ (see e.g. Chap. 11 of [15]). The cuts[2] of $\mathscr{I}$ of constant u thus become metric spheres. The coordinate u is generally known as Bondi parameter or Bondi time. The conformal gauge (2) and the coordinates (4) prove very useful in the analysis of the geometry in a neighbourhood of $\mathscr{I}$ – in particular for the extraction of radiation. The existence of a natural time coordinate (at least up to affine transformations along *each* generator) is very interesting for numerical applications, where at least asymptotically one can get rid of much of the slicing arbitrariness of the interior region. It is nontrivial but rather straightforward to actually (numerically) find this gauge of $\mathscr{I}^+$, which is also required by the standard formulas to compute the energy-momentum at $\mathscr{I}^+$ and the emitted radiation – to be given below.

Before discussing how to compute the radiation, it is useful to idealize a detector (here I will follow the discussion in [24]). In physical space – far away from the sources – we could think of a detector as a triad of spacelike unit vectors attached to the worldline of some (timelike) observer. Let us further assume for simplicity that the observer moves along a timelike geodesic parametrized by proper time and that the triad is transported by Fermi-Walker transport. It is not hard to show – see Frauendiener [24], that taking the appropriate limit in the compactified spacetime, the observer worldline converges to a null geodesic

[1] The structure of i^0 is however quite subtle, significant progress toward its understanding in terms of the field equations has recently been achieved by Friedrich [8,23]

[2] A cut of $\mathscr{I}$ is a two-dimensional spacelike cross section of $\mathscr{I}$ which meets every generator once.

generator of $\mathscr{I}^+$. Taking the limit along a Cauchy surface it converges to the point i^0, where one could naively expect an observer to end up when shifted to larger and larger distances (this limit is however not appropriate in the context of computing the radiation). Furthermore, the proper time parameter of the observer converges to Bondi time. The arbitrariness of boosting the observers is reflected in the affine freedom of choosing the Bondi parameter at $\mathscr{I}$. The description of $\mathscr{I}^+$ thus could be condensed into the statement that it idealizes *us* – the observers of astrophysical phenomena happening far away. By working with the idealization, the approximations and ambiguities associated with detectors at a finite distance have transformed into a surprisingly simple geometric picture! Note that this simplification has to be taken with the typical care required in the treatment of idealizations in (theoretical) physics: under practical circumstances, e.g. computing the actual signal at a gravitational wave detector, $\mathscr{I}$ more realistically corresponds to an observer that is sufficiently far way from the source to treat the radiation linearly, but not so far away that cosmological effects have to be taken into account. In order to compute the detected signal in a realistic application, cosmological data and the fact that an earthbound detector moves in a complicated way relative to the source have to be considered.

We will next discuss a "detector-frame" adapted to $\mathscr{I}^+$ – the commonly used Bondi frame. For a much more complete discussion of Bondi-systems, see e.g. the excellent review by Newman and Tod [14]. There a characteristic framework is used to set up the Bondi frame in a whole neighbourhood of $\mathscr{I}^+$, which is necessary to compute derivatives, entering e.g. the definition of the spin coefficient σ defined below in (5). In the current approach, the Bondi system is only defined at $\mathscr{I}^+$: initial data can be set up, such that all necessary quantities can be propagated along the generators of $\mathscr{I}^+$ [25].

With $\mathscr{I}^+$ being a null surface, it is most natural to use a null frame, consisting of 2 null vectors and 2 spacelike vectors x^a, y^a, which can be considered as the idealizations of the arms of an interferometric gravitational wave detector. The vectors x^a, y^a are commonly treated in the form of two complex null vectors $m^a, \bar{m}^a$, with

$$m^a = x^a + iy^a, \qquad m^a\bar{m}_a = 1 \,,$$

where x^a and y^a are real vectors tangent to the cuts of $\mathscr{I}^+$. The null vectors are taken as the affine tangent n^a and $l_a = \hat{\nabla}_a u$, which satisfy

$$n^a l_a = -1 \,.$$

The tetrad vectors l_a, m^a and $\bar{m}^a$ are parallely propagated along the generators, which yields transport equations that define them on all of $\mathscr{I}^+$ once initial values are chosen.

The Bondi-mass can then be computed in terms of the spin-coefficient σ and the rescaled Weyl tensor components ψ_2 and ψ_4:

$$\sigma \;= \hat{g}_{ab} l^a m^c \hat{\nabla}_c m^b \,, \tag{5}$$

$$\psi_2 = \hat{d}_{abcd} l^a m^b \bar{m}^c n^d \,, \tag{6}$$

$$\psi_4 = \hat{d}_{abcd} n^a \bar{m}^b n^c \bar{m}^d \,. \tag{7}$$

In terms of these quantities the Bondi mass can be defined as

$$M_{\mathrm{B}} = -\frac{\sqrt{A}}{\sqrt{4\pi}^3} \int \left(\psi_2 + \sigma\dot{\bar{\sigma}}\right) \mathrm{d}A \,, \tag{8}$$

the outgoing radiation can be computed to be

$$\dot{M}_{\mathrm{B}} = -\frac{\sqrt{A}}{\sqrt{4\pi}^3} \int \left(\dot{\sigma}\dot{\bar{\sigma}}\right) \mathrm{d}A \,, \tag{9}$$

where A is the area of the cuts of $\mathscr{I}^+$ and $\dot{f} = n^a \hat{\nabla}_a f = \partial_u f$. Furthermore,

$$\ddot{\sigma} = -\bar{\psi}_4$$

can be used to evolve σ, where both σ and $\dot{\sigma}$ can be computed on the initial slice.

This procedure has been implemented by Hübner and Weaver [25] for 2D codes and the 3D code used to obtain the results in Sect. 5 and has been tested and proven accurate for several types of spacetimes [25]. Frauendiener describes his implementation and some results in [26]. There are two essential problems in these implementations: first of all, the gauge conditions will not usually result in a slicing of $\mathscr{I}^+$ by cuts of constant Bondi time u. This means that interpolation has to be used between different slices of the numerical evolution. Second, in those formulations of the conformal field equations that have so far been used in numerical implementations, the conformal factor Ω is an evolution variable and not specified a priori, $\mathscr{I}$ will in general not be aligned with grid points. This results in further technical complications and an additional need for interpolation. When dealing with the physically interesting case of a $\mathscr{I}^+$ of spherical topology, at least two patches have to be used to represent the Bondi tetrad $(l_a, n_a, m_a, \bar{m}_a)$. Frauendiener has achieved to control the movement of $\mathscr{I}^+$ through the grid by the gauge choice for his formulation [26], in particular the shift vector can be chosen such that $\mathscr{I}$ does not change its coordinate location.

2.3 The Conformal Field Equations

Several formulations of the conformal field equations are available, the main difference being whether the conformal factor Ω can be specified a priori or is determined as a variable by the equations. In the original formulation [5,6] and its descendants [7,10,27] Ω (and derivatives) are evolved as dependent variables. All existent numerical codes are based on equations of this type. A later version of the equations allows to fix Ω a priori and has been used to develop a new treatment of spatial infinity i^0 [8,23,9]. However, the formulation and treatment of these equations is more involved, and its numerical solution has not yet been attempted.

In the following we discuss a metric based formulation of the "original" version of the conformal field equations, which forms the basis for Hübner's codes [10–13].

When deriving the conformal field equations, it turns out to be useful to start with the splitting of the Riemann tensor into its trace-free (the Weyl tensor) and trace (Ricci tensor and scalar) parts. Additionally we define the tracefree Ricci tensor $\hat{R}_{ab} = R_{ab} - g_{ab}\, R/4$ and the rescaled Weyl tensor

$$d_{abc}{}^{d} = \Omega^{-1}\, C_{abc}{}^{d} \,. \tag{10}$$

The requirement that the physical scalar curvature $\tilde{R}$ vanishes implies

$$6\,\Omega\, \nabla^a \nabla_a \Omega = 12\,(\nabla^a \Omega)\,(\nabla_a \Omega) - \Omega^2 R \,, \tag{11}$$

Note that this equation is *not* manifestly regular at $\Omega = 0$, but it is actually possible to show that if (11) is satisfied at one point, then by virtue of (12,13,14,17,18), to be given below, it has to be satisfied everywhere. The whole system (11,12,13,14, 17,18) is then regular in the sense that this point does not have to be located at $\mathscr{I}^+$. The vacuum Einstein equations $\tilde{R}_{ab} = 0$ then yield

$$\nabla_a \nabla_b \Omega = \frac{1}{4} g_{ab}\, \nabla^c \nabla_c \Omega - \frac{1}{2}\, \hat{R}_{ab}\, \Omega \,. \tag{12}$$

Finally, commuting covariant derivatives in the expression

$$g^{bc} \nabla_c \nabla_b \nabla_a \Omega$$

and then using (12) again yields

$$\frac{1}{4} \nabla_a \left(\nabla^b \nabla_b \Omega \right) = -\frac{1}{2}\, \hat{R}_{ab}\, \nabla^b \Omega - \frac{1}{24}\, \Omega\, \nabla_a R - \frac{1}{12}\, \nabla_a \Omega\, R \,. \tag{13}$$

Equations for the metric can be obtained by the identity

$$R_{abc}{}^{d} = \Omega d_{abc}{}^{d} + \left(g_{ca} \hat{R}_b{}^{d} - g_{cb} \hat{R}_a{}^{d} - g^{d}{}_{a} \hat{R}_{bc} + g^{d}{}_{b} \hat{R}_{ac} \right)/2 \\ + \left(g_{ca} g_b{}^{d} - g_{cb} g_a{}^{d} \right) \frac{R}{12} \,, \tag{14}$$

which defines the Weyl tensor. Expressing the Riemann tensor $R_{abc}{}^{d}$ in terms of the metric and its derivatives (or the Christoffel quantities in a first order formalism) yields the desired equations. Note that for the physical Riemann tensor the vacuum Einstein equations imply $\tilde{R}_{abc}{}^{d} = \tilde{C}_{abc}{}^{d}$.

We still miss differential equations for $d_{abc}{}^{d}$ and $\hat{R}_{ab}$. These can be obtained from the Bianchi identities $\nabla_{[a} R_{bc]d}{}^{e}$, which in terms of the Weyl and tracefree Ricci tensors imply

$$\nabla_d \tilde{C}_{abc}{}^{d} = 0 \tag{15}$$

for the Weyl tensor of a vacuum spacetime ($\tilde{R}_{ab} = 0$) and

$$\nabla_b \hat{R}_a{}^{b} = \frac{1}{4}\, \nabla_a R \,. \tag{16}$$

While the Weyl tensor is conformally invariant,

$$\tilde{C}_{abc}{}^{d} = C_{abc}{}^{d} \,,$$

this invariance does not hold for (15). Instead however one can show that

$$\tilde{\nabla}_d \tilde{C}_{abc}{}^d = \Omega \nabla_d \left(d_{abc}{}^d \right) ,$$

which implies

$$\nabla_e d_{abc}{}^e = 0 , \tag{17}$$

if the vacuum Einstein equations hold in the physical spacetime.

The Bianchi identity combined with the splitting (14) implies

$$\nabla_a \hat{R}_{bc} - \nabla_b \hat{R}_{ac} = -\frac{1}{12}\left((\nabla_a R)\, g_{bc} - (\nabla_b R)\, g_{ac} \right) - 2\,(\nabla_d \Omega)\, d_{abc}{}^d . \tag{18}$$

Then (11,12,13,14,17,18) constitute the conformal field equations for vacuum general relativity. Here the Ricci scalar R of g_{ab} is considered a given function of the coordinates. For any solution $(g_{ab}, \hat{R}_{ab}, d_{abc}{}^d, \Omega)$, $\hat{R}_{ab}$ is the traceless part of the Ricci tensor and $\Omega\, d_{abc}{}^d$ the Weyl tensor of g_{ab}. Note that the equations are regular even for $\Omega = 0$.

The 3+1 decomposition of the conformal geometry can be carried out as usual in general relativity, e.g.

$$g_{ab} = h_{ab} - n_a n_b = \Omega^2 (\tilde{h}_{ab} - \tilde{n}_a \tilde{n}_b) ,$$

where h_{ab} and $\tilde{h}_{ab}$ are the Riemannian 3-metrics induced by g_{ab}, respectively $\tilde{g}_{ab}$, on a spacelike hypersurface with unit normals n_a and equivalently $n_a = \Omega\, \tilde{n}_a$ (our signature is $(-,+,+,+)$). The relation between the extrinsic curvatures ($\tilde{k}_{ab} = \mathcal{L}_{\tilde{n}} \tilde{h}_{ab}/2$, $k_{ab} = \mathcal{L}_n h_{ab}/2$) is then easily derived as $k_{ab} = \Omega(\tilde{k}_{ab} + \Omega_0 \tilde{h}_{ab})$, where $\Omega_0 = n^a \nabla_a \Omega$.

The additional variables $\hat{R}_{ab}$ and d^d_{abc} can be decomposed into spatial objects by

$$ {}^{(0,1)}\hat{R}_a = n^b h_a{}^c \hat{R}_{bc} , \quad {}^{(0,1)}\hat{R}_{ab} = h_a{}^c h_b{}^d \hat{R}_{cd} ,$$

$$E_{ab} = d_{efcd} h^e{}_a n^f h^c{}_b n^d , \quad B_{ab} = d^*_{efcd} h^e{}_a n^f h^c{}_b n^d ,$$

where E_{ab} and B_{ab} are called the electric and magnetic components of the rescaled Weyl tensor $d_{abc}{}^d$.

Note that for regular components of h_{ab} and k_{ab}, the corresponding components of $\tilde{h}_{ab}$ and $\tilde{k}_{ab}$ with respect to the same coordinate system will in general diverge due to the compactification effect. However for the coordinate independent traces $k = h^{ab} k_{ab}$, $\tilde{k} = \tilde{h}^{ab} \tilde{k}_{ab}$ of the extrinsic curvatures we get

$$\Omega k = (\tilde{k} + 3\Omega_0) ,$$

which can be assumed regular everywhere. Note that at $\mathscr{I}$, $\tilde{k} = -3\Omega_0$. Since $\mathscr{I}^+$ is an ingoing null surface (with $(\nabla_a \Omega)(\nabla^a \Omega) = 0$ but $\nabla_a \Omega \neq 0$), we have that $\Omega_0 < 0$ at $\mathscr{I}^+$. It follows that $\tilde{k} > 0$ at $\mathscr{I}^+$. We will thus call regular spacelike hypersurfaces in $\mathcal{M}$ hyperboloidal hypersurfaces, since in $\tilde{\mathcal{M}}$ they are analogous to the standard hyperboloids $t^2 - x^2 - y^2 - z^2 = 3/\tilde{k}^2$ in Minkowski space, which provide the standard example. Since such hypersurfaces cross $\mathscr{I}$

but are everywhere spacelike in $\mathcal{M}$, they allow to access $\mathscr{I}$ and radiation quantities defined there by solving a Cauchy problem (in contrast to a characteristic initial value problem which utilizes a null surface slicing). Note that in a globally hyperbolic physical spacetime, hyperboloidal hypersurfaces will determine the future of the physical spacetime, but not all of its past, and therefore we call our studies *semiglobal.*

The timelike vector $t^a = (\partial/\partial t)^a$ is decomposed in the standard way into a normal and a tangential component:

$$t^a = Nn^a + N^a, \quad N^a n_a = 0\,. \tag{19}$$

N is called the lapse function, because it determines how fast the time evolution is pushed forward in the direction normal to S and thus determines "how fast time elapses". The tangential component N^a, $N^a n_a = 0$, shifts spatial coordinate points with time evolution, accordingly N^a is called *shift* vector. The lapse N and shift N^a are *not* dynamical quantities, they can be specified freely and correspond to the arbitrary choice of coordinates: the lapse determines the slicing of spacetime, the choice of shift vector determines the spatial coordinates.

We will not discuss the full 3 + 1 equations here for brevity, but rather refer to [10]. Their most essential feature is that they split into constraints plus symmetric hyperbolic evolution equations [10]. The evolution variables are h_{ab}, k_{ab}, the connection coefficients $\gamma^a{}_{bc}$, ${}^{(0,1)}\hat{R}_a$, ${}^{(0,1)}\hat{R}_{ab}$, E_{ab}, B_{ab}, as well as Ω, Ω_0, $\nabla_a\Omega$, $\nabla^a\nabla_a\Omega$ – in total this makes 57 quantities. In addition the gauge source functions q, R and N^a have to be specified. In order to guarantee symmetric hyperbolicity, they are given as functions of the coordinates. Here q determines the lapse as $N = \mathrm{e}^q\sqrt{\det h}$ and N^a is the shift vector. The Ricci scalar R can be thought of as implicitly steering the conformal factor Ω.

The constraints of the conformal field equations (see (14) of [10]) are regular equations on the whole conformal spacetime $(\mathcal{M}, g_{ab})$, but they have not yet been cast into a standard type of PDE system, such as a system of elliptic PDEs (recently however, some progress in this direction has been achieved by Butscher [28]). Therefore some remarks on how to proceed in this situation are in order. A possible resolution is to resort to a 3-step method [11,29,30]:

1. Obtain data for the Einstein equations: the first and second fundamental forms $\tilde{h}_{ab}$ and $\tilde{k}_{ab}$ induced on $\tilde{\Sigma}$ by $\tilde{g}_{ab}$, corresponding in the compactified picture to h_{ab}, k_{ab} and Ω and Ω_0. This yields so-called "minimal data".
2. Complete the minimal data on $\bar{\Sigma}$ to data for *all* variables using the conformal constraints – *in principle* this is mere algebra and differentiation.
3. Extend the data from $\bar{\Sigma}$ to Σ in some ad hoc but sufficiently smooth and "well-behaved" way.

In order to simplify the first step, numerical implementations [11,12,30] so far have been restricted to a subclass of hyperboloidal slices where initially $\tilde{k}_{ab}$ is pure trace, $\tilde{k}_{ab} = \tilde{h}_{ab}\tilde{k}/3$. The momentum constraint

$$\tilde{\nabla}^b\tilde{k}_{ab} - \tilde{\nabla}_a\tilde{k} = 0 \tag{20}$$

then implies $\tilde{k}$ = const. $\neq 0$. We always set $\tilde{k} > 0$. In order to reduce the Hamiltonian constraint

$$^{(3)}\tilde{R} + \tilde{k}^2 = \tilde{k}_{ab}\tilde{k}^{ab}$$

to *one* elliptic equation of second order, we use a modified Lichnerowicz ansatz

$$\tilde{h}_{ab} = \bar{\Omega}^{-2}\phi^4 h_{ab}$$

with *two* conformal factors $\bar{\Omega}$ and ϕ. The principal idea is to choose h_{ab} and $\bar{\Omega}$ and solve for ϕ, as we will describe now. First, the "boundary defining" function $\bar{\Omega}$ is chosen to vanish on a 2-surface $\mathcal{S}$ – the boundary of $\bar{\Sigma}$ and initial cut of $\mathscr{I}$ – with non-vanishing gradient on $\mathcal{S}$. The topology of $\mathcal{S}$ is chosen as spherical for asymptotically Minkowski spacetimes. Then we choose h_{ab} to be a Riemannian metric on Σ, with the only restriction that the extrinsic 2-curvature induced by h_{ab} on $\mathcal{S}$ is pure trace, which is required as a smoothness condition [29]. With this ansatz $\tilde{h}_{ab}$ is singular at $\mathcal{S}$, indicating that $\mathcal{S}$ represents an infinity. The Hamiltonian constraint then reduces to the Yamabe equation for the conformal factor ϕ:

$$4\,\bar{\Omega}^2 \Delta\phi - 4\,\bar{\Omega}(\nabla^a\bar{\Omega})(\nabla_a\phi) - \left(\frac{1}{2}{}^{(3)}R\,\bar{\Omega}^2 + 2\bar{\Omega}\Delta\bar{\Omega} - 3(\nabla^a\bar{\Omega})(\nabla_a\bar{\Omega})\right)\phi = \frac{1}{3}\tilde{k}^2\phi^5\,.$$

This is a semilinear elliptic equation – except at $\mathcal{S}$, where the principal part vanishes for a regular solution. This however determines the boundary values as

$$\phi^4 = \frac{9}{\tilde{k}^2}(\nabla^a\bar{\Omega})(\nabla_a\bar{\Omega})\,. \tag{21}$$

Existence and uniqueness of a positive solution to the Yamabe equation and the corresponding existence and uniqueness of regular data for the conformal field equations using the approach outlined above (assuming the "pure trace smoothness condition) have been proven by Andersson, Chruściel and Friedrich [29].

If the Yamabe equation is solved numerically, the boundary has to be chosen at $\mathcal{S}$, the initial cut of $\mathscr{I}$, with boundary values satisfying (21). If the equation were solved on a larger grid (conveniently chosen to be Cartesian), boundary conditions would have to be invented, which generically would cause the solution to lack sufficient differentiability at $\mathcal{S}$, see Hübner's discussion in [11]. This problem is due to the degeneracy of the Yamabe equation at $\mathcal{S}$. Unfortunately, this means that we have to solve an elliptic problem with *spherical boundary.*

The constraints needed to complete minimal initial data to data for all evolution variables split into two groups: those that require divisions by the conformal factor Ω to solve for the unknown variable, and those which do not. The latter do not cause any problems and can be solved without taking special care at $\Omega = 0$. The first group, needed to compute $^{(1,1)}\hat{R}$, E_{ab} and B_{ab}, however does require special numerical techniques to carry out the division and furthermore it is not known whether solving them on the whole Cartesian time evolution grid actually allows solutions which are sufficiently smooth across $\mathscr{I}$. Thus, at least for these we have to find some ad-hoc extension. There are however also examples of analytically known initial data, e.g. for the Minkowski and Kruskal spacetimes, where all constraints are solved on the whole Cartesian time evolution grid.

3 Examples: Different Ways to Compactify Minkowski Spacetime

The examples presented in this section help to illustrate the compactification procedure – in particular its inherent gauge freedom. They yield interesting numerical tests, some of which will be presented in Sect. 5.

3.1 Almost Static Compactification of Minkowski Spacetime

From the perspective of hyperboloidal initial data, the simplest way to compactify Minkowski spacetime is to choose the initial conformal three-metric as the flat metric, $h_{ab} = \delta_{ab}$, to set $k_{ab} = h_{ab}$, which solves the momentum constraint (20) and to choose the conformal curvature scalar R_g[3] as spherically symmetric, $R_g = R_g(x^2 + y^2 + z^2)$. We know from [29] that a unique solution to the constraints exists. It is not hard to see that it has to be spherically symmetric. Furthermore, it is topologically trivial. From Birkhoff's theorem we can thus conclude that we are dealing with Minkowski spacetime. Choosing the simplest gauge $q = 0$, $N^a = 0$, $R_g = 0$, the resulting unphysical spacetime is actually Minkowski spacetime in standard coordinates:

$$\mathrm{d}s^2 = -\mathrm{d}t + \mathrm{d}\Sigma^2 = \Omega^2\big(- \mathrm{d}T^2 + \mathrm{d}R^2 + R^2 \left(\mathrm{d}\theta^2 + \sin^2\theta \mathrm{d}\phi^2\right) \big) ,$$

where $\mathrm{d}\Sigma^2$ is the standard metric on $\mathbb{R}^3$, $\mathrm{d}\Sigma^2 = \mathrm{d}r^2 + r^2 \left(\mathrm{d}\theta^2 + \sin^2\theta \mathrm{d}\phi^2\right)$, and the conformal factor is

$$\Omega = \left(R^2 - T^2\right)^{-1} = \left(r^2 - t^2\right) , \tag{22}$$

where

$$r = \frac{R}{R^2 - T^2} , \qquad t = \frac{T}{R^2 - T^2} .$$

This setup has been chosen as the basis of Hübner's numerical study of weak data evolutions [13]. With the initial cut of $\mathscr{I}^+$ chosen at $x^2 + y^2 + z^2 = 1$, i^+ is located at coordinate time $t = 1$, the generators of $\mathscr{I}^+$ being straight lines at an angle of 45°.

This conformal representation of Minkowski spacetime is an "almost static" gauge – since the spatial geometry is time-independent, so are all evolution variables *except* for the conformal factor Ω. The physical region inside of $\mathscr{I}^+$ contracts to the regular point i^+ within finite time. This feature is shared with the standard "textbook" example of conformally compactifying Minkowski spacetime, which takes the form of a map into part of the Einstein static universe with $R_g = 6$,

$$\mathrm{d}s^2 = -\mathrm{d}t^2 + \mathrm{d}\Sigma^2 = \Omega^2 \left(-\mathrm{d}T^2 + \mathrm{d}R^2 + R^2 \left(\mathrm{d}\theta^2 + \sin^2\theta \mathrm{d}\phi^2\right)\right) , \tag{23}$$

[3] We change notation from R to R_g for this section to avoid confusion with a coordinate R we will introduce below.

where $\mathrm{d}\Sigma^2$ is the standard metric on S^3, $\mathrm{d}\Sigma^2 = \mathrm{d}\varrho^2 + \sin^2\varrho\left(\mathrm{d}\theta^2 + \sin^2\theta \mathrm{d}\phi^2\right)$, and the conformal factor is

$$\Omega^2 = 4\left(1 + (T-R)^2\right)^{-1}\left(1 + (T+R)^2\right)^{-1} = 4\cos^2\frac{t-\varrho}{2}\cos^2\frac{t+\varrho}{2}\,.$$

Here the coordinate transformations are

$$\varrho = \arctan(T+R) - \arctan(T-R)\,, \tag{24}$$
$$t = \arctan(T+R) + \arctan(T-R)\,. \tag{25}$$

In these coordinates Minkowski spacetime corresponds to the coordinate ranges

$$-\pi < t + \varrho < \pi\,, \tag{26}$$
$$-\pi < t - \varrho < \pi\,, \tag{27}$$
$$\varrho \geq 0\,. \tag{28}$$

For details and pictures of this mapping see the discussions by [4], [15] or [16].

Alternatively, we can choose stereographic spatial coordinates such that

$$\mathrm{d}\Sigma^2 = \omega^2\left(\mathrm{d}r^2 + r^2\left(\mathrm{d}\theta^2 + \sin^2\theta\mathrm{d}\phi^2\right)\right)\,, \qquad \omega = \frac{2}{(1+r^2)}\,.$$

or we may absorb the spatial conformal factor into the spacetime conformal factor by rescaling to

$$\mathrm{d}s'^2 = -\omega^{-2}\mathrm{d}t^2 + \mathrm{d}r^2 + r^2\left(\mathrm{d}\theta^2 + \sin^2\theta\mathrm{d}\phi^2\right)\,, \tag{29}$$

which yields the lapse to be $N = 1$ ($q = -3\log\omega$), respectively $N = \omega^{-2}$ ($q = -2\log\omega$). Note that in the numerical code we use Cartesian coordinates $x = r\sin\theta\cos\phi$, $y = r\sin\theta\sin\phi$, $z = r\cos\theta$.

The conformal transformation leading to (29) changes the scalar curvature from $R_g = 6$ to $R_g = -12\,(1+r^2)^{-2}$. We will see below in Sect. 5 that these simple variations in gauge source functions and conformal rescaling lead to numerical representations which are quite different, e.g. with regard to accuracy and robustness.

3.2 A Static Hyperboloidal Gauge for Minkowski Spacetime

By translating a standard hyperboloid in Minkowski spacetime along the trajectories of the $\partial/\partial t$ Killing vector, one can obtain a gauge where not only the conformal spacetime is static, but also the conformal factor is time-independent – thus also the physical geometry and all evolution variables of the conformal field equations can be made time independent (this has been pointed out to me by M. Weaver and I essentially follow her notes below). See also a talk given by V. Moncrief [31], which we have become aware of after starting to work with this gauge.

In this gauge the point i^+ is not brought into a finite distance and remains in the infinite future. This conformal gauge is particularly useful for stability tests.

To derive this static metric, we start with spherical coordinates (T, R, θ, ϕ) on Minkowski space, where the metric is

$$\mathrm{d}\tilde{s}^2 = -\mathrm{d}T^2 + \mathrm{d}R^2 + R^2\left(\mathrm{d}\theta^2 + \sin^2\theta \mathrm{d}\phi^2\right) . \tag{30}$$

A family of standard hyperboloids with time translation parameter t is given by

$$(T-t)^2 - R^2 = 1 .$$

We transform now to new coordinates $(t, \varrho, \theta, \phi)$, where the level surfaces of t are the standard hyperboloids and $\varrho(R)$ is chosen as a new radial parameter on the hyperboloids. Setting $T = t + \cosh\varrho$ and $R = \sinh\varrho$, the physical metric becomes

$$\mathrm{d}\tilde{s}^2 = -\mathrm{d}t^2 - 2\sinh\varrho\,\mathrm{d}\varrho\,\mathrm{d}t + \mathrm{d}\varrho^2 + \sinh^2\varrho\left(\mathrm{d}\theta^2 + \sin^2\theta\mathrm{d}\phi^2\right) . \tag{31}$$

For simplicity we choose the conformal three-metric to be flat and introduce new spherical coordinates (r, θ, ϕ) such that

$$\mathrm{d}s^2|_{t=\mathrm{const.}} = \mathrm{d}r^2 + r^2\left(\mathrm{d}\theta^2 + \sin^2\theta\mathrm{d}\phi^2\right). \tag{32}$$

Since $h_{ab} = \Omega^2\tilde{h}_{ab}$ we get $\Omega = \mathrm{d}r/\mathrm{d}\varrho$ and

$$\int_\varrho^\infty \frac{\mathrm{d}\varrho'}{\sinh\varrho} = \int_r^1 \frac{\mathrm{d}r'}{r'} . \tag{33}$$

The limits of integration are given by the fact that $\lim_{\varrho\to\infty} r = 1$. Performing the integrals one finds that

$$r = \frac{\mathrm{e}^\varrho - 1}{\mathrm{e}^\varrho + 1} = \frac{1}{R}(\sqrt{1+R^2} - 1) \tag{34}$$

and

$$\Omega = \frac{1-r^2}{2} , \tag{35}$$

our choice thus maps $\mathscr{I}^+$ to the timelike cylinder $r = 1$.

The computer time coordinate t is a Bondi time coordinate on $\mathscr{I}$. In coordinates (t, r, θ, ϕ), the conformal metric reads

$$\mathrm{d}s^2 = -\Omega^2\mathrm{d}t^2 - 2\,r\,\mathrm{d}r\,\mathrm{d}t + \mathrm{d}r^2 + r^2\left(\mathrm{d}\theta^2 + \sin^2\theta\mathrm{d}\phi^2\right) , \tag{36}$$

or

$$\mathrm{d}s^2 = -\Omega^2\mathrm{d}t^2 - 2\,\mathrm{d}t\,\left(x\,\mathrm{d}x + y\,\mathrm{d}y + z\,\mathrm{d}z\right) + \mathrm{d}x^2 + \mathrm{d}y^2 + \mathrm{d}z^2 \tag{37}$$

in Cartesian coordinates, $x = r\sin\theta\cos\phi$, $y = r\sin\theta\sin\phi$, $z = r\cos\theta$, which are used in the numerical code. The shift vector is thus given by $N^i = -x^i$ and the lapse can be computed from $-N^2 + h_{ab}N^aN^b = g_{tt}$ (as implied by (19)) as

$N = (1+r^2)/2$. The three metric has unit determinant, so $q = \ln N$. Note that the shift vector does not become "superluminal" beyond $\mathscr{I}^+$, because the lapse is growing faster than the shift; g_{tt} is nonnegative everywhere and zero only at $\mathscr{I}^+$. The conformal Ricci scalar is

$$R_g = 12\frac{(1-r^2)(3+r^2)}{(1+r^2)^3}\ , \tag{38}$$

which vanishes at $\mathscr{I}^+$.

For a numerical calculation one needs the minimal initial data set,

$$(h_{ab}, \Omega, k_{ab}, \Omega_0) \tag{39}$$

and the gauge source functions, (R, N, N^a). In a numerical calculation in which the Yamabe equation is solved to find Ω, one gives $(h_{ab}, \bar{\Omega}, \operatorname{tr} k)$. In a test case such as this, which is an explicitly known solution, one can just take $\bar{\Omega} = \Omega$. It remains therefore to calculate k_{ab} and Ω_0. From

$$k_{ab} = \frac{1}{2N}(\partial_t h_{ab} - \mathcal{L}_N h_{ab}) \tag{40}$$

we find that the components of the extrinsic curvature are $k_{ij} = \delta_{ij}/N$ and $k = 3/N$. From the identity $\tilde{k} = \Omega\, k - 3\,\Omega_0$ we find that

$$\Omega_0 = -\frac{2r^2}{1+r^2}\ . \tag{41}$$

4 History

This section tries to give a broad overview of what has been achieved so far in the field of numerical treatment of the conformal field equations. Historically, this field was started by Peter Hübner by studying a scalar field coupled to gravity in spherical symmetry in his PhD thesis [32] finished in 1993. His subsequent work has lead to the development of both a 2D and a 3D evolution code, formulated in "metric" variables. Jörg Frauendiener has also developed an independent 2D code, formulated in frame variables.

4.1 Early Work on Spherical Symmetry

The first numerical implementation of the conformal field equations is due to Peter Hübner, who has studied the spherically symmetric collapse of scalar fields in his PhD thesis [32] and subsequently in [33]. In his gauge both future null infinity ($\mathscr{I}^+$) and future timelike infinity (i^+) are compactified and the whole spacetime is covered in finite coordinate time. Hübner studies the global structure of the spacetime, including the appearance of singularities and the localization of the event horizon. To handle the latter, floating point exceptions are caught and grid points are flagged as "singular", grid points whose values depend on information from singular grid points are correspondingly flagged as singular as well. Even though this method does not allow to actually trace the singularity in a strict sense (computers cannot actually deal with infinite values), the method traces the singularity as tightly as possible. In contrast to typical black hole excision schemes, which are based on locating the apparent horizon, this scheme could thus be termed "tight excision". The method has not yet been implemented in higher dimensions, where one has to face more intricate technical problems and where the structure of the singularity is likely to be much more complex as well. The paper also studies critical collapse. Hübner's results are consistent with the black hole mass power-law scaling with the correct exponent, however no echoing related to discrete self-similarity has been seen in his results. This has created some discussion, whether the results of other authors are numerical artefacts, or artefacts of boundary conditions at finite distance. However numerical critical collapse simulations in a compactified characteristic framework have recently shown both the correct power-law scaling and discretely self-similar echoing [34].

The coordinates in this approach are based on the geometric structure of double null-coordinates that is available in spherical symmetry. Unfortunately this choice does not generalize in the absence of spherical symmetry. Finding a gauge that would allow to run, say, the Kruskal spacetime in a 3D code for "arbitrarily long" Bondi times is an open problem, where significant insight could be gained from studying more general gauges in a manifestly spherically symmetric code.

4.2 Axially Symmetric Spacetimes with Toroidal $\mathscr{I}$ in the Frame Formulation

Following Hübner's encouraging results for spherically symmetric simulations [32,33], numerical codes have been developed by Frauendiener and Hübner to study axially symmetric spacetimes. For simplicity, e.g. to avoid numerical stability problems at the axis of symmetry and to avoid problems associated with a $\mathscr{I}$ of spherical topology – which does not align with Cartesian coordinates – both Hübner and Frauendiener considered the asymptotically A3-spacetimes [35,36], which do not possess an axis of symmetry and where $\mathscr{I}$ has toroidal topology. These spacetimes are modelled after the A3-metric in the Ehlers-Kundt classification [37], which provides an analogue of the Schwarzschild metric in plane symmetry. These spacetimes are not physical, but they contain a large class of nontrivial radiative vacuum spacetimes, which make them an interesting toy model to study numerical techniques, gauges and the extraction of radiation.[4] These axisymmetric codes thus have been designed to treat the vacuum case, and matter couplings have not yet been implemented. An advantage for code-testing is that exact solutions are known [35,38].

In the first [27] of a series of papers [24,26,27,30] on his axisymmetric code, Frauendiener gives a nice overview of the motivation for using the conformal

[4] One of the notable differences with the Minkowski case is that one can only define a Bondi-mass but no Bondi four-momentum.

field equations of numerical simulations of isolated systems. He discusses the conformal field equations in the space spinor formalism [39], which is chosen because of compactness of notation, and because it allows a very straightforward 3+1 decomposition of the equations, rendering the equations in symmetric hyperbolic form. His formalism contains 8 free functions which determine the gauge: the harmonicity $F := \nabla_c \nabla^c t$ determines the choice of time coordinate t, the shift is given in terms of frame coefficients, the scalar curvature R (Λ in his notation) of the compactified spacetime and an imaginary and symmetric space spinor field F_{AB} (i.e. three numbers), which determines rotations of the spatial frame (for $F_{AB} = 0$ the frame is transported via Fermi-Walker transport). He also discusses the implications of the assumptions of the toroidal symmetry, in particular for the choice of gauge – e.g. the adoption of the frame.

In the second paper [26] of the series, Frauendiener discusses his numerical methods and gauge choices and presents results for evolutions of initial data corresponding to the exact solution presented in [35]. Here one of the two Killing vectors is disguised by a coordinate transformation. The numerical evolution proceeds via a generalization of the Lax-Wendroff scheme to 2D, which Frauendiener proves to be stable and second order accurate. The time step is such that the numerical domain of dependence is contained in the domain of dependence as defined by the equations. An essential difficulty – as usual – is posed by the treatment of the boundary. Well-posedness of the associated initial-boundary-value problem has not yet been proven and numerical analysis can only provide rough guidelines to work out stable algorithms [40]. Frauendiener's boundary treatment is based on the identification of ingoing and outgoing modes at the boundary, as determined from the symmetric hyperbolic character of the equations. He sets boundary values for inward-propagating quantities (e.g. motivated by the exact solution) and sets values for the outward propagating quantities by extrapolation from the interior. This method can be applied just a few grid-points outside of $\mathscr{I}$ and is found to be stable as long as the gauge source functions do not depend on the evolution variables – which would change the characteristics. Note that the constraints will in general not be satisfied on the boundary, which may trigger constraint-violating modes of the equations.

Frauendiener gives a detailed discussion of the problems associated with the choice of the gauge and performs a number of numerical experiments in this respect, evolving data corresponding to exact solutions [35,38] with singular i^+. One of the problems is that if the gauge source functions are allowed to depend on the evolution variables, this will change the characteristics of the system and will in general spoil the symmetric hyperbolic character of the system. Experiments in this direction, where $F = F(N, K)$, indeed exhibited a boundary instability. Regarding the choice of time coordinate, that is, the harmonicity function F, several choices are tested: a "natural" gauge, which is taken from the exact solution, the Gauss gauge (where the lapse N is spatially constant), the harmonic gauge, $F = 0$ and a family of gauges that interpolates between the "natural" and harmonic gauge.

The "natural" gauge is found to provide the best performance and the approach to the singularity is found to be essentially limited by machine precision. The harmonic gauge leads to a coordinate singularity before reaching the singularity. This feature is shared by most of the gauges that interpolate between natural and harmonic gauge. For the "Gauss" gauge with $N = \text{const.}$, caustics (coordinate shocks) develop quickly and crash the simulations.

Regarding the choice of shift vector, a prescription for $\mathscr{I}$ fixing – that is, steering the evolution of the surface $\Omega = 0$ – is discussed, which can be easily implemented in Frauendiener's formulation. This however relies on the specific form of the frame equations and does not carry over to equations as those used in Hübner's codes [10]. In particular he studies the case of "$\mathscr{I}$ freezing" – holding the coordinate position of $\mathscr{I}$ in place such that no loss of resolution occurs in the physical domain.

Finally, he discusses the extraction of gravitational radiation, e.g. by computing the Bondi mass and shows some results.

In order to study more general spacetimes, Frauendiener has implemented a numerical scheme for determining hyperboloidal initial data sets for the conformal field equations by using pseudo-spectral methods as described in [30]. He uses the implicit approach of first solving the Yamabe equation and then carrying out the division by the conformal factor for certain fields which vanish on $\mathscr{I}$. The challenge there is to numerically obtain a smooth quotient. The division problem is treated by a transformation to the coefficient space, where a QR-factorization of a suitable matrix is used and then transforming back.

In [24] Frauendiener gives a pedagogical discussion of the issue of radiation extraction in asymptotically flat space-times within the framework of conformal methods for numerical relativity. The aim is to show that there exists a well defined and accurate extraction procedure which mimics the physical measurement process and operates entirely intrinsically within $\mathscr{I}^+$. The notion of a detector at infinity is defined by idealizing local observers in Minkowski space. A detailed discussion is presented for Maxwell fields and the generalization to linearized and full gravity is performed by way of the similar structure of the asymptotic fields.

Recently, Hein has written a 2D axisymmetric code that allows for an axis [41], i.e. can treat the physical situation with a $\mathscr{I}^+$ of spherical topology. The usual problem of the coordinate singularity at the axis in adapted coordinates is solved by using Cartesian coordinates, following a method developed by Alcubierre et al. [42]. The code has so far been tested by evolving Minkowski spacetime in various gauges, further tests with nontrivial spacetimes are currently underway.

4.3 Metric-Based 2D and 3D Codes

The basic design of Hübner's approach is outlined in [10], where he presents the first order time evolution equations as obtained from a 3+1 split of the conformal field equations. The evolution equations can be brought into symmetric hyperbolic form by a change of variables. He discusses his motivation of avoiding artificial boundaries and how the conformal field equations formally allow placement of the grid boundaries outside the physical spacetime.

A particularly subtle part of the evolution is usually the boundary treatment. In the conformal approach we are in the situation that the boundary can actually be placed outside of the physical region of the grid – this is one of its essential advantages! In typical explicit time evolution algorithms, such as our Runge-Kutta method of lines, the numerical propagation speed is larger than the speed of all the characteristics (in our case the speed of light). Thus $\mathscr{I}$ does *not* shield the physical region from the influence of the boundary – but this influence has to converge to zero with the convergence order of the algorithm – fourth order in our case. In principle one therefore does not have to choose a "physical" boundary condition. The only requirements are stability and "practicality" – e.g. the boundary condition should avoid, if possible, the development of large gradients in the unphysical region to reduce the numerical "spill over" into the physical region, or even code crashes. It seems likely however, that this practicality requirement will eventually lead to a treatment of the boundary which satisfies the constraints at the boundary.

Hübner develops the idea of modifying the equations near the grid boundaries to obtain a consistent and stable discretization. The current implementation of the boundary treatment relies on this introduction of a "transition layer" in the unphysical region, which is used to transform the rescaled Einstein equations to trivial evolution equations, which are stable with a trivial copy operation at outermost gridpoint as a boundary condition (see [10] for details and references). He thus replaces

$$\partial_t f + A^i \partial_i f - b = 0$$

by

$$\partial_t f + \alpha(\Omega) \left(A^i \partial_i f - b \right) = 0 \, ,$$

where α is chosen as $\alpha(\Omega) = 0$ for $\Omega \leq \Omega_0 < \Omega_1 < 0$ and 1 for $\Omega \geq \Omega_1$. One potential problem is that the region of large constraint violations outside of $\mathscr{I}$ may trigger constraint violating modes of the equations that can grow exponentially. Another problem is that a "thin" transition zone causes large gradients in the coefficients of the equations – thus eventually leading to large gradients in the solution, while a "thick" transition zone means loosing many gridpoints. If no transition zone is used at all and the Cartesian grid boundary touches $\mathscr{I}$, the ratio of the number of grid points in the physical region versus the number of grid points in the physical region is already $\pi/6 \approx 0.52$.

Furthermore he discusses his point of view concerning possible advantages of the conformal approach and discusses potential problems of the Cauchy and Cauchy-Characteristic matching approaches to numerical relativity. He outlines the geometric scenario of his approach and stresses that these techniques allow, in principle, to calculate the complete future of scenarios such as initial data for N black holes.

The second paper [11] of the series deals with the technical details of construction of initial data and of time-evolution of such data. The second and fourth order discretizations, which are used for the construction of the complete data set and for the numerical integration of the time evolution equations, are described and their efficiencies compared. Results from tests for $A3$ and disguised Minkowski spacetimes confirm convergence for the 2D and 3D codes.

The simplest approach to the division by Ω would be implementing l'Hôpital's rule. However this leads to non-smooth errors and consequently to a loss of convergence [11]. Instead, Hübner [11] has developed a technique to replace a division $g = f/\Omega$ by solving an elliptic equation of the type

$$\nabla^a \nabla_a (\Omega^2 g - \Omega f) = 0$$

for g (actually some additional terms added for technical reasons are omitted here for simplicity). For the boundary values $\Omega^2 g - \Omega f = 0$, the unique solution is $g = f/\Omega$. The resulting linear elliptic equations for g are solved by the same numerical techniques as the Yamabe equation. For technical details see Hübner [12].

Finally, we have to extend the initial data to the full Cartesian spatial grid in some way. Since solving all constraints also outside of $\mathscr{I}$ will in general not be possible in a sufficiently smooth way [11], we have to find an ad hoc extension, which violates the constraints outside of $\mathscr{I}$ but is sufficiently well behaved to serve as initial data. The resulting constraint violation is not necessarily harmful for the evolution, since $\mathscr{I}$ causally disconnects the physical region from the region of constraint violation. On the numerical level, errors from the constraint violating region *will* in general propagate into the physical region, but if our scheme is consistent, these errors have to converge to zero with the convergence order of the numerical scheme (fourth order in our case). There may of course still be practical problems that prevent us from reaching this aim: making the ad-hoc extension well behaved is actually quite difficult, the initial constraint violation may trigger constraint violating modes in the equations, which take us away from the true solution, singularities may form in the unphysical region, etc.

Since the time evolution grid is Cartesian, its grid points will in general not coincide with the collocation points of the pseudo-spectral grid. Thus fast Fourier transformations cannot be used for transformation to the time evolution grid. The current implementation instead uses standard discrete ("slow") Fourier transformations, which typically take up the major part of the computational effort of producing initial data.

It turns out that the combined procedure works reasonably well for certain data sets. For other data sets the division by Ω is not yet solved in a satisfactory way and constraint violations are of order unity for the highest available resolutions. In particular this concerns the constraint $\nabla_b E_a{}^b = -{}^{(3)}\varepsilon_{abc} k^{bd} B_d{}^c$ ((14d) in [10]), since E_{ab} is computed last in the hierarchy of variables and requires two divisions by Ω. Further research is required to analyze the problems and either improve the current implementation or apply alternative algorithms. Ultimately, it seems desirable to change the algorithm for obtaining initial data to a method that solves the conformal constraints directly and therefore does not suffer the current problems. This approach may of course introduce new problems like having an elliptic system too large to be handled in practice.

The time evolution algorithm is an implementation of a standard fourth order method of lines (see e.g. [43]). In the method of lines we formally write

$$\partial_t f = B(f, \partial_i f) \;, \tag{42}$$

where $B(f, \partial_i f) = -A^i(f)\partial_i f + b(f)$. Discretizing the spatial derivatives parametrizes the ordinary differential equations by grid point index. For the present code, fourth order accurate centered spatial differences have been implemented, e.g. for the x-derivative:

$$\partial_x f \to \frac{1}{12\Delta x}\left(-f_{i+2,j,k} + 8f_{i+1,j,k} - 8f_{i-1,j,k} + f_{i-1,j,k}\right) .$$

The numerical integration of the ordinary differential equations proceeds via the standard fourth order Runge-Kutta scheme:

$$f^{l+1}_{i,j,k} = f^l_{i,j,k} + \frac{1}{6}\left(k^l_{i,j,k} + 2k^{l+1/4}_{i,j,k} + 2k^{l+1/2}_{i,j,k} + k^{l+3/4}_{i,j,k}\right) , \tag{43}$$

where

$$\begin{aligned}
k^l_{i,j,k} &= \Delta t\, B(f^l_{i,j,k}, \partial_i f^l_{i,j,k}) \;, \\
k^{l+1/4}_{i,j,k} &= \Delta t\, B(f^{l+1/4}_{i,j,k}, \partial_i f^{l+1/4}_{i,j,k}) \;, \quad & f^{l+1/4}_{i,j,k} &= f^l_{i,j,k} + \frac{1}{2} k^l_{i,j,k} \;, \\
k^{l+1/2}_{i,j,k} &= \Delta t\, B(f^{l+1/2}_{i,j,k}, \partial_i f^{l+1/2}_{i,j,k}) \;, \quad & f^{l+1/2}_{i,j,k} &= f^l_{i,j,k} + \frac{1}{2} k^{l+1/4}_{i,j,k} \;, \\
k^{l+3/4}_{i,j,k} &= \Delta t\, B(f^{l+3/4}_{i,j,k}, \partial_i f^{l+3/4}_{i,j,k}) \;, \quad & f^{l+3/4}_{i,j,k} &= f^l_{i,j,k} + k^{l+1/2}_{i,j,k} \;.
\end{aligned}$$

Additionally, a dissipation term of the type discussed in theorems 6.7.1 and 6.7.2 of Gustafsson, Kreiss and Oliger [43] is added to the right-hand-sides to damp out high frequency oscillations and keep the code numerically stable. The dissipation term used is

$$\sigma Q_2 := \frac{\sigma}{64\,N}(\Delta x)^5 \sum_{i=1}^{N} \partial_i{}^6 f \;,$$

where the spatial derivatives are discretized as

$$\begin{aligned}
\partial_x{}^6 f^l_{i,j,k} \to \frac{1}{(\Delta x)^6}\Big(& f^l_{i-3,j,k} - 6f^l_{i-2,j,k} + 15f^l_{i-1,j,k} \\
& - 20f^l_{i,j,k} + 15f^l_{i+1,j,k} - 6f^l_{i+2,j,k} + f^l_{i+3,j,k}\Big) .
\end{aligned}$$

Numerical experiments show that usually small amounts of dissipation (σ of the order of unity or smaller) are sufficient and do not change the results in any significant manner. Numerical tests for Minkowski spacetime with disguised symmetries and an explicitly known A3-like solution with radiation [38] are described in [11].

Further extensive tests of the 2D code have been performed by Weaver [44]. She studied the choice of gauge source functions for an A3-like solution, solving the Yamabe equation for the conformal factor. She found that for this solution it is quite simple to prescribe a shift so that $\mathscr{I}$ is fixed to a very good approximation. She also studied the use of the gauge source function q to prolong the numerical simulation inside physical spacetime. In cases where $q = 0$ results in a "singularity" developing outside physical spacetime (which causes the code to crash), prescription of q so that the evolution inside physical spacetime is prolonged compared to the outside allows the simulation to essentially cover the physical spacetime to the future of the initial data surface. She thus found that in this context the ad hoc prescription of gauge source functions was sufficient to achieve desired effects, and caused no instabilities. Also she explored the effect of turning off the transition zone, while still simply copying data at the outer grid boundary into the ghost zone, along with prescription of q, so that the evolution is slowed down at the outer boundary. In the A3-like 2D runs this alternative boundary treatment was successful and avoided problems created by the transition zone.

In the third part of the series [12], a pseudospectral solver for the constraints is described. Since the implementation depends on the topology, it discusses both the asymptotically A3 and asymptotically Minkowski cases. At the end also some remarks are made about a possible extension to the multi-blackhole case, using a multi-patch scheme (the Schwarz alternating procedure).

In the fourth part of the series [13] Hübner presents results of 3D calculations for initial data which evolve into a regular point i^+ and which thus could be called "weak data". The initial conformal metric is chosen in Cartesian coordinates as

$$\mathrm{d}s^2 = \left(1 + \frac{A}{3}\bar{\Omega}^2\left(x^2 + 2y^2\right)\right)\mathrm{d}x^2 + \mathrm{d}y^2 + \mathrm{d}z^2 \,. \tag{44}$$

We choose $\bar{\Omega} = \left(1 - \left(x^2 + y^2 + z^2\right)\right)/2$ as the boundary defining function $\bar{\Omega}$ appearing in this ansatz. It is used to satisfy the smoothness condition for the conformal metric at $\mathscr{I}$. For the gauge source functions, Hübner has made the "trivial" choice: $R = 0$, $N^a = 0$, $q = 0$, i.e. the conformal spacetime has vanishing scalar curvature, the shift vanishes and the lapse is given by $N = \mathrm{e}^q\sqrt{\det h} = \sqrt{\det h}$. This simplest choice of gauge is completely sufficient for $A = 1$ data and has lead to a milestone result of the conformal approach – the evolution of weak data which evolve into a regular point i^+ of $\mathcal{M}$, which is resolved as a single grid cell. With this result Hübner has illustrated a theorem by Friedrich, who has shown that for sufficiently weak initial data there exists a regular point i^+ of $\mathcal{M}$ [45]. The complete future of (the physical part of) the initial slice can thus be reconstructed in a finite number of computational time steps. This calculation is an example of a situation for which the usage of the conformal field equations is ideally suited: main difficulties of the problem are directly addressed and solved by using the conformal field equations.

A natural next question to ask is: what happens if one increases the amplitude A? To answer this question, I have performed and analyzed runs for integer values of A up to $A = 20$. Preliminary results have been presented in [46]. While for $A = 1, 2$ the code was found to be able to continue beyond i^+ without problems, for all higher amplitudes the "trivial" gauge leads to code crashes before reaching i^+. While the physical data still decay quickly in time, a sharp peak of the lapse develops outside of $\mathscr{I}$ and crashes the code after Bondi time ~ 8 (320 M) for $A = 3$ and ~ 1.5 (3 M) for $A = 20$ (here M is the initial Bondi mass). A partial cure of the problem was obtained using a modified gauge source function $q = -r^2/a$ ($N = \mathrm{e}^{-r^2/a}\sqrt{\det h}$), where a is tuned such that one gets a smooth lapse and smooth metric components. For $A = 5$, for example, a value of $a = 1$ was found by moderate tuning of a (significantly decreasing or increasing a crashes the code before the regular i^+ is reached). Unfortunately, this modification of the lapse is not sufficient to achieve much higher amplitudes. As A is increased, the parameter a requires more fine tuning, which was only achieved for $A \leq 8$. For higher amplitudes the code crashes with significant differences in the maximal and minimal Bondi time achieved, while the radiation still decays very rapidly. Furthermore, the curvature quantities do not show excessive growth – it is thus natural to assume that we are still in the weak-field regime and the crash is not connected to the formation of an apparent horizon or singularity. While some improvement is obviously possible through simple non-trivial models for the lapse (or other gauge source functions), this approach seems quite limited and more understanding will be necessary to find practicable gauges. An interesting line of research would be to follow the lines of [47] in order to find evolution equations for the gauge source functions which avoid the development of pathologies.

Schmidt has presented hyperboloidal initial data for the Kruskal spacetime, a hyperboloidal foliation for the future of these hyperboloidal initial data [48] and results from numerical simulations evolving these initial data with different gauges, which have been performed by Weaver with Hübner's 3D code. The explicit hyperboloidal version of the Kruskal spacetime is very useful for numerically testing the conformal approach in the treatment of black hole spacetimes. These runs have been performed in octant mode. The runs typically proceed until the determinant of the three metric becomes negative [44], caused by some feature in the exact solution which is no longer adequately resolved and which is growing, leading to large narrow spikes in the numerical data. Future work will have to be directed toward improving the choice of gauge source functions such that rapidly growing sharp features are avoided.

In the next section, I will present new results obtained with the 3D code for asymptotically Minkowski spacetimes, which will illustrate some of the current problems. One of these is the presence of exponentially growing constraint violating modes. The problem of controlling the growth of the constraints for the conformal field equations has first been addressed by Florian Siebel in a diploma thesis [49] and subsequently by Hübner and Siebel in [50]. The key idea in this work is to develop a λ–system [51] for the conformal field equations in 1+1 dimensions (with toroidal $\mathscr{I}$'s). A λ–system is an enlarged evolution system, where evolution equations for the constraints are added in, consistently with symmetric hyperbolicity. One then has a large parameter space of coefficient functions available in which to find choices such that the new system has the constraint surface as an attractor. The main conclusion of this work is that it was not possible to significantly improve the fidelity of the numerical calculations. In those cases where moderate improvements regarding the constraints could be achieved, the deviation from the known exact solution would get larger.

5 Results from 3D Calculations

All the results presented in this section have been performed with 121^3 grids on 32 processors of the AEI's SGI origin 2000. The outer boundary has been placed at a radius of $r = 1.15$ in these runs ($\mathscr{I}^+$ is initially located at a radius $r = 1$).

5.1 Minkowski Data

We will first discuss some results for Minkowski spacetime, which in spite of its simplicity provides some nontrivial numerical tests. As has been first demonstrated by Hübner in [13], for weak data – in particular Minkowski space – it is possible via the conformal approach to cover the whole domain of dependence of initial data reaching out to $\mathscr{I}^+$ with a finite number of time steps. Let us thus first consider the gauges of Sect. 3.1, where the compactified geometry is time-independent, but a time-dependent conformal factor Ω is responsible for contracting the cuts of $\mathscr{I}^+$ to a point within finite coordinate time.

We have compared the gauges where the conformal spacetime is Minkowski, (22), the Einstein static universe (23), or the spacetime given by (29). Essentially, the result is that the Minkowski case yields the highest accuracy, the Einstein universe case works in principle and in the case (29) the code crashes before reaching i^+. In Fig. 1 the Minkowski and Einstein universe cases are compared by plotting $h_{xx} - 1$ and the value of the constraint $\nabla_x \Omega = \Omega_x$ at the center versus coordinate time (where t is scaled such that $t(i^+) = 1$. The Minkowski case – denoted by the unbroken line – clearly yields better accuracy, although the growth of $h_{xx} - 1$ is faster and approximately exponential during the later stage of "physical" evolution. Note that the constraint grows very fast in both cases.

Figures 2 – 5 show a comparison of the less optimal Einstein universe case with the case (29) to illustrate some of the problems one expects in the evolution of nontrivial spacetimes. Figure 2 shows the time evolution of h_{xx} along the positive x–axis versus coordinate time for the Einstein universe case and for the case of (29). Figure 2 compares the corresponding contour lines. While no deviation from staticity is visible for the Einstein universe case, the other case shows a rapidly growing peak in h_{xx} and the lapse (shown in Fig. 4) (and thus of $\det h$), which is located in the transition zone outside of $\mathscr{I}^+$. Eventually this feature cannot be resolved any more and the code crashes. In the Einstein

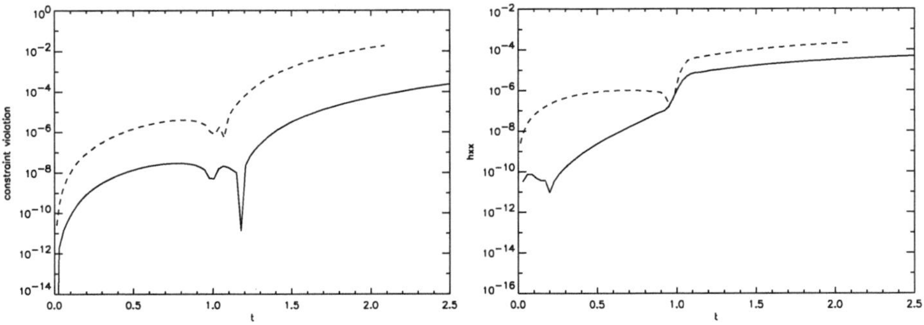

Fig. 1. Comparing the Minkowski (solid line) and Einstein universe (dashed line) cases: left, the value of the constraint $\nabla_x \Omega = \Omega_x$ at the center is plotted versus coordinate time. In the right image $h_{xx} - 1$ is plotted vs. coordinate time (where t is scaled such that $t(i^+) = 1$)

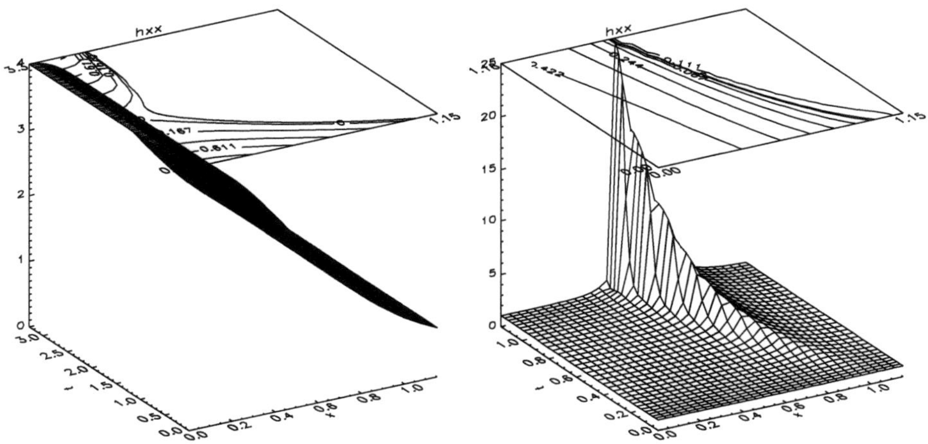

Fig. 2. The value of the metric component h_{xx} for $x \geq 0$ is plotted versus coordinate time. The left image shows the Einstein universe case. The right image shows case (29). There the maximum of h_{xx} in the region where $\Omega > 0$ is approximately at the value 5

static case the code was simply stopped by running out of time in the queue. Fig. 5 shows the sum over the L^2–norms (taken in the physical region) of all the constraints versus time. While in the Einstein static case the constraints show a rapid decrease in the physical region, followed by a steep growth after passing through i^+, the case (29) exhibits roughly exponential overall growth almost from the start.

Results for the completely static gauge given by (37) are shown in Figs. 6 – 9. This gauge poses a harder challenge than the previous ones, where i^+ is reached in finite time. Now the goal is to maintain an indefinite stable evolution. However, the evolution shows exponential growth, illustrated in Figs. 8 and 9 by the values of h_{xx} and constraints $\nabla_x h_{xx}$ and $\nabla_x \Omega = \Omega_x$. It is interesting, however, that the curvature invariants I and J are *decreasing* during the evolution as shown

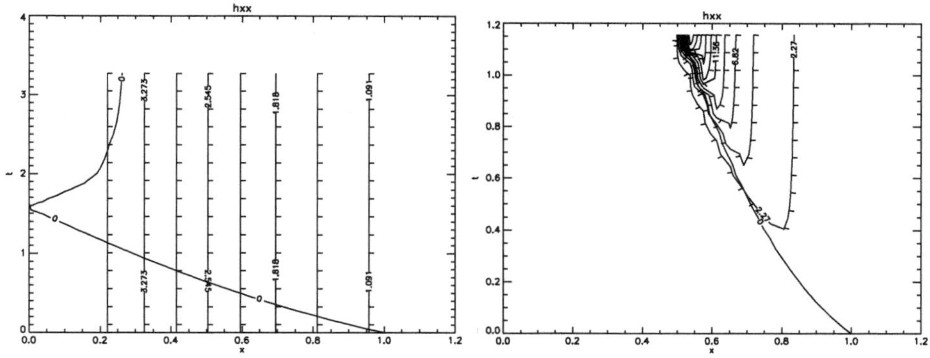

Fig. 3. Contour lines of the metric component h_{xx} for $x \geq 0$ are plotted versus coordinate time. The left image shows the Einstein universe case. The right image shows case (29). The thicker line marks $\Omega = 0$, i.e. $\mathscr{I}^+$

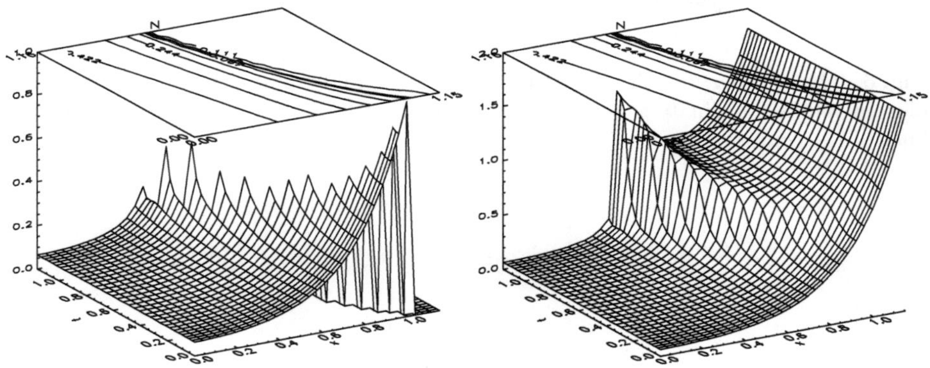

Fig. 4. The value of the lapse N for $x \geq 0$ is plotted for the case (29) versus coordinate time. The left image shows the points where $\Omega > 0$. The right image shows all points

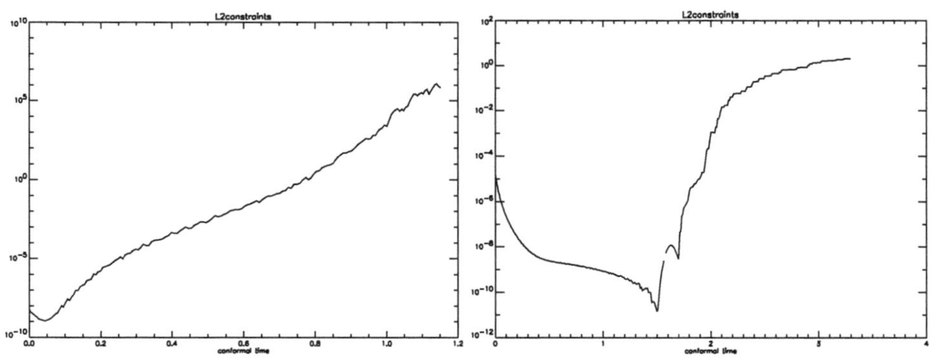

Fig. 5. The sum over the L^2–norms (taken in the physical region) of all the constraints is plotted versus coordinate time for the Einstein universe (left) and case (29) (right)

in Figs. 6 and 7. The exponential blowup crashes the code at $t \sim 5.1$. This time seems to be roughly independent of resolution, size of time step, amount of dissipation, location of the boundary and location of the transition zone. A possible explanation is exponentially growing constraint violating modes on the continuum level.

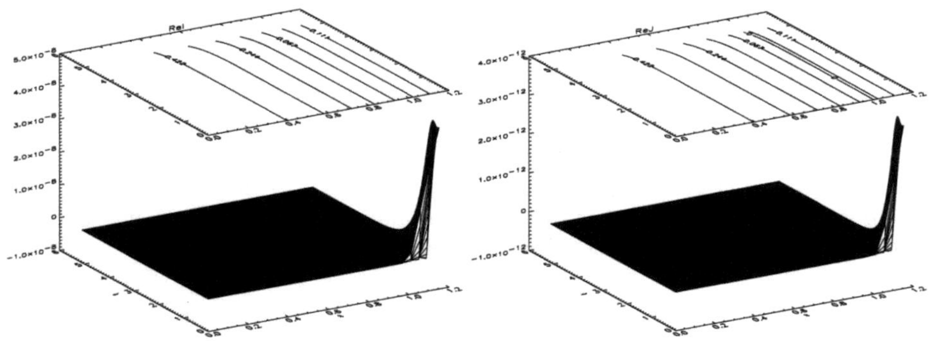

Fig. 6. The real parts of the curvature invariants I (left) and J (right) for $x \geq 0$ are plotted versus coordinate time for the static gauge of (37). Superimposed are contour lines of the conformal factor Ω

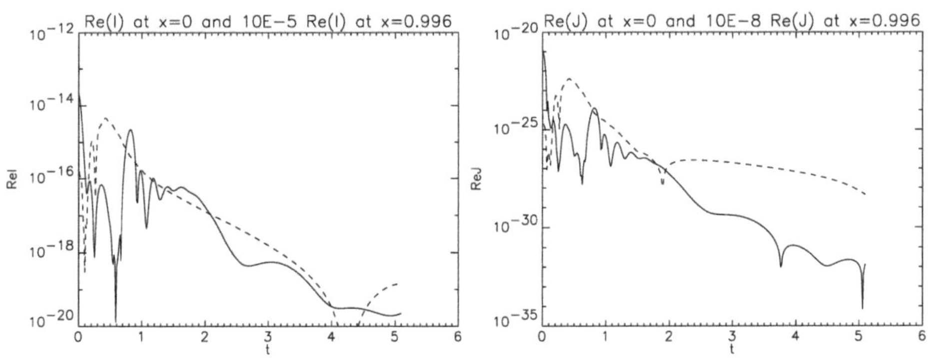

Fig. 7. The real parts of the curvature invariants I (left) and J (right) are plotted versus coordinate time for the static gauge of (37). The solid line is for the gridpoint at the center of the grid, the dashed line for a grid point at $x = 0.996$, $y = z = 0$, multiplied by a factor of 10^{-5} for I and 10^{-8} for J

5.2 "Brill" Data

We use an axisymmetric Brill–wave type ansatz to look at initial data that contain radiation and set

$$ds^2 = \omega^2 \left(e^{2Q}(d\varrho^2 + dz^2) + \varrho^2 d\varphi^2 \right) , \tag{45}$$

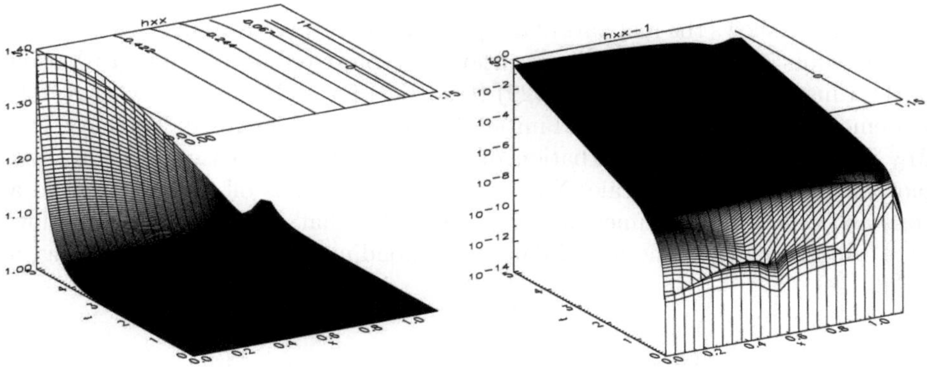

Fig. 8. The value of the metric component h_{xx} for $x \geq 0$ is plotted versus coordinate time with linear (left) and logarithmic (right) scaling for the static gauge of (37). Approximately exponential growth is obvious. The largest amplitude of the growth is in the center

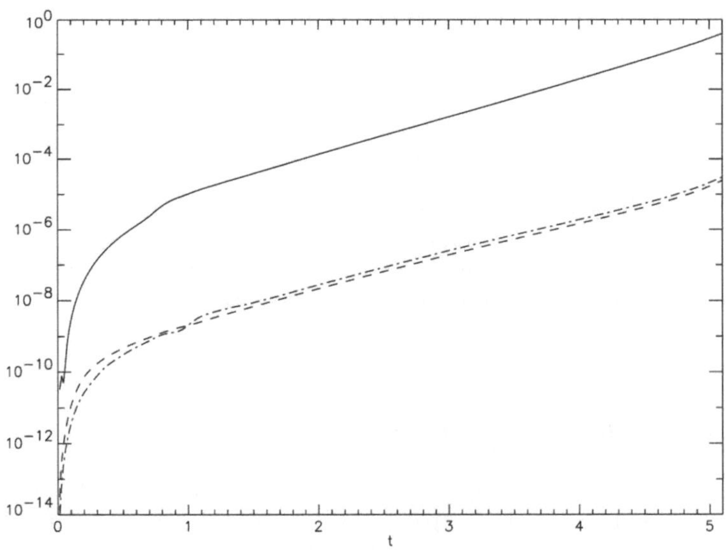

Fig. 9. The values of h_{xx} (solid line) and the constraints $\nabla_x h_{xx}$ (dot–dashed) and $\nabla_x \Omega = \Omega_x$ (dashed) are plotted versus coordinate time for the static gauge of (37)

where $\varrho^2 = x^2 + y^2$. With $Q = \ln(1 + A\bar{\Omega}^2 \varrho^2 f(\varrho^2))/2$, in Cartesian coordinates the conformal three-metric becomes

$$h_{\mathrm{B}} = \omega^2 \begin{pmatrix} 1 + A\,x^2\,\bar{\Omega}^2\,f & A\,xy\,\bar{\Omega}^2\,f & 0 \\ A\,xy\,\bar{\Omega}^2\,f & A\,y^2\,\bar{\Omega}^2\,f & 0 \\ 0 & 0 & 1 + A\,\bar{\Omega}^2\,f \end{pmatrix} .$$

The axial symmetry makes it easier to analyze the data and choose the gauges. Here we set $\omega = f = 1$ and $A = 1$.

Figure 10 shows the real part of the physical curvature invariant $\tilde{I} = \Omega^6 I$ and the mass loss $\dot{M}_{\rm B}$. The curvature invariant $\tilde{I}$ is computed both as a perturbation of the Einstein universe and case (29) (triangles) for a "Brill wave" with $A = 1$, to demonstrate that the physical initial data are indeed identical. The mass loss $\dot{M}_{\rm B}$ is computed as a perturbation of the Einstein static case ($R_g = 6$) and plotted in a logarithmic scale. Note that the falloff levels off at late times to a constant value due to numerical error. Note also that oscillations, as we show here, are absent from the initial data corresponding to (44) as shown in Fig. 5 of [46].

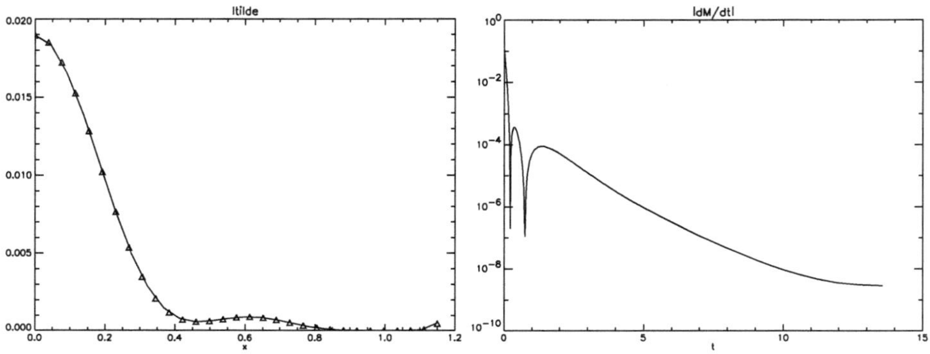

Fig. 10. The left image shows the real part of the physical curvature invariant $\tilde{I} = \Omega^6 I$, computed as a perturbation of the Einstein universe (line) and case (29) (triangles) for a "Brill wave" with $A = 1$. The right image shows the corresponding mass loss function $\dot{M}_{\rm B}$, computed as as a perturbation of the Einstein static case ($R_g = 6$)

6 Conclusions and Outlook

Bringing the conformal approach to numerical relativity to full fruition such that it can be used as a tool to explore new physics – in particular in black hole spacetimes – will be a long term effort. In order to contemplate the scope of this project, let us give a drastically oversimplified definition of the art of numerical relativity as a procedural recipe:

1. Find a well posed formulation of the initial(-boundary) value and initial data (constraint) problems for general relativity (optimally, well-posedness should be a theorem but good numerical evidence may be considered sufficient).
2. Without destroying well-posedness, modify your equations and choose your gauges, such that your problem actually becomes well-conditioned[5].
3. Construct a solid numerical implementation, flexible enough to handle experiments as required by science and by finding solutions to the problems associated with point two.

[5] Ill conditioned problems are those where a result depends very strongly on input, i.e. on initial data, see e.g. Sect. 1.6 and 6.1 of [52].

4. Discover (new) results in physics.
5. Explain what you achieved (and how) to fellow numerical relativists and others, such as mathematical relativists, astrophysicists, cosmologists, or mathematicians.

Even without considering the last point (which the present article humbly tries to serve), numerical relativity is a challenging enterprise.

The conformal approach complies with point one in the optimal sense: the equations are regular in the whole spacetime, including the asymptotic region, there are no ambiguities associated with ad-hoc cutoffs at finite distance and the evolution equations are symmetric hyperbolic, which guarantees well-posedness of the initial value problem and allows well-posed initial-value-boundary problems.

Point two, however, already poses a significant challenge: well-defined is not well-conditioned, well-posed problems may still be hopelessly ill-conditioned for numerical simulation. A simple example is provided by any chaotic dynamical system (in the sense of ordinary differential equations). When it comes to solving the Einstein equations, the gauge freedom of the theory results in having more equations (constraints and evolution equations) than variables and more variables than physical degrees of freedom. This redundancy can easily lead to spurious approximate solutions. Different ways to write the equations are only equivalent with regard to exact solutions, but approximations will tend to exhibit constraint violating or gauge modes that may grow very fast (e.g. exponentially). This is perfectly consistent with well-posedness but not acceptable numerically. Even without triggering instabilities, the choice of a bad gauge is likely to create features in the solution which are in practice impossible to resolve. The "good news" is that many of the problems encountered with the conformal field equations have counterparts in traditional approaches to numerical relativity. The way toward solving these problems usually takes the form of gaining insight from simplifications and analytical studies, which then have to be tested in full numerical simulations. This requires a flexible code that is geared toward performing the necessary experiments, which leads to point three – another hard task for classical relativists, because it requires an engineering attitude many relativists are not familiar with. The gauge freedom of general relativity and absence of a natural background creates an additional twist when it comes to point four, which leads to numerous technical and conceptual subtleties.

What is the roadmap for the future? In order to comply with points two and three of the above recipe, preliminary work is carried out toward a new 3D code that will be flexible enough to carry out a range of numerical experiments in order to come up with well-conditioned algorithms for the conformal field equations. One major issue in the improvement of algorithms is to implement a better boundary condition which does not require a transition zone, allows the boundary to be closer to $\mathscr{I}^+$ and minimize constraint violations generated at the boundary or outside $\mathscr{I}$. Here an essential problem is that $\mathscr{I}^+$ has spherical cuts and algorithms based on Cartesian grids are probably not optimal. Certainly, a lot of energy will have to be devoted to the question of finding appropriate gauge conditions. Particularly hard seems to be the question of how to choose the Ricci scalar R of the unphysical spacetime. Since R steers the conformal factor implicitly through nonlinear PDEs, it seems very hard to influence the conformal factor in any desired way.

An important role in improving the analytical understanding and in setting up numerical experiments will be played by the utilization of simplifications. Particularly important are spacetime symmetries and perturbative studies. Minkowski and Kruskal spacetimes provide particularly important cases to be studied in this context. An alternative route to simplification, which has been very successful in numerical relativity, is perturbative analysis, e.g. with Minkowski or Schwarzschild backgrounds. In the context of compactification this has been carried out numerically with characteristic codes in [53,54] (using appropriate variables in the Teukolsky equations, the perturbation equations are made regular at $\mathscr{I}^+$). Some of the problems that showed up there are likely to be relevant also for the conformal approach.

The theory of general relativity is known as a never drying out source for subtle questions in physics and mathematics. Numerical relativity is hoped to help answer some important questions – but at the same time poses many new ones. Without a thorough understanding of how to obtain approximate solutions, our insight into the theory seems incomplete. For isolated systems, the mastering of compactification techniques promises reliability and precision. The next years are expected to see some significant progress in this direction.

Acknowledgments

The author thanks R. Beig, J. Frauendiener, H. Friedrich, B. Schmidt, and M. Weaver for helpful discussions, C. Lechner and J. Valiente Kroon for a careful reading of the manuscript, and P. Hübner and M. Weaver for letting him use their codes, explaining their results and providing general support in order to take over this project.

References

1. M. Choptuik: Phys. Rev. Lett. **70**, 9 (1993)
2. B. Kleihaus, J. Kunz: Phys. Rev. Lett. **78**, 2527 (1997), Phys. Rev. Lett. **79** 1595 (1997)
3. A. Rendall, M. Weaver: Class. Quantum Grav. **18**, 2959 (2001)
4. R. Penrose: Phys. Rev. Lett. **10**, 66 (1963)
5. H. Friedrich, Proc. R. Soc. **A375**, 169 (1981)
6. H. Friedrich, Proc. R. Soc. **A378**, 401 (1981)
7. H. Friedrich, Commun. Math. Phys. **91**, 445 (1993)
8. H. Friedrich, Journal of Geometry and Physics **24**, 83 (1998)
9. H. Friedrich: gr-qc/0201006
10. P. Hübner, Black Class. Quantum Grav. **16**, 2145 (1999)
11. P. Hübner, Equation Class. Quantum Grav. **16**, 2823 (1999)
12. P. Hübner, Class. Quantum Grav. **18**, 1421 (2001)

13. P. Hübner, Class. Quantum Grav. **18**, 1871 (2001)
14. E. T. Newman, K. P. Tod: 'Asymptotically Flat Space-Times'. In: *General Relativity and Gravitation.* ed. by A. Held (Plenum Press, New York, 1980) pp. 1–36.
15. R. Wald: *General Relativity* (University of Chicago Press, Chicago, 1984)
16. J. Stewart: *Advanced General Relativity* (Cambridge University Press, Cambridge, 1991)
17. J. Frauendiener: Living Rev. Rel. **4**, 1 (2000)
18. R. Penrose: Proc. R. Soc. **A284** 159 (1965)
19. R. Arnowitt, S. Deser, C. W. Misner: 'The dynamics of general relativity'. In: *Gravitation: An Introduction to Current Research.* ed. by L. Witten (Wiley, New York, 1962) pp. 227–265
20. H. Bondi, M. van der Burg, A. Metzner: Proc. R. Soc. **A269**, 21 (1962)
21. R. Schoen, S. T. Yau: Commun. Math. Phys. **65**, 54 (1979)
22. G. Horowitz, M. Perry: Phys. Rev. Lett. **48**, 371 (1982)
23. H. Friedrich, 'Einstein's Equation and Geometric Asymptotics'. In: *Proceedings of the GR-15 conference.* ed. by N. Dadhich, J. Narlikar (IUCAA, Pune, 1998)
24. J. Frauendiener: Class. Quantum Grav. **17**, 373 (2000)
25. P. Hübner, M. Weaver: unpublished
26. J. Frauendiener: Phys. Rev. **D58**, 064003 (1998)
27. J. Frauendiener: Phys. Rev. **D58**, 064002 (1998)
28. A. Butscher: In: *Proceedings of the International Workshop on the Conformal Structure of Spacetime.* ed. by J. Frauendiener, H. Friedrich (Springer Verlag, Lecture Notes of Physics series, to be published)
29. L. Andersson, P. T. Chruściel, H. Friedrich: *Comm. Math. Phys.* **149**, 587 (1992)
30. J. Frauendiener: J. Comput. Appl. Math. **109**, 475 (1999)
31. V. Moncrief: In: *Workshop on Mathematical Issues in Numerical Relativity* http://online.itp.ucsb.edu/online/numrel00/moncrief/ (ITP, Santa Barbara January 10 – 28th, 2000)
32. P. Hübner: Numerische und analytische Untersuchungen von (singulären) asymptotisch flachen Raumzeiten mit konformen Techniken. PhD thesis, Ludwig-Maximilians-Universität, München, (1993)
33. P. Hübner, Phys. Rev. **D53**, 701 (1996)
34. M. Pürrer. Diploma thesis, University of Vienna, Vienna, in preparation
35. B. Schmidt: Class. Quantum Grav. **13** 2811 (1996)
36. T. Foertsch: Spacetimes Admitting Null Infinities with Toroidal Sections. Diploma thesis, Technical University of Berlin, Berlin (1997)
37. J. Ehlers, W. Kundt, 'Exact solutions of the gravitational field equations' In: *Gravitation: an introduction to current research.* ed. by L. Witten (J. Wiley, New York, 1962)
38. P. Hübner, Class. Quantum Grav. **15**, L21 (1998)
39. R. Penrose, W. Rindler: *Spinors and Spacetime*, volume 2 (Cambridge University Press, Cambridge, 1986)
40. L. N. Trefethen: SIAM Review, **24**, 113 (1982)
41. M. Hein: Numerische Simulation axialsymmetrischer, isolierter Systeme in der Allgemeinen Relativitätstheorie. Diploma thesis, University of Tübingen, Tübingen (2002)
42. M. Alcubierre, S. Brandt, B. Bruegmann, D. Holz, E. Seidel, R. Takahashi, J. Thornburg: Int. J. Mod. Phys. **D10**, 273 (2001)
43. B. Gustafsson, H.-O. Kreiss, J. Oliger: *Time Dependent Problems and Difference Methods*, (Wiley, New York, 1995)

44. M. Weaver: private communication
45. H. Friedrich: Commun. Math. Phys. **107**, 587 (1986)
46. S. Husa: *Proceedings of the International Workshop on the Conformal Structure of Spacetime.* ed. by J. Frauendiener, H. Friedrich (Springer Verlag, Lecture Notes of Physics series, to be published)
47. M. Alcubierre, B. Brügmann: Phys. Rev. **D63**, 104006 (2001)
48. B. Schmidt: 'Numerical Evolution of the Kruskal Spacetime Using the Conformal Field Equations'. In: *Proceedings of the 20th Texas Symposium on Relativistic Astrophysics.* ed. by J. Craig Wheeler, H. Martel (American Institute of Physics, 2001), pp. 729–733
49. F. Siebel: Simultanes Numerisches Lösen von Zeitentwicklungsgleichungen und Zwangsbedingungen der Einsteinschen Feldgleichungen. Diploma thesis, LMU, Munich (1999)
50. P. Hübner, F. Siebel: Phys. Rev. **D64**, 024021 (2001)
51. O. Brodbeck, S. Frittelli, P. Hübner, O. Reula: J. Math. Phys. **40**, 909 (1999)
52. K. E. Atkinson: *An introduction to numerical analysis*, 2nd edn. (Wiley, New York, 1989)
53. M. Campanelli, R. Gomez, S. Husa, J. Winicour, Y. Zlochower: Phys. Rev. **D63**, 124013 (2001)
54. S. Husa, Y. Zlochower, R. Gomez, J. Winicour: Phys. Rev. **D** (to be published)

Part III

Gravitational Waves: Production and Detection

Binary Black Holes and Gravitational Wave Production: Post-Newtonian Analytic Treatment

Gerhard Schäfer

Friedrich-Schiller-Universität Jena, Theoretisch-Physikalisches Institut,
Max-Wien-Platz 1, D-07743 Jena, Germany

Abstract. Within a post-Newtonian approximation scheme the Einstein field equations are solved for the dynamics of the decaying orbits of binary black holes and the related gravitational wave emission. The black holes are modelled by two Dirac delta functions in conformally related d-dimensional euclidean space. The limit from the Einstein field equations in (d+1)-dimensional spacetime is applied to achieve well-defined mathematical expressions in 4-dimensional spacetime. The conservative orbital dynamics is presented up to the third post-Newtonian order of approximation and the decaying orbital phase up to the four-and-a-half post-Newtonian order. The gravitational waveform is given to second post-Newtonian order.

1 Introduction

In General Relativity (GR), the objects with the least (internal) degrees of freedom are black holes. Correspondingly, the simplest two-body problem is the one containing two black holes. As simple as the problem seems to be, nonstationarity because of gravitational wave emission makes the full-analytic solution of the problem unsolvably complicated. Even using most advanced supercomputer techniques in numerical relativity, the problem is still far from being solved. It is thus of great importance to construct approximate analytic solutions which allow the description of the motion of binary black holes with high accuracy. Here the problem immediately arises that black holes are strongly self-gravitating objects whereas their relative orbital dynamics can be as weak as one likes, as weak as to fulfil the Newtonian dynamics very well. An approximation scheme has therefore to be applied which at the same time covers as well the weak-field motion as the strong-field self-energy regimes.

Knowing that the Brill-Lindquist binary black-hole initial value solution [1] can be deduced from two Dirac delta sources living in conformally related euclidean 3-space [2], we adopt that approach to model our binary black-hole system. At the same time we generalize the 3-dimensional space to d-dimensional space. In this way we formally obtain weakly self-gravitating black holes. This procedure delivers a mathematically consistent framework regarding the distributional calculus, at least as far as explicit calculations have shown [3]. Furthermore, in going over to (d+1)-dimensional spacetime, neither basic principles of the theory like the principle of coordinate covariance are destroyed nor specific symmetry properties like the isometry property of the analytically continuated Schwarzschild 3-metric. In the limit where the dimension d approaches 3, we get a well-defined binary black-hole structure in 4-dimensional physical spacetime.

Besides the Brill-Lindquist initial value solution there exists the Misner-Lindquist one [4], [5]. The latter solution is defined by the isometry property of the analytically continuated 3-metric. The solution can be represented by two sets of infinite many Dirac delta functions in conformally related flat space [2]. In the electrostatics, the Brill-Lindquist and Misner-Lindquist solutions correspond to the electric fields of two point-charges and two conducting charged spheres, respectively. The imaging method introduces an infinite number of point-charges in the virtual insides of both spheres. As seen from a technical point of view, the binary black-hole configuration we are going to work out in this paper is the simpler one. Finally we point out that the Brill-Lindquist and Misner-Lindquist black holes are *eternal* black holes, the configurations of which are shown up in the non-radiative parts of the Hamiltonians.

2 Binary Black-Hole Model

In this paper we will treat the Einstein field equations using the ADM (d+1)-decomposition of spacetime in which spacetime is foliated into d-dimensional spacelike hypersurfaces depending on time, t, parametrically [6], [7]. On the hypersurfaces there are living the four constraint equations

$$g^{1/2}\mathrm{R} = \frac{1}{g^{1/2}}\left(\pi^i_j\pi^j_i - \frac{1}{d-1}\pi^2\right) + \frac{16\pi G}{c^3}\sum_a \left(m_a^2c^2 + g^{ij}p_{ai}p_{aj}\right)^{1/2}\delta_a \quad (1)$$

and

$$-2\partial_j\pi^j_i + \pi^{kl}\partial_i g_{kl} = \frac{16\pi G}{c^3}\sum_a p_{ai}\delta_a \ , \quad (2)$$

where g_{ij} denotes the (symmetric) metric of a d-dimensional hypersurface and R its curvature scalar. g, g^{ij}, $\pi^{ij}c^3/16\pi G$ are, respectively, the determinant, the inverse metric, and the canonical conjugate of the d-dimensional metric, whereby the conjugate is a density of weight one; $\pi = \pi^i_i$, where $\pi^i_j = g_{jk}\pi^{ik}$. The linear momentum of point-mass a (a=1,2) is denoted by p_{ai} and its bare mass by m_a. c is the speed of light and G the Newtonian gravitational constant generalized to d-dimensional space. $\partial_i = \partial/\partial x^i$ is the partial derivative with respect to the hypersurface coordinates x^i ($i = 1, 2, ..., d$). The Dirac delta function $\delta_a = \delta(x^i - x^i_a(t))$ is defined by $\delta(x^i - x^i_a(t)) = 0$ if $x^i \neq x^i_a(t)$ and $\int \mathrm{d}^d x\ \delta(x^i - x^i_a(t)) = 1$, where $x^i_a(t)$ is the space coordinate of point-mass a at time t.

The ADM coordinate conditions which generalize the isotropic Schwarzschild metric to arbitrary $(d+1)$-dimensional spacetimes read, e.g. see [3],

$$g_{ij} = \left(1 + \frac{d-2}{4(d-1)}\phi\right)^{4/(d-2)}\delta_{ij} + h^{\mathrm{TT}}_{ij} \quad (3)$$

and

$$\pi^{ii} = 0 \ , \quad (4)$$

where h_{ij}^{TT} denotes the tranverse-traceless (TT) part of the metric g_{ij} with respect to the euclidean d-metric δ_{ij}. The latter metric is the conformal metric mentioned in the introduction; it is directly related to the isotropic part of the metric g_{ij} under the adapted coordinate conditions. Notice the interesting property that for $d = 2$, $g_{ij} = \mathrm{e}^{\phi}\delta_{ij} + h_{ij}^{\mathrm{TT}}$ holds, as well as $g^{1/2}\mathrm{R} = -\Delta\phi$ if additionally $h_{ij}^{\mathrm{TT}} = 0$ is valid (see (17) below).

Taking into account the gauge condition for π^{ij}, the following decomposition can be applied,

$$\pi^{ij} = \tilde{\pi}^{ij} + \pi^{ij}_{\mathrm{TT}} \,, \tag{5}$$

where

$$\tilde{\pi}^{ij} = \partial_i \pi^j + \partial_j \pi^i - \frac{2}{d}\delta_{ij}\partial_k \pi^k \tag{6}$$

and where $\pi^{ij}_{\mathrm{TT}}c^3/16\pi G$ denotes the canonical conjugate to h_{ij}^{TT}.

The Hamilton functional of the binary black hole system is given by

$$H\left[x_a^i, p_{ai}, h_{ij}^{\mathrm{TT}}, \pi^{ij}_{\mathrm{TT}}\right] = -\frac{c^4}{16\pi G}\int \mathrm{d}^d x \; \Delta\phi\left[x_a^i, p_{ai}, h_{ij}^{\mathrm{TT}}, \pi^{ij}_{\mathrm{TT}}\right] \,, \tag{7}$$

where Δ is the Laplacian in d-dimensional flat space. For the derivation of the dynamics of the binary black-hole system it has turned out to be most advantageous to introduce a Routh functional of the form [8],

$$R\left[x_a^i, p_{ai}, h_{ij}^{\mathrm{TT}}, \partial_t h_{ij}^{\mathrm{TT}}\right] = H - \frac{c^3}{16\pi G}\int \mathrm{d}^d x \; \pi^{ij}_{\mathrm{TT}}\partial_t h_{ij}^{\mathrm{TT}} \,. \tag{8}$$

This functional is a Hamiltonian for the particle (point-mass) degrees of freedom, and a Lagrangian for the independent gravitational field degrees of freedom. The evolution equation for the latter reads

$$\frac{\delta \int R(t')\mathrm{d}t'}{\delta h_{ij}^{\mathrm{TT}}(x^k, t)} = 0 \,, \tag{9}$$

where the δ-symbol denotes the standard functional (Fréchet) derivative. The solution of this equation will be constructed iteratively, through iterative inversion of the d'Alembertian in (d+1)-dimensional flat spacetime with no-incoming radiation condition applied at each step [see e.g. [9] for the retarded Green's function in (d+1)-dimensional spacetime]. Although the calculations are done in a fictitious spacetime with fictitious past lightlike infinity, the propagation property in the true metric may be realized at each step checking the physically to be imposed no-incoming boundary condition with respect to true lightlike past infinity, cf. (27) below.

The Hamilton equations of motion for the two black holes take the form

$$\dot{p}_{ai} = -\frac{\partial R}{\partial x_a^i} \,, \qquad \dot{x}_a^i = \frac{\partial R}{\partial p_{ai}} \,. \tag{10}$$

The *conservative* part R_c of the binary dynamics is given by

$$R_c(t) = \frac{1}{2}[R_{on}(t) + R_{on}(-t)] , \tag{11}$$

where

$$R_{on}(t) = R\left[x_a^i, p_{ai}, h_{ij}^{TT}[x_a^k, p_{ak}], \partial_t h_{ij}^{TT}[x_a^k, p_{ak}]\right] \tag{12}$$

denotes the *on-field-shell* Routh functional. It is easily shown that

$$\int R_c(t)\mathrm{d}t = \int R_{on}(t)\mathrm{d}t$$

holds. Obviously, the non-time-symmetric part of an on-field-shell Routh functional is a total time derivative which does not contribute to equations of motion for matter. The conservative equations of motion read

$$\dot{p}_{ai}(t) = -\frac{\delta \int R_c(t')\mathrm{d}t'}{\delta x_a^i(t)} , \qquad \dot{x}_a^i(t) = \frac{\delta \int R_c(t')\mathrm{d}t'}{\delta p_{ai}(t)} , \tag{13}$$

where the functional derivatives read

$$\frac{\delta \int R_c(t')\mathrm{d}t'}{\delta z(t)} = \frac{\partial R_c}{\partial z(t)} - \frac{\mathrm{d}}{\mathrm{d}t}\frac{\partial R_c}{\partial \dot{z}(t)} + \dots , \tag{14}$$

with $z = x_a^i$ or $z = p_{ai}$.

By the aid of a higher order contact transformation new phase-space coordinates $\bar{x}_a^i, \bar{p}_{ai}$ can be introduced in such a way that an *ordinary* Hamiltonian $H_{co}(\bar{x}_a^i, \bar{p}_{ai})$ results [10], also see [11]. This Hamiltonian can be obtained by simply putting

$$H_{co}(x,p) = R_c[x, p, \dot{x}(x,p), \dot{p}(x,p), \dots] \tag{15}$$

and by recognizing that a phase-space coordinate transformation has occurred implicitly. There is no need to know the new phase-space coordinates in terms of the former ones as long as coordinate (gauge) invariant quantities are considered like in Sects. 4.1, 5.1, and 5.2 below.

The ordinary Hamilton equations read

$$\dot{\bar{p}}_{ai} = -\frac{\partial H_{co}}{\partial \bar{x}_a^i} , \qquad \dot{\bar{x}}_a^i = \frac{\partial H_{co}}{\partial \bar{p}_{ai}} . \tag{16}$$

The Hamilton function of Sect. 4 below is of H_{co}-type.

2.1 Brill-Lindquist Initial Value Solution

To explicitly prove that the introduced source model produces binary black-hole spacetimes we will derive the Brill-Lindquist initial value solution for uncharged black holes. Putting $h_{ij}^{\mathrm{TT}} = 0$ and $p_{ai} = 0$, and hereof $\pi^{ij} = 0$, the constraint equation (1) takes the simple form

$$-\left(1 + \frac{d-2}{4(d-1)}\,\phi\right)\Delta\phi = \frac{16\pi G}{c^2}\sum_a m_a\delta_a\,. \tag{17}$$

The exact solution of this equation, in the limit of 3-dimensional space, uniquely results in, cf. [2], [3],

$$\phi = \frac{4G}{c^2}\left(\frac{\alpha_1}{r_1} + \frac{\alpha_2}{r_2}\right), \tag{18}$$

where r_a denotes the euclidean distance between x^i and x_a^i. The solution (18) has been achieved by the aid of the inverse Laplacian in d-dimensional space [9]

$$-\Delta^{-1}\delta_a = \frac{\Gamma((d-2)/2)}{4\pi^{d/2}}r_a^{2-d} \tag{19}$$

and the related ansatz

$$\phi = \frac{2G}{c^2}\frac{d-1}{d-2}\left(\frac{\alpha_1}{r_1^{d-2}} + \frac{\alpha_2}{r_2^{d-2}}\right). \tag{20}$$

The insertion of this ansatz into (17), for $1 < d < 2$, yields a well-defined finite solution using the property $r_a^{2-d}\delta_a = 0$. The analytic continuation of the obtained solution to 3-dimensional space results in the final solution (18). The coefficients α_a, for $d = 3$, read [12]

$$\alpha_a = \frac{m_a - m_b}{2} + \frac{c^2 r_{ab}}{G}\left(\sqrt{1 + \frac{m_a + m_b}{c^2 r_{ab}/G} + \left(\frac{m_a - m_b}{2c^2 r_{ab}/G}\right)^2} - 1\right), \tag{21}$$

where r_{ab} denotes the euclidean distance between x_a^i and x_b^i.

The Brill-Lindquist Hamilton function results in, see (7) and (17),

$$H_{\mathrm{BL}} = (\alpha_1 + \alpha_2)\,c^2 = (m_1 + m_2)\,c^2 - G\,\frac{\alpha_1\alpha_2}{r_{12}}\,. \tag{22}$$

It describes the total (initial) energy between two uncharged Brill-Lindquist black holes. Notice that the Hamiltonian is well-behaved in the limit $r_{12} \to 0$. Therefore, full GR shows up no infinite self-energies. The dimensional regularization approach from fictitious d-dimensional space parallels the finiteness of black-hole self-energies in physical space in the sense that it also produces finite, in latter case zero, self-energies (the naked masses in [2] are identical with the bare masses when dimensional regularization is applied).

3 Post-Newtonian Expansion

The expansion of the Brill-Lindquist Hamilton function in powers of $1/c^2$, i.e.

$$H_{\mathrm{BL}} - mc^2 = \sum_{n=0}^{\infty}\left(\frac{1}{c^2}\right)^n H_n(r_{12})\,, \tag{23}$$

is well-defined if $Gm/r_{12} < c^2$ is valid with $m = m_1 + m_2$. H_0 is the Newtonian energy and H_n is called the nth-Post-Newtonian (nPN) one. The generalization of such a post-Newtonian expansion to arbitrary Hamiltonians, respectively Routh functionals, reads

$$R\left[x_a^i, p_{ai}, h_{ij}^{\mathrm{TT}}, \partial_t h_{ij}^{\mathrm{TT}}\right] - mc^2 = \sum_{n=0}^{\infty} \left(\frac{1}{c^2}\right)^n R_n\left[x_a^i, p_{ai}, \hat{h}_{ij}^{\mathrm{TT}}, \partial_t \hat{h}_{ij}^{\mathrm{TT}}\right], \quad (24)$$

where $G\hat{h}_{ij}^{\mathrm{TT}} = c^4 h_{ij}^{\mathrm{TT}}$, with G being introduced for convenience. The expansion (24) makes sense only if the occurring velocities and gravitational potentials are small compared to the speed of light, respectively the squared of it. In the expansion, $\hat{h}_{ij}^{\mathrm{TT}}$ has to be treated as independent from $1/c$. In this way the expansion guarantees that the gravitational field generated by the Newtonian black-hole orbital dynamics starts with the correct power in $1/c^2$, namely $(1/c^2)^2$, see (26). The insertion of the expansion (24) into (9) results in an inhomogeneous wave equation of the form (also see [13])

$$\left(\Delta - \frac{\partial_t^2}{c^2}\right) h^{\mathrm{TT}} = \frac{G}{c^4} \sum_{n=0}^{\infty} \left(\frac{1}{c^2}\right)^n D_n^{\mathrm{TT}}[x, x_a(t), p_a(t), \hat{h}^{\mathrm{TT}}(t), \partial_t \hat{h}^{\mathrm{TT}}(t)] \,. \quad (25)$$

This equation is solved order by order in the way explained above where the ordering is defined by the orders of the source terms D_n^{TT}. As the speed of light appears in the d'Alembertian too, the solution will not have the same PN-structure as the source part of the wave equation has. This poses however no problems as a quantitative ordering is still achievable, cf. (26) below.

3.1 The Gravitational Radiation Field

The gravitational radiation field, in the asymptotic rest frame of the source, can be put in the wave zone (w.z.) into the form (after repeated use of the (10)), given here for 4-dimensional spacetime only [14],

$$h_{ij}^{\mathrm{TT}}(\mathbf{x}, t) = \frac{G}{c^4} \frac{P_{ijkm}(\mathbf{n})}{r} \sum_{l=2}^{\infty} \left\{ \left(\frac{1}{c^2}\right)^{(l-2)/2} \frac{4}{l!} \, \mathrm{M}_{kmi_3\ldots i_l}^{(l)} \left(t - \frac{r_*}{c}\right) N_{i_3\ldots i_l} \right.$$
$$\left. + \left(\frac{1}{c^2}\right)^{(l-1)/2} \frac{8l}{(l+1)!} \, \varepsilon_{pq(k} \, S_{m)pi_3\ldots i_l}^{(l)} \left(t - \frac{r_*}{c}\right) n_q \, N_{i_3\ldots i_l} \right\}, \quad (26)$$

where M_{A_l} and S_{A_l} are symmetric and tracefree (STF) radiative mass and current multipole moments which parametrize the radiation field in a cartesian basis. A_l denotes a multi-index of length l, i.e. $A_l = A_{i_1 i_2 \ldots i_l}$, where the is run over 1,2,3. $N_{A_l} = n_{i_1} n_{i_2} \ldots n_{i_l}$, where n_i is the unit normal in the direction of the radial vector $\mathbf{x} = r\mathbf{n}$, pointing from the source to the observer. The parentheses around indices mean to take the symmetric part of the corresponding tensor. $\mathrm{M}^{(l)} = \mathrm{d}^l\mathrm{M}/\mathrm{d}t^l$ and $\mathrm{S}^{(l)} = \mathrm{d}^l\mathrm{S}/\mathrm{d}t^l$, and $P_{ijkl}(\mathbf{n})$ denotes the transverse-traceless projection orthogonal to $\mathbf{x}$ acting on symmetric tensors:

$P_{ijkl}(\mathbf{n}) = (\delta_{ik} - n_i n_k)(\delta_{jl} - n_j n_l) - (\delta_{ij} - n_i n_j)(\delta_{kl} - n_k n_l)/2$. ε_{ijk} is the usual antisymmetric tensor of Levi-Civita. r_* is the radial coordinate such that $t - r_* = \text{const.}$ describes the physical light propagation. The multipole moments are given by space integrals over local expressions in time as well as by time integrals of those expressions. The latter non-local-in-time expressions are called *tail* terms. The mass-quadrupole moment e.g., to 1.5PN order, is given by [15], [16],

$$\mathrm{M}_{ij}\left(t - \frac{r_*}{c}\right) = \widehat{\mathrm{M}}_{ij}\left(t - \frac{r_*}{c}\right)$$

$$+\frac{2Gm}{c^3}\int_0^\infty \mathrm{d}v \, \ln\left(\frac{v}{2b}\right) \widehat{\mathrm{M}}^{(2)}_{ij}\left(t - \frac{r_*}{c} - v\right) + O(1/c^4)\,, \tag{27a}$$

$$r_* = r + \frac{2Gm}{c^2}\ln\left(\frac{r}{cb}\right) + O(1/c^3)\,, \tag{27b}$$

where $\widehat{\mathrm{M}}_{ij}$ is a local-in-time expression which has to be calculated to $1/c^2$ accuracy. The parameter b is purely of gauge type. It drops out from all observations. As seen from (26), the lowest gravitational multipole moments are the quadrupolar ones ($l = 2$).

The gravitational energy flux is given by [14],

$$\mathcal{L}(t) = \frac{c^3}{32\pi G}\oint_{w.z.} (\partial_t h^{\mathrm{TT}}_{ij})^2 r^2 \mathrm{d}\Omega\,. \tag{28}$$

Its decomposition in a power series in terms of $1/c^2$ may be written as

$$\mathcal{L} = \frac{G}{5c^5}\sum_{n=0}^{\infty}\left(\frac{1}{c^2}\right)^n \hat{\mathcal{L}}_n\,. \tag{29}$$

The term $G\hat{\mathcal{L}}_n/5c^{5+2n}$ is called the energy flux at the nPN order. Correspondingly, the wave field which contributes to this power is called the wave field of the nPN order. Assuming the well-established balance between energy flux and (binding) energy loss of the source system [13], denoted by $d\mathcal{E}/dt$, the following equation holds

$$-\left\langle \frac{d\mathcal{E}(t)}{dt} \right\rangle = \langle \mathcal{L}(t) \rangle\,, \tag{30}$$

where the brackets mean time-averaging over the longest orbital period. The reason for this averaging procedure is the general radiation-emission property that the local-in-time energy flux and the local-in-time energy loss are not necessarily identical at a specific instant of time because of field-energy that is floating forth and back in the induction zone.

In the following we often will denote post-Newtonian orders of wave expressions by nPN(W) and those of particle-motion expressions by nPN(M). Using this convention, the nPN(W) energy flux then corresponds to the (n+5/2)PN(M) energy loss. The 2PN(W) energy flux takes the form

$$\mathcal{L} = \frac{G}{5c^5} \left\{ \mathrm{M}_{ij}^{(3)} \mathrm{M}_{ij}^{(3)} + \frac{1}{c^2} \left[\frac{5}{189} \mathrm{M}_{ijk}^{(4)} \mathrm{M}_{ijk}^{(4)} + \frac{16}{9} \mathrm{S}_{ij}^{(3)} \mathrm{S}_{ij}^{(3)} \right] \right.$$

$$\left. + \frac{1}{c^4} \left[\frac{5}{9072} \mathrm{M}_{ijkm}^{(5)} \mathrm{M}_{ijkm}^{(5)} + \frac{5}{84} \mathrm{S}_{ijk}^{(4)} \mathrm{S}_{ijk}^{(4)} \right] \right\} . \tag{31}$$

4 Conservative Hamiltonian to 3PN Order

The unique conservative 3PN(M) Hamiltonian of our binary black-hole model in 4-dimensional spacetime has been derived only recently using dimensional regularization [3]. From all the applied methods only this one has turned out to be fully satisfactory. In the following we will only give the result, in the center-of-mass frame and in reduced variables, for details see [17] and references therein; for the precise identification of the Brill-Lindquist case, see [18]. By the aid of the following definitions

$$\widehat{H} = \frac{H_{co}}{\mu} \tag{32a}$$

$$\mathbf{p} = \frac{\bar{\mathbf{p}}_1}{\mu} = -\frac{\bar{\mathbf{p}}_2}{\mu} , \quad \mathbf{q} = \frac{1}{Gm}(\bar{\mathbf{x}}_1 - \bar{\mathbf{x}}_2) , \quad \mathbf{q} = q\bar{\mathbf{n}} , \ \bar{\mathbf{n}} \cdot \bar{\mathbf{n}} = 1 \tag{32b}$$

$$\mu = \frac{m_1 m_2}{m} , \quad \nu = \frac{\mu}{m} , \quad 0 \le \nu \le \frac{1}{4} \tag{32c}$$

one gets the *reduced* Hamiltonian in the form,

$$\widehat{H}(\mathbf{q},\mathbf{p}) = \widehat{H}_{\mathrm{N}}(\mathbf{q},\mathbf{p}) + \frac{1}{c^2}\widehat{H}_{1\mathrm{PN}}(\mathbf{q},\mathbf{p}) + \frac{1}{c^4}\widehat{H}_{2\mathrm{PN}}(\mathbf{q},\mathbf{p}) + \frac{1}{c^6}\widehat{H}_{3\mathrm{PN}}(\mathbf{q},\mathbf{p}) , \tag{33}$$

where

$$\widehat{H}_{\mathrm{N}}(\mathbf{q},\mathbf{p}) = \frac{\mathbf{p}^2}{2} - \frac{1}{q} , \tag{34a}$$

$$\widehat{H}_{1\mathrm{PN}}(\mathbf{q},\mathbf{p}) = \frac{1}{8}(3\nu - 1)(\mathbf{p}^2)^2 - \frac{1}{2}\left[(3+\nu)\mathbf{p}^2 + \nu(\bar{\mathbf{n}}\cdot\mathbf{p})^2\right]\frac{1}{q} + \frac{1}{2q^2} , \tag{34b}$$

$$\begin{aligned}\widehat{H}_{2\mathrm{PN}}(\mathbf{q},\mathbf{p}) = {} & \frac{1}{16}\left(1 - 5\nu + 5\nu^2\right)(\mathbf{p}^2)^3 \\ & + \frac{1}{8}[(5 - 20\nu - 3\nu^2(\mathbf{p}^2)^2 - 2\nu^2(\bar{\mathbf{n}}\cdot\mathbf{p})^2\mathbf{p}^2 - 3\nu^2(\bar{\mathbf{n}}\cdot\mathbf{p})^4]\frac{1}{q} \\ & + \frac{1}{2}[(5+8\nu)\mathbf{p}^2 + 3\nu(\bar{\mathbf{n}}\cdot\mathbf{p})^2]\frac{1}{q^2} - \frac{1}{4}(1+3\nu)\frac{1}{q^3} ,\end{aligned} \tag{34c}$$

$$\widehat{H}_{3\mathrm{PN}}(\mathbf{q},\mathbf{p}) = \frac{1}{128}\left(-5 + 35\nu - 70\nu^2 + 35\nu^3\right)(\mathbf{p}^2)^4$$

$$
\begin{aligned}
&+ \frac{1}{16}\left[\left(-7+42\nu-53\nu^2-5\nu^3\right)(\mathbf{p}^2)^3+(2-3\nu)\nu^2(\bar{\mathbf{n}}\cdot\mathbf{p})^2(\mathbf{p}^2)^2\right.\\
&\left.+\,3(1-\nu)\nu^2(\bar{\mathbf{n}}\cdot\mathbf{p})^4\mathbf{p}^2-5\nu^3(\bar{\mathbf{n}}\cdot\mathbf{p})^6\right]\frac{1}{q}\\
&+\left[\frac{1}{16}\left(-27+136\nu+109\nu^2\right)(\mathbf{p}^2)^2+\frac{1}{16}(17+30\nu)\nu(\bar{\mathbf{n}}\cdot\mathbf{p})^2\mathbf{p}^2\right.\\
&\left.+\,\frac{1}{12}(5+43\nu)\nu(\bar{\mathbf{n}}\cdot\mathbf{p})^4\right]\frac{1}{q^2}\\
&+\left\{\left[-\frac{25}{8}+\left(\frac{1}{64}\pi^2-\frac{335}{48}\right)\nu-\frac{23}{8}\nu^2\right]\mathbf{p}^2\right.\\
&\left.+\left(-\frac{85}{16}-\frac{3}{64}\pi^2-\frac{7}{4}\nu\right)\nu(\bar{\mathbf{n}}\cdot\mathbf{p})^2\right\}\frac{1}{q^3}\\
&+\left[\frac{1}{8}+\left(\frac{109}{12}-\frac{21}{32}\pi^2\right)\nu\right]\frac{1}{q^4}\,. \qquad (34\text{d})
\end{aligned}
$$

In the next Sect., orbital observables will be derived from this Hamiltonian. The test body case is achieved by $\nu = 0$, the equal-mass case by $\nu = 1/4$.

4.1 Dynamical Invariants

The easiest way to calculate dynamical invariants is by the aid of the Hamilton-Jacobi theory [10]. The knowledge of the (reduced) radial action $i_r(E, j)$, where E is the numerical value of $\widehat{H}(\mathbf{q}, \mathbf{p})$ and where j is the absolute value of the reduced angular momentum $\mathbf{q}\times\mathbf{p}$, immediately leads to the fractional periastron advance per orbital revolution $k = (\Phi - 2\pi)/2\pi$ and to the orbital period P,

$$\frac{\Phi}{2\pi} = 1 + k = -\frac{\partial}{\partial j} i_r(E, j)\,, \qquad (35)$$

$$\frac{P}{2\pi G m} = \frac{\partial}{\partial E} i_r(E, j)\,. \qquad (36)$$

The explicit expressions read

$$
\begin{aligned}
k = \frac{1}{c^2}\frac{3}{j^2}&\left\{1+\frac{1}{c^2}\left[\frac{5}{4}(7-2\nu)\frac{1}{j^2}+\frac{1}{2}(5-2\nu)\,E\right]\right.\\
&\left.+\,\frac{1}{c^4}\left[a_1(\nu)\frac{1}{j^4}+a_2(\nu)\frac{E}{j^2}+a_3(\nu)\,E^2\right]\right\} \qquad (37)
\end{aligned}
$$

and

$$
\begin{aligned}
\frac{P}{2\pi G m} = \frac{1}{(-2E)^{3/2}}&\left\{1-\frac{1}{c^2}\frac{1}{4}(15-\nu)E\right.\\
&+\frac{1}{c^4}\left[\frac{3}{2}(5-2\nu)\frac{(-2E)^{3/2}}{j}-\frac{3}{32}(35+30\nu+3\nu^2)\,E^2\right]
\end{aligned}
$$

$$+ \frac{1}{c^6}\left[a_2(\nu)\frac{(-2E)^{3/2}}{j^3} - 3a_3(\nu)\frac{(-2E)^{5/2}}{j} + a_4(\nu)\,E^3\right]\Bigg\} , \tag{38}$$

where

$$a_1(\nu) = \frac{5}{2}\left(\frac{77}{2} + \left(\frac{41}{64}\pi^2 - \frac{125}{3}\right)\nu + \frac{7}{4}\nu^2\right) , \tag{39a}$$

$$a_2(\nu) = \frac{105}{2} + \left(\frac{41}{64}\pi^2 - \frac{218}{3}\right)\nu + \frac{45}{6}\nu^2, \tag{39b}$$

$$a_3(\nu) = \frac{1}{4}(5 - 5\nu + 4\nu^2) , \tag{39c}$$

$$a_4(\nu) = \frac{5}{128}(21 - 105\nu + 15\nu^2 + 5\nu^3) . \tag{39d}$$

The years-long measurements of the orbital motion of the Hulse-Taylor binary pulsar PSR1913+16 have become such accurate that the 2PN(M) structure of k and P has been reached observationally [19], [20]. The applicability of the binary point-mass 2PN(M) equations of motion to neutron stars has been explored in several publications [21], [22], [23], [24], [25].

In the case of *circular* orbits, the angular frequency is a straightforward observable of the orbital motion. It relates to the radial and periastron-advance frequencies in the following way,

$$\omega_{\rm circ} = \omega_{\rm radial} + \omega_{\rm periastron} = 2\pi\frac{1+k}{P} . \tag{40}$$

In terms of the dimensionless quantity x, defined by

$$x = \left(\frac{Gm\omega_{\rm circ}}{c^3}\right)^{2/3} , \tag{41}$$

the binding (reduced) energy for circular orbits is given by

$$E_{\rm circ}(x;\nu) = -\frac{1}{2}\,x\left[1 + E_1(\nu)\,x + E_2(\nu)\,x^2 + E_3(\nu)\,x^3 + O(x^4)\right] , \tag{42}$$

where

$$E_1(\nu) = -\frac{1}{12}(9+\nu) , \tag{43a}$$

$$E_2(\nu) = -\frac{1}{24}(81 - 57\nu + \nu^2) , \tag{43b}$$

$$E_3(\nu) = -\frac{10}{3}\left(\frac{405}{128} + \frac{1}{64}\left(41\pi^2 - \frac{6889}{6}\right)\nu + \frac{31}{64}\nu^2 + \frac{7}{3456}\nu^3\right) . \tag{43c}$$

In the test body case (small black hole in the field of a heavy one), the binding energy is known exactly, reading

$$E_{\rm circ}(x;\nu=0) = \frac{1-2x}{\sqrt{1-3x}} - 1 . \tag{44}$$

The innermost stable circular orbit, in the literature also called last stable orbit (LSO), is given by the minimum of the function $E_{\rm circ}(x)$, respectively the function $j_{\rm circ}(x)$ because of the relation $\mathrm{d}E_{\rm circ} = x^{3/2}\mathrm{d}j_{\rm circ}$, e.g. see [17]. At the minimum, $\mathrm{d}E_{\rm circ}(x)/\mathrm{d}x = 0$ holds. The test body case yields $x_{\rm LSO} = 1/6$ (Schwarzschild value). For equal-mass black holes ($\nu = 1/4$), $x_{\rm LSO}$ turns out to be 0.255 taking into account (42) and (43). An a-priori-improved 3PN-calculation, using an effective and Padé-improved one-body approach, yielded the value 0.198, see [17]. Obviously, further investigations are needed for better insight into the location of the LSO for binary black holes.

5 Energy Loss and Gravitational Wave Emission

In Sect. 3.1 we have seen that the energy flux to nPN(W) order implies energy loss to (n+5/2)PN(M) order. Hereof it follows that energy-loss calculations are quite efficient via energy-flux calculations. Because of this we will apply the balance property between emitted and lost energies to easily derive the energy loss from the energy flux. In general, only after averaging over orbital periods both expressions coincide (see (30)). In case of circular orbits, however, this averaging procedure is not needed.

5.1 Orbital Decay to 4.5PN Order

The binding energy of our binary system is given by $\mu E_{\rm circ}$. Therefore, for the energy loss we get

$$-\mu\frac{dE_{\rm circ}}{dt} = \mathcal{L} = \frac{32c^5}{5G}\nu^2x^5\left[1+\left(-\frac{1247}{336}-\frac{35}{12}\nu\right)x + 4\pi x^{3/2}\right.$$
$$\left. + \left(-\frac{44711}{9072}+\frac{9271}{504}\nu+\frac{65}{18}\nu^2\right)x^2\right] , \tag{45}$$

where the 2PN(W) energy flux is from [26]; for the 3.5PN(W) energy flux see [27], [28].

Taking into account (32) we obtain a differential equation for x which is easily solved with accuracy $1/c^9$. In terms of the dimensionless time variable

$$\tau = \frac{\nu c^3}{5Gm}(t_{\rm c}-t) , \tag{46}$$

where $t_{\rm c}$ denotes the coalescence time, the solution reads [26],

$$x = \frac{1}{4}\tau^{-1/4}\left[1+\left(\frac{743}{4032}+\frac{11}{48}\nu\right)\tau^{-1/4} - \frac{1}{5}\pi\tau^{-3/8}\right.$$
$$\left. + \left(\frac{19583}{254016}+\frac{24401}{193536}\nu+\frac{31}{288}\nu^2\right)\tau^{-1/2}\right] . \tag{47}$$

Taking into account the relation between phase and frequency $\mathrm{d}\phi/\mathrm{d}t = \omega$, respectively $\mathrm{d}\phi/\mathrm{d}\tau = -5x^{3/2}/\nu$, the phase evolution results in

$$\phi = \phi_c - \frac{1}{\nu}\tau^{5/8}\left[1 + \left(\frac{3715}{8064} + \frac{55}{96}\nu\right)\tau^{-1/4} - \frac{3}{4}\pi\tau^{-3/8}\right.$$
$$\left. + \left(\frac{9275495}{14450688} + \frac{284875}{258048}\nu + \frac{1855}{2048}\nu^2\right)\tau^{-1/2}\right] . \tag{48}$$

5.2 Gravitational Waveform to 2PN Order

The radiation field can be decomposed into two orthogonal polarization states. The polarization states h_+ and $h_\times$ are defined by

$$h_+ = \frac{1}{2}(u_i u_j - v_i v_j)h_{ij}^{\mathrm{TT}} , \tag{49a}$$
$$h_\times = \frac{1}{2}(u_i v_j + v_i u_j)h_{ij}^{\mathrm{TT}} , \tag{49b}$$

where $\mathbf{u}$ and $\mathbf{u}$ denote two vectors in the polarization plane forming an orthogonal right-handed triad with the direction $\mathbf{n}$ from the source to the detector. The detector is directly sensitive to a linear combination of the polarization waveforms h_+ and $h_\times$, namely

$$h(t) = F_+ h_+(t) + F_\times h_\times(t) , \tag{50}$$

where F_+ and $F_\times$ are the so-called beam-pattern functions of the detector depending on two angles giving the direction $-\mathbf{n}$ of the source as seen from the detector and a polarization angle specifying the orientation of the vectors $\mathbf{u}$ and $\mathbf{v}$ around that direction (for explicit expressions, see [29]).

For our binary system, the two polarizations h_+ and $h_\times$ are chosen such that the polarization vectors $\mathbf{u}$ and $\mathbf{v}$ lie respectively along the major and minor axis of the projection onto the plane of the sky of the circular orbit, with $\mathbf{u}$ oriented toward the ascending node, the point at which black hole 1 crosses the plane of the sky moving towards the observer. The result, to 2PN(W) order, reads [30]

$$h_{+,\times} = \frac{2G\mu x}{c^2 r}\left[H^{[0]}_{+,\times} + x^{1/2}H^{[1/2]}_{+,\times} + xH^{[1]}_{+,\times} + x^{3/2}H^{[3/2]}_{+,\times} + x^2 H^{[2]}_{+,\times}\right] , \tag{51}$$

where the plus polarization is given by

$$H^{[0]}_+ = -(1 + c_i^2)\cos 2\psi , \tag{52a}$$

$$H^{[1/2]}_+ = -\frac{s_i}{8}\frac{\delta m}{m}[(5 + c_i^2)\cos\psi - 9(1 + c_i^2)\cos 3\psi] , \tag{52b}$$

$$H^{[1]}_+ = \frac{1}{6}[19 + 19c_i^2 - 2c_i^4 - \nu(19 - 11c_i^2 - 6c_i^4)]\cos 2\psi$$

$$- \frac{4}{3}s_i^2(1+c_i^2)(1-3\nu)\cos 4\psi\ , \tag{52c}$$

$$\begin{aligned} H_+^{[3/2]} &= \frac{s_i}{192}\frac{\delta m}{m}\{[57+60c_i^2-c_i^4-2\nu(49-12c_i^2-c_i^4)]\cos\psi \\ &- \frac{27}{2}[73+40c_i^2-9c_i^4-2\nu(25-8c_i^2-9c_i^4)]\cos 3\psi \\ &+ \frac{625}{2}(1-2\nu)s_i^2(1+c_i^2)\cos 5\psi\} - 2\pi(1+c_i^2)\cos 2\psi\ , \end{aligned} \tag{52d}$$

$$\begin{aligned} H_+^{[2]} &= \frac{1}{120}[22+396c_i^2+145c_i^4-5c_i^6+\frac{5}{3}\nu(706-216c_i^2-251c_i^4+15c_i^6) \\ &- 5\nu^2(98-108c_i^2+7c_i^4+5c_i^6)]\cos 2\psi \\ &+ \frac{2}{15}s_i^2[59+35c_i^2-8c_i^4-\frac{5}{3}\nu(131+59c_i^2-24c_i^4) \\ &+ 5\nu^2(21-3c_i^2-8c_i^4)]\cos 4\psi \\ &- \frac{81}{40}(1-5\nu+5\nu^2)s_i^4(1+c_i^2)\cos 6\psi \\ &+ \frac{s_i}{40}\frac{\delta m}{m}\{[11+7c_i^2+10(5+c_i^2)\ln 2]\sin\psi - 5\pi(5+c_i^2)\cos\psi \\ &- 27[7-10\ln(3/2)](1+c_i^2)\sin 3\psi + 135\pi(1+c_i^2)\cos 3\psi\}\ , \end{aligned} \tag{52e}$$

and the cross polarization by

$$H_\times^{[0]} = -2c_i\sin 2\psi\ , \tag{53a}$$

$$H_\times^{[1/2]} = -\frac{3}{4}s_ic_i\frac{\delta m}{m}[\sin\psi - 3\sin 3\psi]\ , \tag{53b}$$

$$\begin{aligned} H_\times^{[1]} &= \frac{c_i}{3}[17-4c_i^2-\nu(13-12c_i^2)]\sin 2\psi \\ &- \frac{8}{3}c_is_i^2(1-3\nu)\sin 4\psi\ , \end{aligned} \tag{53c}$$

$$\begin{aligned} H_\times^{[3/2]} &= \frac{s_ic_i}{96}\frac{\delta m}{m}\{[63-5c_i^2-2\nu(23-c_i^2)]\sin\psi \\ &- \frac{27}{2}[67-15c_i^2-2\nu(19-15c_i^2)]\sin 3\psi \\ &+ \frac{625}{2}(1-2\nu)s_i^2\sin 5\psi\} - 4\pi c_i\sin 2\psi\ , \end{aligned} \tag{53d}$$

$$\begin{aligned} H_\times^{[2]} &= \frac{c_i}{60}[68+226c_i^2-15c_i^4+\frac{5}{3}\nu(572-490c_i^2+45c_i^4) \\ &- 5\nu^2(56-70c_i^2+15c_i^4)]\sin 2\psi \\ &+ \frac{4}{15}c_is_i^2[55-12c_i^2-\frac{5}{3}\nu(119-36c_i^2) \\ &+ 5\nu^2(17-12c_i^2)]\sin 4\psi \end{aligned}$$

$$
\begin{aligned}
&- \frac{81}{20}(1-5\nu+5\nu^2)c_i s_i^4 \sin 6\psi \\
&- \frac{3}{20} s_i c_i \frac{\delta m}{m} \{[3+10\ln 2]\cos\psi + 5\pi \sin\psi \\
&- 9[7-10\ln(3/2)]\cos 3\psi - 45\pi \sin 3\psi\} \,, \qquad \text{(53e)}
\end{aligned}
$$

where $c_i = \cos i$ and $s_i = \sin i$ and i denotes the inclination angle between the direction of the detector, as seen from the binary's center-of-mass, and the normal to the orbital plane which is assumed to be right-handed with respect to the sense of motion so that $0 \leq i \leq \pi$. $\delta m = m_1 - m_2$, and the phase variable ψ is given by

$$
\psi = \phi - 3x^{3/2} \ln \left(\frac{x}{x_0} \right) \,, \qquad (54)
$$

where ϕ is the actual orbital phase of the binary, namely the angle oriented in the sense of motion between the ascending node and the direction of black hole 1 ($\phi = 0$ mod 2π when the two black holes lie along $\mathbf{u}$, with black hole 1 at the ascending node). The logarithmic phase modulation originates from the propagation of tails in the wave zone. The constant scale x_0 can be chosen arbitrarily; it relates to the arbitrary constant b in the (27).

6 Concluding Remarks

In this paper we have presented the conservative dynamics of the orbital motion of Brill-Lindquist binary black holes up to the 3PN(M) order and the emitted gravitational waves to 2PN(W) order. Corresponding to the latter, the radiation damping has been given to 4.5PN(M) order.

On the level of motion of binary black holes, known are the equations of motion to the 3.5PN(M) order [3], [31], [25] as well as the radiation damping (energy loss) to the 6PN(M) order apart from a few terms at the 5.5PN(M) order [28]. When no eternal black holes are considered but forming ones, a fluid matter approach is appropriate. Here, the derivation of the 3PN(M) equations of motion is progressing [24], [25].

On the level of gravitational waves from binary black holes, the 3.5PN(W) waveform is known apart from a few terms at the 3PN(W) order [28]. All the unknown terms mentioned above may probably be calculable uniquely using dimensional regularization.

Acknowledgments

I thank the organizers of the Spanish Relativity Meeting for their kind invitation and generous financial support.

References

1. D.R. Brill, R.W. Lindquist: Phys. Rev. **131**, 471 (1963)
2. P. Jaranowski, G. Schäfer: Phys. Rev. **D61**, 064008 (2000)
3. T. Damour, P. Jaranowski, G. Schäfer: Phys. Lett. **B513**, 147 (2001)
4. C.W. Misner: Ann. Phys. (N.Y.) **24**, 102 (1963)
5. R.W. Lindquist: J. Math. Phys. **4**, 938 (1963)
6. R. Arnowitt, S. Deser, C.W. Misner: 'The Dynamics of General Relativity'. In: *Gravitation: An Introduction to Current Research*, ed. by L. Witten (John Wiley, New York 1962) pp. 227–265
7. D.D. Holm: Physica **17D**, 1 (1985)
8. P. Jaranowski, G. Schäfer: Phys. Rev. **D57**, 7274 (1998)
9. M. Riesz: Acta Mathematica **81**, 1 (1949)
10. T. Damour, P. Jaranowski, G. Schäfer: Phys. Rev. **D62**, 044024 (2000)
11. T. Damour, G. Schäfer: J. Math. Phys. **32**, 127 (1991)
12. P. Jaranowski, G. Schäfer: Phys. Rev. **D60**, 124003 (1999)
13. G. Schäfer: 'Motion of Binary Systems and Higher Order Post-Newtonian Approximations'. In: *Analytical and Numerical Approaches to Relativity: Sources of Gravitational Radiation, Spain, September 1997*, ed. by C. Bona, J. Carot, L. Mas, J. Stela (Universitat de les Illes Balears, 1998) pp. 3–13
14. K.S. Thorne: Rev. Mod. Phys. **52**, 299 (1980)
15. G. Schäfer: Astron. Nachr. **311**, 4 (1990)
16. L. Blanchet, G. Schäfer: Class. Quantum Grav. **10**, 2699 (1993)
17. T. Damour, P. Jaranowski, G. Schäfer: Phys. Rev. **D62**, 084011 (2000)
18. P. Jaranowski, G. Schäfer: Ann. Phys. (Leipzig) **9**, 378 (2000)
19. J.H. Taylor: Rev. Mod. Phys. **66**, 711 (1994)
20. T. Damour, G. Schäfer: Nuovo Cimento **B101**, 127 (1988)
21. S.M. Kopeikin: Sov. Astron. **29**, 516 (1985)
22. L.P. Grishschuk, S.M. Kopeikin: 'Equations of Motion for Isolated Bodies with Relativistic Corrections Including the Radiation Reaction Force'. In: *Relativity in Celestial Mechanics and Astronomy, IAU Symposium 114, Leningrad 1985*, ed. by J. Kovalevsky, V.A. Brumberg (Reidel, Dordrecht 1986) pp. 19–34
23. T. Damour: 'The Problem of Motion in Newtonian and Einsteinian Gravity'. In: *Three Hundred Years of Gravitation*, ed. by S. Hawking, W. Israel (Cambridge University Press, Cambridge 1987) pp. 128–198
24. Y. Itoh, T. Futamase, H. Asada: Phys. Rev. **D63**, 064038 (2001)
25. M.E. Pati, C.M. Will: Phys. Rev. **D65**, 104008 (2002)
26. L. Blanchet, T. Damour, B.R. Iyer: Phys. Rev. **D51**, 5360 (1995)
27. L. Blanchet, G. Faye, B.R. Iyer, B. Joguet: Phys. Rev. **D65**, 061501 (2002)
28. L. Blanchet, B.R. Iyer, B. Joguet: Phys. Rev. **D65**, 064005 (2002)
29. J.A. Lobo, this volume
30. L. Blanchet, B.R. Iyer, C.M. Will, A.G. Wiseman: Class. Quantum Grav. **13**, 575 (1996)
31. L. Blanchet, G. Faye: Phys. Rev. **D63**, 062005 (2001)

The Detection of Gravitational Waves

J. Alberto Lobo

Departament de Física Fonamental, Universitat de Barcelona
Diagonal 647, E-08028 Barcelona, Spain

Abstract. This chapter is concerned with the question: how do gravitational waves (GWs) interact with their detectors? It is intended to be a *theoretical review* of the fundamental concepts involved in interferometric and acoustic (Weber bar) GW antennas. In particular, the type of signal the GW deposits in the detector in each case will be assessed, as well as its intensity and deconvolution. Brief reference will also be made to detector sensitivity characterisation, including very summary data on current state of the art of GW detectors.

1 Introduction

Gravitational waves (GW), on very general grounds, seem to be a largely unavoidable consequence of the well established fact that no known interaction propagates instantly from source to observer: gravitation would be the first exception to this rule, should it be described by Newton's theory. Indeed, the Newtonian gravitational potential $\phi(\boldsymbol{x}, t)$ satisfies Poisson's equation

$$\nabla^2\phi(\boldsymbol{x}, t) = -4\pi G\, \varrho(\boldsymbol{x}, t) \;, \tag{1}$$

where $\varrho(\boldsymbol{x}, t)$ is the density of matter in the sources of gravitational fields, and G is Newton's constant. But, since (1) contains no time derivatives, the time dependence of $\phi(\boldsymbol{x}, t)$ is purely *parametric*, i.e., time variations in $\varrho(\boldsymbol{x}, t)$ *instantly* carry over to $\phi(\boldsymbol{x}, t)$, irrespective of the value of $\boldsymbol{x}$. So, for example, non-spherically symmetric fluctuations in the mass distribution of the Sun (such as e.g. those caused by solar storms) would instantly *and* simultaneously be felt both in the nearby Mercury and in the remote Pluto...

Quite independently of the *quantitative* relevance of such instant propagation effect in this particular example – which is none in practice –, its very existence is *conceptually* distressing. In addition, the asymmetry between the space and time variables in (1) does not even comply with the basic requirements of Special Relativity.

Einstein's solution to the problem of gravity, General Relativity (GR), does indeed *predict* the existence of radiation of gravitational waves. As early as 1918, Einstein himself provided a full description of the polarisation and propagation properties of weak GWs [1]. According to GR, GWs travel across otherwise flat empty space at the speed of light, and have *two* independent and *transverse* polarisation amplitudes, often denoted $h_+(\boldsymbol{x}, t)$ and $h_\times(\boldsymbol{x}, t)$, respectively [2]. In a more general framework of so called *metric theories* of gravity, GWs are allowed to have up to a maximum 6 amplitudes, some of them transverse and some longitudinal [3].

The theoretically predicted existence of GWs poses of course the experimental challenge to *measure* them. Historically, it took a long while even to attempt the construction of a *gravitational telescope*: it was not until the decade of the 1960's that J. Weber first took up the initiative, and developed the first gravitational antennas. These were elastic cylinders of aluminum, most sensitive to *short bursts* (a few milliseconds) of GWs. After analysing the data generated by *two* independent instruments, and looking for events in coincidence in both, he reported evidence that a considerably large number of GW flares had been sighted [4].

Even though Weber never gave up his claims of *real* GW detection [5], his contentions eventually proved untenable. For example, the rate and intensity of the reported events would imply the happening of several supernova explosions per week in our galaxy [6], which is astrophysically very unlikely.

It became clear that more sensitive detectors were necessary, whose design and development began shortly afterwards. In the mid 1970's and early eighties, the new concept of *interferometric* GW detector started to develop [7,8], which would later lead to the larger *LIGO* and *VIRGO* projects, as well as others of more reduced dimensions (*GEO-600* and *TAMA*), and to the future *space* antenna *LISA*. In parallel, *cryogenic* resonant detectors were designed and constructed in several laboratories, and towards mid 1990's the next generation of *ultracryogenic* antennas, *NAUTILUS* (Rome), *AURIGA* (Padua), *ALLEGRO* (Baton Rouge, Louisiana) and *NIOBE* (Australia), began *taking data*. More recently, data exchange protocols have been signed up for multiple detector coincidence analysis [9]. Based on analogous physical principles, new generation *spherical* GW detectors are being programmed in Brazil, Holland and Italy [10–12].

In spite of many years of endeavours and hard work, GWs have proved elusive to all dedicated detectors constructed so far. However, the discovery of the binary pulsar *PSR 1913+16* by R. Hulse and J. Taylor in 1974 [13], and the subsequent long term detailed monitoring of its orbital motion, brought a breeze of fresh air into GW science: the measured decay of the orbital period of the binary system due to *gravitational bremsstrahlung* accurately conforms to the predictions of General Relativity. Hulse and Taylor were awarded the 1993 Nobel Prize in Physics for their remarkable work. As of 1994 [14], the accumulated binary pulsar data confirm GR to a high precision of a tenth of a percent[1].

The binary pulsar certainly provides the most compelling evidence to date of the GW phenomenon as such, yet it does so thanks to the observation of a *back action* effect on the source. Even though I do not consider accurate the statement, at times made by various people, that this is only *indirect* evidence of GWs, it is definitely a matter of fact that there is more to GWs than revealed by the binary pulsar... For example, amplitude, phase and polarisation parameters

[1] It seems that priorities in pulsar observations have since shifted to other topics of astrophysical interest, so it is difficult to find more recent information on *PSR 1913+16*.

of a GW can only be measured, according to current lore, with dedicated GW antennas.

But how do GW telescopes interact with the radiation they are supposed to detect? This is of course a fundamental question, and is also the subject of the present contribution, where I intend to review the *theoretical foundations* of this problem, and its solutions as presently understood. In Sect. 2, I summarise the essential properties of GWs within a rather large class of possible theories of the gravitational interaction; Sect. 3 briefly bridges the way to Sects. 4 and 5 where interferometric and acoustic detector concepts are respectively analysed in some detail. Sect. 5 also includes aboundant reference to new generation spherical detectors in its various variants (solid, hollow, dual). For the sake of completeness, I have added a section (Sect. 6) with a very short summary of detector characterisation concepts, so that the interested reader gets a flavour of how sensitivities are defined, what do they express and how do GW signals compare with local noise disturbances in currently conceived detectors. Sect. 7 closes the paper with a few general remarks.

2 The Nature of Gravitational Waves

Quite generally, a time varying mass-energy distribution creates in its surroundings a time varying gravitational field (*curvature*). As already stressed in Sect. 1, we do not expect such time variations to travel instantly to distant places, but rather that they travel as *gravity waves* across the intervening space.

Now, how are these waves "seen"? A *single* observer may of course not feel any variations in the gravitational field where he/she is immersed, if he/she is in *free fall* in that field – this is a consequence of the *Equivalence Principle* [15]. *Two* nearby observers have instead the capacity to do so: for, both being in free fall, they can take each other as a reference to measure any relative *accelerations* caused by a non-uniform gravitational field, in particular those caused by a *gravitational wave field.* We can rephrase this argument saying that gravitational waves show up as *local tides*, or gradients of the local gravitational field at the observatory.

In the language of Differential Geometry, tides are identified as *geodesic deviations*, i.e., variations in the four-vector connecting nearby geodesic lines. It is shown in textbooks, e.g. [2], that the geodesic deviation equation is

$$\frac{D^2\xi_\mu}{ds^2} = R_{\mu\nu\varrho\sigma}\,\dot{x}^\nu\dot{x}^\sigma\xi^\varrho\ , \tag{2}$$

where "D" means *covariant derivative*, s is proper time for *either* geodesic, $R_{\mu\nu\varrho\sigma}$ is the Riemann tensor, $\dot{x}^\nu$ is a unit tangent vector (again to either geodesic), and ξ^μ is the vector connecting corresponding points of the two geodesic lines.

The GW fields considered in this paper will be restricted to a class of *perturbations* of the geometry of otherwise flat space-time, with the additional assumptions that they be

– small
– time-dependent
– vacuum perturbations.

This is certainly not the most general definition of a GW yet it will suffice to our purposes here: any GWs generated in astrophysical sources and reaching a man-made detector definitely satisfy the above requirements. The interested reader is referred to [16] for a thorough treatment of more general GWs.

Following the above assumptions, a GW can be described by perturbations $h_{\mu\nu}(\boldsymbol{x},t)$ of a *flat* Lorentzian metric $\eta_{\mu\nu}$, i.e., there exist *quasi-Lorentzian* coordinates $(\boldsymbol{x},t)$ in which the space-time metric $g_{\mu\nu}(\boldsymbol{x},t)$ can be written

$$g_{\mu\nu}(\boldsymbol{x},t) = \eta_{\mu\nu} + h_{\mu\nu}(\boldsymbol{x},t) \tag{3}$$

with

$$\eta_{\mu\nu} = \mathrm{diag}\,(-1,1,1,1)\ , \qquad |h_{\mu\nu}(\boldsymbol{x},t)| \ll 1\ . \tag{4}$$

The actual effect of a GW on a pair of test particles is, according to (2), determined by the Riemann tensor $R_{\mu\nu\varrho\sigma}$, and this in turn is determined by the functions $h_{\mu\nu}$. I now review briefly the different possibilities in terms of which the theory underlying the physics of gravity waves is, i.e., which are the field equations which the $h_{\mu\nu}$ satisfy.

2.1 Plane GWs According to General Relativity

The vacuum field equations of General Relativity are, as is well known [2],

$$R_{\mu\nu} = 0\ , \tag{5}$$

where $R_{\mu\nu}$ is the Ricci tensor of the metric $g_{\mu\nu}$. If quadratic and successively higher order terms in the perturbations $h_{\mu\nu}$ are neglected, then this tensor can be seen to be given by

$$R_{\mu\nu} = \Box\bar{h}_{\mu\nu} - \partial_\mu\partial_\varrho\bar{h}^\varrho_\nu - \partial_\nu\partial_\varrho\bar{h}^\varrho_\mu\ , \tag{6}$$

with

$$\Box \equiv \eta^{\mu\nu}\,\partial_\mu\partial_\nu\ , \qquad \bar{h}_{\mu\nu} \equiv h_{\mu\nu} - \tfrac{1}{2}\,h\,\eta_{\mu\nu}\ , \qquad h \equiv \eta^{\mu\nu}\,h_{\mu\nu}\ . \tag{7}$$

New coordinates $(\boldsymbol{x}',t')$ can be defined by means of transformation equations

$$x'^\mu = x^\mu + \varepsilon^\mu(\boldsymbol{x},t)\ , \tag{8}$$

and these will still be quasi-Lorentzian if the functions $\varepsilon^\mu(\boldsymbol{x},t)$ are *sufficiently small*. More precisely, the GW components are, in the new coordinates,

$$h'_{\mu\nu} = h_{\mu\nu} - \partial_\mu\varepsilon_\nu - \partial_\nu\varepsilon_\mu \tag{9}$$

provided higher order terms in ε^μ are neglected. Thus "sufficiently small" means that the derivatives of ε_μ be of the order of magnitude of the metric perturbations $h_{\mu\nu}$ – so that in the new coordinates x'^μ the metric tensor *also* splits up as

$g'_{\mu\nu} = \eta_{\mu\nu} + h'_{\mu\nu}$. It is now possible, see [2], to choose new coordinates in such a way that the *gauge conditions*

$$\partial_\nu \bar{h}^\nu_\varrho = 0 \tag{10}$$

hold. This being the case, (5) read

$$\Box \bar{h}_{\mu\nu} = 0 \, , \tag{11}$$

which are vacuum *wave equations*. Therefore GWs travel across empty space at the speed of light, according to GR theory. *Plane wave solutions* to (11) satisfying (10) can now be constructed [2] which take the form

$$h_{\mu\nu}(\boldsymbol{x},t) = h^{\rm TT}_{\mu\nu}(\boldsymbol{x},t) = \begin{pmatrix} 0 & 0 & 0 & 0 \\ 0 & h_+(t-z) & h_\times(t-z) & 0 \\ 0 & h_\times(t-z) & -h_+(t-z) & 0 \\ 0 & 0 & 0 & 0 \end{pmatrix} \tag{12}$$

for waves travelling down the z-axis. The label TT is an acronym for *transverse-traceless*, the usual denomination for this particular gauge.

The physical meaning of the polarisation amplitudes in (12) is clarified by looking at the effect of an incoming wave on test particles. Consider e.g. two equal test masses whose center of mass is at the origin of TT coordinates; let ℓ_0 be their distance in the absence of GWs, and (θ, φ) the orientation (relative to the TT axes) of the vector joining both masses. Making use of the geodesic deviation equation (2), with the Riemann tensor associated to (12), it can be seen that the GW only affects the *transverse* projection of the distance relative to the wave propagation direction (the z-axis); in fact, if $\ell(t) \equiv (\xi^\mu \, \xi_\mu)^{1/2}$ is such distance, then some simple algebra leads to the result[2]

$$\ell(t) = \ell_0 \left[1 + \frac{1}{2} \, \{ h_+(t) \, \cos 2\varphi + h_\times(t) \, \sin 2\varphi \} \, \sin^2\theta \right] \, . \tag{13}$$

It is very important at this point to stress that the *wavelength* λ of the incoming GW must be much larger than the distance $\ell(t)$ between the particles for (13) to hold, i.e.,

$$\ell(t) \ll \lambda \, , \tag{14}$$

and this is a condition which must be *added* to the already made assumption that $|h_{\mu\nu}| \ll 1$.

A graphical representation of the result (13) is displayed in Fig. 1: a number of test particles are evenly distributed around a circle perpendicular to the incoming GW, i.e., in the xy plane. When a periodic signal comes in, the distances between those particles change following (13); note that the changes are modulated by the angular factors, i.e., according to the positions of the particles on the circle. The "+" mode is characterised by a vanishing wave amplitude $h_\times$, while in the "×" mode h_+ vanishes.

[2] Note that the Riemann tensor is calculated at the center of mass of the test particles, therefore at $\boldsymbol{x} = 0$. But it can also be calculated at the position of *either* one of them – this would only make up for a negligible *second order* difference.

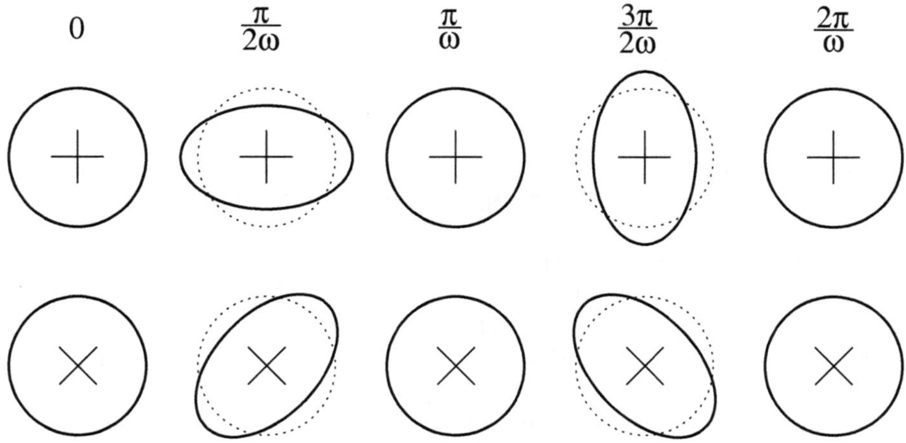

Fig. 1. The "+" and "×" polarisation modes of a GW, according to GR theory. The dotted lines (circles) indicate the position of the test particles in the absence of GW signal. Each step in the graph corresponds to a quarter of the period of the driving GW, as labelled atop

2.2 Plane GWs According to *Metric Theories* of Gravity

Although General Relativity has never been questioned so far by experiment, there are in fact alternative theories, e.g. Brans-Dicke theory [17], which are interesting for a number of reasons, for example *cosmological* reasons [18]. Generally, these theories make their own specific predictions about GWs, and they partly differ from those of GR – just discussed. The term *metric theory* indicates that the gravitational interaction affects the *geometry* of space-time, i.e., the metric tensor $g_{\mu\nu}$ is a fundamental ingredient – though other fields may also be necessary to complete the theoretical scheme. Obviously, General Relativity falls within this class of theories.

The appropriate scheme to assess the physics of such more general class of GWs was provided long ago [19]. The idea is to consider only *plane gravitational waves*, which should be an extremely good approximation for astrophysics, given our great distance even to the nearest conceivable GW source, and to characterise them by their *Newman-Penrose scalars* [20].

It appears that only *six* components of the Riemann tensor out of the usual 20 are independent in a *plane* GW; these are given by the four Newman-Penrose scalars

$$\Psi_2(v), \Psi_3(v), \Psi_4(v) \text{ and } \Phi_{22}(v)\ , \tag{15}$$

of which Ψ_2 and Φ_{22} are real, while Ψ_3 and Ψ_4 are *complex* functions of the null variable v – see [20] for all notation details. If a quasi-Lorentzian coordinate system is chosen such that GWs travel along the z-axis, then $v = t - z$ and one can calculate the scalars (15) to obtain

$$\Psi_2(t-z) = -\tfrac{1}{6}\, R_{tztz}\ , \tag{16a}$$

$$\Psi_3(t-z) = -\tfrac{1}{2}\left(-R_{txtz} + \mathrm{i}R_{tytz}\right) , \tag{16b}$$

$$\Psi_4(t-z) = -R_{txtx} + R_{tyty} + 2\mathrm{i}R_{txty} , \tag{16c}$$

$$\Phi_{22}(t-z) = -R_{txtx} - R_{tyty} . \tag{16d}$$

General Relativity is characterised by $\Psi_4(t-z)$ being the only non-vanishing scalar, while in Brans-Dicke theory $\Phi_{22}(t-z)$ also is different from zero – see [19] for full details.

It is relevant to remark at this stage that the only non-trivial components of the Riemann tensor of a plane GW are the so called "*electric*" components, R_{titj}, as we see in (16a–d) above. These are, incidentally, *also* the only ones which appear in the geodesic deviation equation (2), since one may naturally choose $\dot{x}^\mu = (1;0,0,0)$[3].

This fact helps us make a graphical representation of all *six* possible polarisation states of a general metric GW in the same manner as in Fig. 1. The result is represented in Fig. 2 – whose source is [3]. The idea is to take a ring of test particles, let a GW pass by, and analyse the results of the displacements it causes in the distributions of those particles, just as done in Sect. 2.2. Note that the first three modes are *transverse*, while the other three are longitudinal – see the caption to the figure. As already stressed, GR only gives rise to the two Ψ_4 modes.

3 Gravitational Wave Detection Concepts

We are now ready to discuss the objectives and procedures to detect GWs: knowing their physical structure, one can design systems whose interaction with the GWs be sufficiently well understood and under control; suitable monitoring of the dynamics of such systems will be the source of information on whatever GW parameters may show up in a given observation experiment.

There are *two* major detection concepts: *interferometric* and *acoustic* detection. Historically, the latter came first through the pioneering work of J. Weber, but interferometric GW antennas are at present attracting the larger stake of the investment in this research field, both in hardware and in human commitment. This is because much hope has been deposited in their capabilities to reach sufficient sensitivity to *measure* GWs for the first time.

Interferometric detectors aim to measure phase shifts between light rays shone along two different (straight) lines, whose ends are defined by *freely suspended* test masses. This is done in a Michelson layout, using mirrors, beam splitters and photodiodes. Acoustic detectors are instead based on elastically linked test masses – rather than freely suspended – which *resonantly* respond to GW excitations.

[3] Note that, with this choice, $R_{\mu\varrho\nu\sigma}\,\dot{x}^\mu\dot{x}^\nu = R_{t\varrho t\sigma}$ but, because of the symmetries of the Riemann tensor, only values of ϱ and σ different from zero, i.e., $\{\varrho\sigma\} = \{ij\}$, give non-zero contributions.

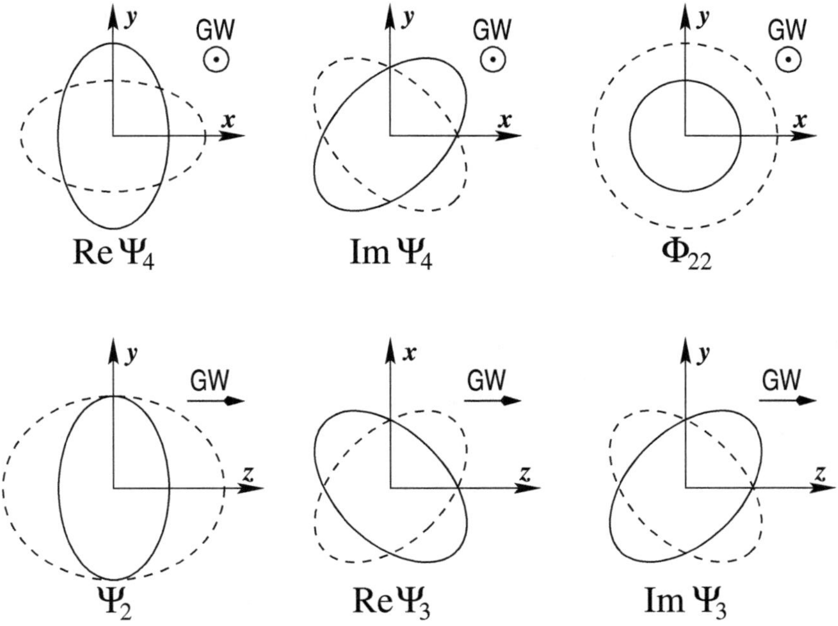

Fig. 2. The six polarisation modes of a plane, metric GW. The dotted and solid lines correspond to distributions separated by a half period of the incoming wave, and it is assumed that the particles lie on a circle when no GW is acting. Note the indicators that the modes in the upper row are excited by a wave which arrives *perpendicular* to the plane of the particles, while those in the lower row correspond to GWs which are in the *same* plane as the particles. It must however be clarified that the mode Φ_{22} has spherical symmetry, so it includes a combination of longitudinal and transverse excitations in like proportions

These *qualitative* ideas can be made quantitatively precise, but the process is a non-trivial one and has important subtleties which must be properly understood for a thorough assessment of the detector workings and readout. The next sections are devoted to explain with some detail which are the *theoretical principles* governing the behaviour and response of both kinds of GW antennas.

4 *Interferometric* GW Detectors

A rather *naïve* idea to measure the effect of an incoming GW is provided by the following argument – see also Fig. 3 for reference: let a GW having a "+" polarisation (assume GR for simplicity at this stage) come in perpendicular to the local horizontal at a given observatory; if three masses are laid down on the vertices of an ideally oriented isosceles right triangle then, as we saw (Fig. 1), the catets shrink and stretch with opposite phases. If a beam of laser light is now shone into the system, and a beam-splitter attached to mass M_0 and mirrors attached to M_1 and M_2, then one can think of measuring the distance changes between the masses by simple interferometry.

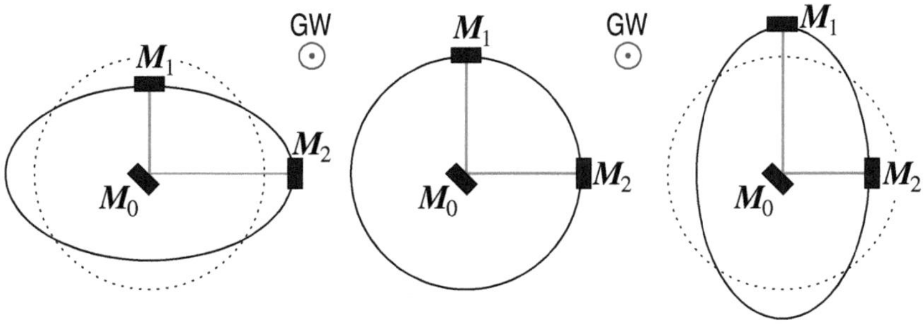

Fig. 3. The "naïve" interferometric detector concept: a GW coming perpendicular to the plane of the sheet, "+" polarised relative to the x and y axes, causes the end masses M_1 and M_2 to oscillate in phase opposition relative to the central mass M_0. Light is shone into the system, and suitable beam splitters and mirrors are attached to the masses; length changes are then measured interferometrically, which directly lead to determine the GW amplitude

This may look like a very reasonable proposal for a detector yet the following criticism readily suggests itself: gravitation is concerned with *geometry*, i.e., gravitational fields alter lengths and angles; therefore GWs will affect *identically* both distances between the masses of Fig. 3 *and* the wavelength of the light travelling between them – thus leading to a *cancellation* of the conjectured interferometric effect...

While the criticism is certainly correct, the conclusion is *not*. The reason is that it overlooks the fact that gravitation is concerned with the geometry of *space-time* – not just space. In the case of the above GW it so happens that, in TT coordinates, the time dimension of space-time is not *warped* in the GW geometry – see the form of the metric tensor in (12) – while the transverse space dimensions *are*. Consequently an electromagnetic wave travelling in the xy plane experiences *wavelength* changes depending on the propagation direction, but it does *not* experience *frequency* changes. The net result of this is that the *phase* of the electromagnetic wave differs from direction to direction of the xy plane, and *this* makes a GW amenable to detection by interferometric principles.

Looked at in this way, the masses represented in Fig. 3 only play a *passive* role in the detector, in the sense that they simply make the interference between the two light beams possible by providing physical support for the mirrors and beam splitters. In other words, the physical principles underlying the working of an interferometric GW detector must have to do with the *interaction between GWs and electromagnetic waves* rather than with geodesic deviations of the *masses*.

Admittedly, this is not the most common point of view [21]. It can however be made precise by the following considerations, which are studied in depth in [22] and [23].

4.1 Test Light Beams in a GW-Warped Space-Time

According to the above, it seems that we must address the question of how GW-induced fluctuations in the geometry of a background space-time affect the properties (amplitude and phase) of a plane electromagnetic wave – a *light beam* – which travels through such space-time. It will be sufficient to consider that the light beam is a *test* beam, i.e., its back action on the surrounding geometry is negligible.

Let then $A_\mu(\boldsymbol{x}, t)$ be the vector potential which describes an electromagnetic wave travelling in vacuum; A_μ thus satisfies Maxwell's equations

$$\Box A_\mu = 0 \,, \tag{17}$$

where $\Box$ stands for the generalised d'Alembert operator:

$$\Box \equiv g^{\varrho\sigma}\, \nabla_\varrho \nabla_\sigma \,. \tag{18}$$

We need only retain *first order* terms in $h_{\mu\nu}$ in the covariant derivatives, so that (17) reads

$$\Box A_\mu \simeq \eta^{\varrho\sigma}\,(\partial_\varrho\partial_\sigma A_\mu - 2\Gamma^\nu_{\mu\sigma}\,\partial_\varrho A_\nu - \Gamma^\nu_{\varrho\sigma}\,\partial_\nu A_\mu - A_\nu\,\partial_\varrho \Gamma^\nu_{\mu\sigma}) = 0 \,, \tag{19}$$

To this equation, *gauge conditions* must be added. We shall conventionally adopt the usual Lorentz conditions, $\nabla_\mu A^\mu = 0$, which, to lowest order in the gravitational perturbations, read

$$\nabla_\mu A^\mu \simeq -\partial_t A_t + \partial_x A_x + \partial_y A_y + \partial_z A_z - h^{\mu\nu}\,\partial_\mu A_\nu = 0 \,. \tag{20}$$

In addition to the weakness of the GW perturbation, it is also the case in actual practice that:

- The GW typical frequencies, ω, are much smaller than the frequency of the light, Ω: $\omega \ll \Omega$.
- The wavelength of the GW is much larger than the cross sectional dimensions of the light beam.

Wave front distortions and beam curvature are effects which can also be safely neglected in first order calculations [23]. Finally, I shall make the simplifying assumption of perpendicular incidence, i.e., the incoming GW propagates in a direction orthogonal to the arms of the interferometer[4].

We shall thus consider one of the interferometer arms aligned with the x-direction, and the other with the y-direction, while the incoming GW will be assumed to approach the detector down the z-axis. The interaction GW-light beam will thus occur in the $z = 0$ plane, hence the GW perturbations can be suitably described by a function of time alone, i.e.,

$$h_{\mu\nu}(\boldsymbol{x}, t) \longrightarrow h_{\mu\nu}(\omega t) \,, \tag{21}$$

[4] This condition can be easily relaxed, but it complicates the equations to an extent which is inconvenient for the purposes of the present review. Details are fully given in [22].

where ω is the frequency of the GW, which we can also safely assume to be plane-fronted, since its source will in all cases of interest be far removed from the observatory. In addition, for a beam running along the x-axis, the electromagnetic vector potential will only depend on the space variable x and on time, i.e.,

$$A_\mu(\boldsymbol{x},t) \longrightarrow A_\mu(x,t)\ . \tag{22}$$

The following *ansatz* suggests itself as a solution to (19):

$$A_t\,(t,x) = \varepsilon_0(\omega t)\ , \tag{23a}$$
$$A_x\,(t,x) = \varepsilon_1(\omega t)\ , \tag{23b}$$
$$A_y\,(t,x) = a_2\ \exp[\mathrm{i}\Omega\,(t-x)]\,\mathrm{e}^{\mathrm{i}\,\phi_2(\omega t)}\ , \tag{23c}$$
$$A_z\,(t,x) = a_3\ \exp[\mathrm{i}\Omega\,(t-x)]\,\mathrm{e}^{\mathrm{i}\,\phi_3(\omega t)}\ , \tag{23d}$$

where $\varepsilon_0(\omega t)$, $\varepsilon_1(\omega t)$, $\phi_2(\omega t)$ and $\phi_3(\omega t)$ are small quantities of order h, a_2 and a_3 are constants, and Ω is the frequency of the light. Clearly, these expressions reproduce the plane wave solutions to vacuum Maxwell's equations in the limit of flat space-time, i.e., when $h_{\mu\nu}=0$.

Let us now take an incoming GW of the form

$$h_{\{+,\times\}}(t) = \tilde{H}_{\{+,\times\}}\,\mathrm{e}^{\mathrm{i}\omega t}\ , \tag{24}$$

and substitute it into (19) and (20), with the *ansatz* (23a-d), neglecting higher order terms in the ratio ω/Ω. Then [22], *both* ϕ_2 and ϕ_3 are seen to satisfy the approximate differential equation

$$\ddot{\phi}(t) + \frac{2\mathrm{i}\Omega}{\omega}\,\dot{\phi}(t) + \frac{\mathrm{i}\Omega^2}{\omega^2}\,h_+(t) = 0\ . \tag{25}$$

The solution to this equation which is *independent of the initial conditions* is, to the stated level of accuracy,

$$\phi_2(t) \simeq \phi_3(t) \simeq \frac{\Omega}{2\omega}\,\tilde{H}_+\sin\omega t\ . \tag{26}$$

As shown in [22], we need not worry about either $\varepsilon_0(\omega t)$ or $\varepsilon_1(\omega t)$ at this stage because the *longitudinal* component of the electric field (i.e., E_x) is an order of approximation smaller than the transverse components, which are given by

$$E_y = \partial_t A_y\ , \qquad E_z = \partial_t A_z\ , \tag{27}$$

hence

$$E_y(x,t) \simeq \mathrm{i}\Omega\,a_2\,\exp\left[\mathrm{i}\Omega(t-x) + \mathrm{i}\,\frac{\Omega}{2\omega}\,\tilde{H}_+\sin\omega t\right]\ , \tag{28a}$$

$$E_z(x,t) \simeq \mathrm{i}\Omega\,a_3\,\exp\left[\mathrm{i}\Omega(t-x) + \mathrm{i}\,\frac{\Omega}{2\omega}\,\tilde{H}_+\sin\omega t\right]\ . \tag{28b}$$

These expressions beautifully show how the incoming GW causes a *phase shift* in an electromagnetic beam of light. Note that this phase shift is a *periodic* function of time, with the frequency of the GW. If we consider a real interferometer, such as it is very schematically shown in Fig. 3, and call τ the *round trip* time for the light to go from M_0 to M_2 and back, then the accumulated phase shift is, according to these formulas,

$$\delta_x\phi = 2 \times \frac{\Omega}{2\omega}\,\tilde{H}_+ \sin\frac{\omega\tau}{2}\,, \tag{29}$$

since there is an obvious symmetry between light rays travelling to the right and to the left for a GW arriving perpendicularly to them[5].

The arguments leading to (28a-b) can be very easily reproduced, *mutatis mutandi*, to obtain the phase shift experienced by a light ray travelling in the y, rather than the x direction – everything in fact amounts to a simple interchange $\{x \longleftrightarrow y\}$ in the equations, which includes $\{h_+ \longleftrightarrow -h_+\}$ as this is equivalent to $\{h_{xx} \longleftrightarrow h_{yy}\}$, see (12). The result is

$$\delta_y\phi = -2 \times \frac{\Omega}{2\omega}\,\tilde{H}_+ \sin\frac{\omega\tau}{2}\,. \tag{30}$$

In the actual interferometer, provided it has equal arm lengths, the two laser rays recombine in the beam splitter with a net phase difference

$$\delta\phi = \delta_x\phi - \delta_y\phi = \frac{2\Omega}{\omega}\,\tilde{H}_+ \sin\frac{\omega\tau}{2}\,, \tag{31}$$

and this produces an *interference signal*, which is in principle measurable – if the instrumentation is sufficiently sensitive.

The reader may wonder how it is that the detector signal only depends on one of the GW amplitudes, h_+, but not on the other, $h_\times$. The reason is that we have made a very special assumption regarding the orientation of the polarisation axes of the GW relative to the light beam propagation directions. In a realistic case, even if perpendicular GW incidence happens, the arms of the detector will not be aligned with the natural axes of the GW, let alone the most likely case of *oblique* incidence. An important conclusion one should draw from this section is a *conceptual* one, that interferometric detectors are able to measure GW amplitudes and polarisations as a result of the interaction between the electromagnetic field of light rays and the background space-time geometry they travel across.

Beyond this, though, (31) has very relevant *quantitative* consequences, too. For example, as stressed in [22], its range of validity is *not* limited to interferometer arms short compared to the GW wavelength. Therefore, according to the formula, a *null* effect (signal cancellation) happens if the round trip time τ equals the period of the GW, $2\pi/\omega$. Likewise, (31) also tells us that *maximum* detector signal occurs when $\tau = \pi/\omega$. All this happen to be true for arbitrary

---

[5] The reader is warned that this symmetry does *not* happen if the GW and the light beam are not perpendicular, see [23].

incidence and polarisation of the incoming GWs as well. For GW frequencies in the 1 kHz range, the best detector should thus have arm lengths in the range of 150 kilometers – and even longer for lower GW frequencies. No ground based GW antenna has ever been conceived of such dimensions yet there are intelligent ways to *store* the light in shorter arms for suitably tuned GW periods. I shall not go into details of these technical matters, see [8] and [21] for thorough information.

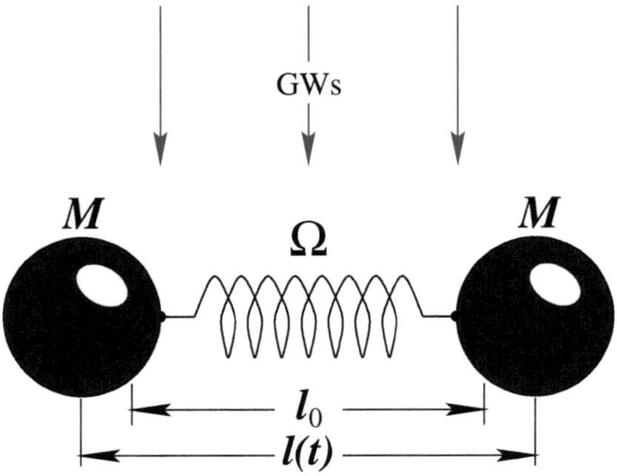

Fig. 4. The acoustic detector concept: a GW coming perpendicular to the spring drives the two masses M at its ends to oscillate at the frequency of the GW. Resonant amplification is obtained when the latter equals the characteristic of the spring frequency, Ω

5 *Acoustic* GW Detectors

Acoustic GW detectors work based on a completely different concept – see Fig. 4: the idea is to set up test masses M linked together by a spring of relaxed length ℓ_0, so that GW *tides* drive their oscillations around the equilibrium position, with significant mechanical amplification at the characteristic frequency of the spring Ω. The spring deformation

$$q(t) \equiv \ell(t) - \ell_0 \tag{32}$$

thus obeys the following equation of motion[6]:

$$\ddot{q}(t) + \Omega^2 q(t) = \frac{1}{2}\,\ell_0 \ddot{h}(t) \;, \tag{33}$$

where

$$h(t) = [h_+(t)\,\cos 2\varphi + h_\times(t)\,\sin 2\varphi]\,\sin^2\theta \;, \tag{34}$$

as follows from (13) – see also [24] for a complete discussion of this case.

[6] I shall not include any *dissipative* terms at this stage, for they do not influence the key points of our present discussion.

This is the main idea, but in practice acoustic GW detectors are *elastic solids* rather than a single spring, i.e., they do not have a single characteristic frequency but a whole spectrum. The response of an elastic solid to a GW excitation is assessed by means of the classical theory of Elasticity, as described for example in [25]. In such theory the deformations of the solid are given by the values of a vector field of displacements, $\boldsymbol{u}(\boldsymbol{x},t)$, which satisfies the evolution equations

$$\varrho\frac{\partial^2\boldsymbol{u}}{\partial t^2} - \mu\nabla^2\boldsymbol{u} - (\lambda+\mu)\,\nabla(\nabla\cdot\boldsymbol{u}) = \boldsymbol{f}(\boldsymbol{x},t)\ , \tag{35}$$

where ϱ is the (undeformed) density of the solid, and λ and μ are its Lamé coefficients, related to the Poisson ratio and Young modulus of the material the solid is made of [25]. The function in the rhs of the equation is the density of *external forces* driving the motion of the system; in the present case, these are the *tides* generated by the sweeping GW, i.e.,

$$f_i(\boldsymbol{x},t) = \varrho\,R_{titj}(t)\,x_j\ , \tag{36}$$

where $R_{titj}(t)$ are components of the Riemann tensor evaluated at a fixed point of the solid, most expediently chosen at its center of mass. As already discussed in Sect. 2.2, plane GWs have at most six degrees of freedom, adequately associated with the six *electric components* of the Riemann tensor, $R_{titj}(t)$. The six components are *one monopole* amplitude and *five quadrupole* amplitudes, and this important structure is made clear by the following expression of the density of GW tidal forces:

$$\boldsymbol{f}(\boldsymbol{x},t) = \boldsymbol{f}^{(0,0)}(\boldsymbol{x})\,g^{(0,0)}(t)\ +\ \sum_{m=-2}^{2}\boldsymbol{f}^{(2,m)}(\boldsymbol{x})\,g^{(2,m)}(t)\ , \tag{37}$$

which is entirely equivalent to (36) – see [26] –, and uses the common (l,m) notation convention to indicate multipole terms. It is very important to stress at this stage that $\boldsymbol{f}^{(l,m)}(\boldsymbol{x})$ are pure *form factors*, simply depending on the fact that tides are monopole-quadrupole quantities, while all the relevant *dynamic* information carried by the GW is encoded in the time dependent coefficients $g^{(l,m)}(t)$. According to these considerations, we see that the ultimate objective of an acoustic GW antenna is to produce values of $g^{(l,m)}(t)$ – or indeed to extract from the readout of the device as much information as possible about those quantities.

Somewhat lengthy algebra permits to write down a *formal solution* to (36) and (37) in terms of an orthogonal series expansion [26]:

$$\boldsymbol{u}(\boldsymbol{x},t) = \sum_N \omega_N^{-1}\,\boldsymbol{u}_N(\boldsymbol{x})\left[\sum_{\substack{l=0\text{ and }2\\ m=-l,\dots,l}} f_N^{(l,m)}\,g_N^{(l,m)}(t)\right]\ , \tag{38}$$

where

$$f_N^{(l,m)} \equiv \frac{1}{M}\int_{\text{Solid}}\boldsymbol{u}_N^*(\boldsymbol{x})\cdot\boldsymbol{f}^{(l,m)}(\boldsymbol{x})\,\mathrm{d}^3x\ , \tag{39a}$$

$$g_N^{(l,m)}(t) \equiv \int_0^t g^{(l,m)}(t')\,\sin\omega_N(t-t')\,\mathrm{d}t' \;, \tag{39b}$$

with M the whole mass of the solid; ω_N is the (possibly *degenerate*) characteristic frequency of the elastic body, and $\boldsymbol{u}_N(\boldsymbol{x})$, the corresponding *wavefunction*, both determined by the solution to the *eigenvalue* problem

$$\mu\nabla^2\boldsymbol{u}_N + (\lambda+\mu)\,\nabla(\nabla\boldsymbol{\cdot}\boldsymbol{u}_N) = -\omega_N^2\,\varrho\,\boldsymbol{u}_N \;, \tag{40}$$

with the boundary conditions that the surface of the solid be free of any tensions and/or tractions – see [26] for full details.

Historically, the first GW antennas were Weber's elastic *cylinders* [4], but more recently, *spherical* detectors have been seriously considered for the next generation of GW antennas, as they show a number of important advantages over cylinders. I shall devote the next sections to a discussion of both types of systems, though clear priority will be given to spheres, due to their much richer capabilities and theoretical interest.

5.1 Cylinders

First thing to study the response of an elastic solid to an incoming GW is, as we have just seen, to determine its characteristic oscillation *modes*, i.e., its frequency spectrum ω_N and associated wavefunctions, $\boldsymbol{u}_N(\boldsymbol{x})$. In the case of a cylinder this is a formidable task; although its formal solution is known [27,28], cylindrical antennas happen in practice to be narrow and long [29,30], and so approximate solutions can be used instead which are much simpler to handle, and sufficiently good – see also [15].

It appears that, in the *long rod* approximation, the most efficiently coupled modes are the *longitudinal* ones, and these have typical sinusoidal profiles, of the type

$$\delta z(z,t) \propto \sin\left(\frac{n\pi z}{L}\right)\,\sin\left(\frac{n\pi v_{\mathrm{s}}t}{L}\right) \;, \qquad n=1,2,3,\ldots \tag{41}$$

for a rod of length L whose end faces are at $z=\pm L/2$, and in which the speed of sound is v_{s}. Figure 5 graphically shows the longitudinal deformations of the cylinder which correspond to (41), including *transverse* distortions which, though not reflected in the simplified equation (41), do happen in practice as a result of the Poisson ratio being different from zero [25]. An important detail to keep in mind is that *odd* n modes have maximum displacements at the end faces, while *even* n modes have nodes there. In fact, the latter do not couple to GWs [24]. It is also interesting to stress that the center of the cylinder is always a node – this is relevant e.g. to suspension design issues [30].

A very useful concept to characterise the sensitivity of an acoustic antenna is its *cross section* for the absorption of GW energy. If an incoming GW flux density of $\Phi(\omega)$ watts per square metre and hertz sweeps the cylinder and sets it

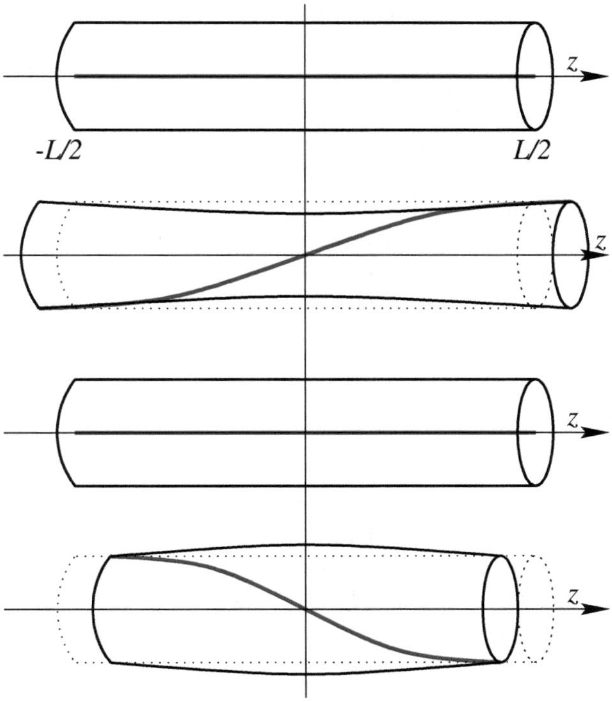

Fig. 5. The first longitudinal oscillation mode of a long cylinder. Note that it has a node at the center and maxima at the end faces. A whole period is represented, and transverse deformations are also shown

to oscillate with energy $E(\omega)$ joules then the cross section is *defined* by the ratio

$$\sigma_{\rm abs}(\omega) = \frac{E(\omega)}{\Phi(\omega)} , \tag{42}$$

which is thus measured in $\mathrm{m}^2\,\mathrm{Hz}$. Simple calculations show that, for optimal antenna orientation (perpendicular to the GW incidence direction) this quantity is given by [24]

$$\sigma_{\rm abs}(\omega_n) = \frac{8}{\pi n^2} \frac{GMv_{\rm s}^2}{c^3} , \qquad n = 1, 3, 5, \dots , \tag{43}$$

where M is the total mass of the cylinder.

It is interesting to get a flavour of the order of magnitude this quantity has: consider a cylinder of Al5056 (an aluminum alloy, for which $v_{\rm s} \simeq 5400\,\mathrm{m/s}$), 3 metres long and 60 centimetres across, which has a mass of 2.3 tons[7]; the above formula tells us that, for the first mode ($n = 1$),

$$\sigma_{\rm abs}(\omega_1) = 4.3 \times 10^{-21}\ \mathrm{cm}^2\,\mathrm{Hz} . \tag{44}$$

[7] These figures correspond to a real antenna, see [29].

This is a *very small* number indeed, and gives an idea of how weak the coupling between GWs and matter is.

The weakness of the coupling gives an indication of how difficult it is to detect GWs. By the same token, though, GWs are very weakly damped as they travel through matter, which means they can produce information about otherwise invisible regions, such as the interior of a supernova, or even the big bang.

Equation (43) is only valid for *perpendicular* GW incidence. If incidence is instead oblique then a significant damping factor of $\sin^4\theta$ comes in, where θ is the angle between the GW direction and the axis of the cylinder [24][8]. This is a severe penalty, and it also happens in interferometric detectors not optimally oriented [22,31].

5.2 Solid Spheres

The first initiatives to construct and operate GW detectors are due to J. Weber, who decided to use elastic cylinders. This philosophy and practice has survived him[9], and still today (February 2002) all GW detectors in continuous data taking regimes are actually Weber bars, though with significant sensitivity improvements [32] derived, amongst other, from ultracryogenic and *SQUID* techniques.

About ten years after Weber began his research, R. Forward published an article [33] where he pondered in a semi-quantitative way the potential virtues of a *spherical*, rather than cylindrical GW detector. Ashby and Dreitlein [34] estimated how the whole Earth, as an *auto-gravitating* system, responds to GWs bathing it, and later Wagoner [35] developed a theoretical model to study the response of an elastic sphere to GW excitations.

Interest in this new theoretical concept then waned to eventually re-emerge in the 1990s. The *ALLEGRO* detector group at Louisiana constructed a room temperature prototype antenna [36,37], which produced sound experimental evidence that it is actually possible to have a working system capable of making *multimode* measurements – I'll come to this in detail shortly –, thus proving that a full-fledged spherical GW detector is within reach of current technological state of the art, as developed for Weber bars. It was apparently the fears to find unsurmountable difficulties in this problem which deterred further research on spherical GW antennas for years [38].

In this section I will give the main principles and results of the *theory* of the spherical GW detector, based on a formalism which has already been partly used in Sect. 5, and for whose complete detail the reader is referred to [26].

As we have seen, first thing we need is the *eigenmodes* and *frequency spectrum* of the spherical solid. This is a classical problem, long known in the literature, the solution to which I will briefly review here, with some added emphasis on the issues of our present concern.

---

[8] Such factor can incidentally be inferred easily from (13), if one notices that the energy of oscillation appearing in the numerator of (43) is proportional to $\dot{\ell}^2$.

[9] Professor Joseph Weber died on 30th September 2000 at the age of 81.

The oscillation eigenmodes of a solid elastic body fall into two families: *spheroidal* and *torsional* modes [26]. Of these, only the former couple to GWs, while torsional modes do not couple at all [39]. Spheroidal wavefunctions have the analytic form

$$\boldsymbol{u}_{nlm}(\boldsymbol{x}) = A_{nl}(r)\,Y_{lm}(\boldsymbol{n})\,\boldsymbol{n} - B_{nl}(r)\,\mathrm{i}\boldsymbol{n}\times\boldsymbol{L}Y_{lm}(\boldsymbol{n})\;, \tag{45}$$

where Y_{lm} are spherical harmonics [40], $\boldsymbol{n} = \boldsymbol{x}/R$ is the outward pointing normal, $\boldsymbol{L}$ is the 'angular momentum operator', $\boldsymbol{L} \equiv -\mathrm{i}\boldsymbol{x} \times \nabla$, $A_{nl}(r)$ and $B_{nl}(r)$ are somewhat complicated combinations of spherical Bessel functions [26], and $\{nlm\}$ are 'quantum numbers' which label the modes. The frequency spectrum appears to be composed of ascending series of *multipole harmonics*, ω_{nl}, i.e., for *each* multipole value l there are an infinite number of frequency harmonics, ordered by increasing values of n. For example, there are *monopole* frequency harmonics ω_{10}, ω_{20}, ω_{30}, etc.; then *dipole* frequencies ω_{11}, ω_{21},..., then *quadrupole* harmonics ω_{12}, ω_{22}, and so on. Each of these frequencies is $(2l+1)$-fold degenerate, and this is a fundamental fact which makes of the sphere a theoretically ideal GW detector, as we shall shortly see.

If the above expressions are substituted into (39a-b), then into (38), one easily obtains the response of the sphere function as

$$\begin{aligned}\boldsymbol{u}(\boldsymbol{x},t) &= \sum_{n=1}^{\infty} \frac{a_{n0}}{\omega_{n0}}\,\boldsymbol{u}_{n00}(\boldsymbol{x})\,g_{n0}^{(0,0)}(t) \\ &+ \sum_{n=1}^{\infty} \frac{a_{n2}}{\omega_{n2}} \left[\sum_{m=-2}^{2} \boldsymbol{u}_{n2m}(\boldsymbol{x})\,g_{n2}^{(2,m)}(t)\right]\;,\end{aligned} \tag{46}$$

where

$$a_{n0} = -\frac{1}{M}\int_0^R A_{n0}(r)\,\varrho\,r^3\,\mathrm{d}r\;, \tag{47a}$$

$$a_{n2} = -\frac{1}{M}\int_0^R \left[A_{n2}(r) + 3\,B_{n2}(r)\right]\,\varrho\,r^3\,\mathrm{d}r\;, \tag{47b}$$

and

$$g_{nl}^{(l,m)}(t) \equiv \int_0^t g^{(l,m)}(t')\,\sin\omega_{nl}(t-t')\,\mathrm{d}t'\;. \tag{48}$$

The series expansion (46) transparently shows that *only* monopole and quadrupole spherical modes can possibly be excited by an incoming GW. The monopole will of course not be excited at all if General Relativity is the true theory of gravitation. A spherical solid is thus seen to be the *best possible* shape for a GW detector. This is because of the optimality of the *overlap coefficients* a_{nl} between the universal form factors $\boldsymbol{f}^{(l,m)}(\boldsymbol{x})$ and the eigenmodes of the sphere $\boldsymbol{u}_{nlm}(\boldsymbol{x})$, which comes about due to the clean *multipole structure* of the latter: only the $l=0$ and $l=2$ *spheroidal* modes couple to GWs, hence *all the GW energy* is deposited into them, and only them. Any other shape of solid, e.g. a cylinder, has eigenmodes most of which have some amount of monopole/quadrupole projections in the form of the coefficients (39a), and this means the incoming GW energy is distributed amongst *many modes*, thus making detection less efficient. We shall assess quantitatively the efficiency of the spherical detector in terms of *cross section* values below.

But, as just stated, quadrupole modes are *degenerate*. More specifically, they are 5-fold degenerate, each degenerate wavefunction corresponding to one of the five integer values m can take between -2 and $+2$. Monopole modes are instead non-degenerate. Figure 6 shows the shapes of all these modes [41] – see the caption to the figure for further details.

Degeneracy is a key concept for the *multimode* capabilities of the spherical detector. For, as explicitly shown by (46), monopole and quadrupole *detector* modes are driven by one and five *GW amplitudes*, respectively, i.e., $g^{(0,0)}(t)$ and $g^{(2,m)}(t)$. Therefore, if one could *measure* the amplitudes of these modes, i.e., the amplitudes of the deformations displayed in Fig. 6, then a *complete deconvolution* of the GW signal would be accomplished. This is a *unique* feature of the spherical antenna, which is not shared by any other GW detector: it enables the determination of *all* the GW amplitudes, not just a combination of them, no matter where the signal comes from. In Sect. 5.3 below I shall give a more detailed review of how the multimode capability can be implemented in practice.

Cross Sections. The general definition (42) applies in this case, too. Since cross section is a *frequency* dependent concept, and since quadrupole modes are degenerate, it is clear that energies deposited in *each* of the five degenerate modes of a given frequency harmonic must be added up to obtain $E(\omega_{n2})$ for that mode. Such energy must be calculated by means of volume integrals – to add up the energies of all differentials of mass throughout the solid –, the details of which I omit here. The final result turns out to be a remarkable one [26]: cross sections factorise in the form

$$\sigma_{\rm abs}(\omega_{nl}) = \mathcal{K}_l(\aleph)\,\frac{GMv_{\rm t}^2}{c^3}\,(k_{nl}a_{nl})^2 \qquad (l = 0 \text{ or } 2)\;, \tag{49}$$

where $GMv_{\rm t}^2/c^3$ is a characteristic of the material of the sphere[10], and $(k_{nl}a_{nl})$ a dimensionless quantity associated with the $\{nl\}$-th frequency harmonic; finally, and this is the stronger theoretical point of this expression, $\mathcal{K}_l(\aleph)$ is a coefficient which is characteristic of the underlying theory of GWs, symbolically indicated

[10] $v_{\rm t}$ is the so called 'transverse speed of sound', and is related to the true speed of sound, $v_{\rm s}$, by the formula

$$v_{\rm t} = (2+2\sigma_{\rm P})^{-1/2}\,v_{\rm s}\;,$$

with $\sigma_{\rm P}$ the Poisson ratio of the material.

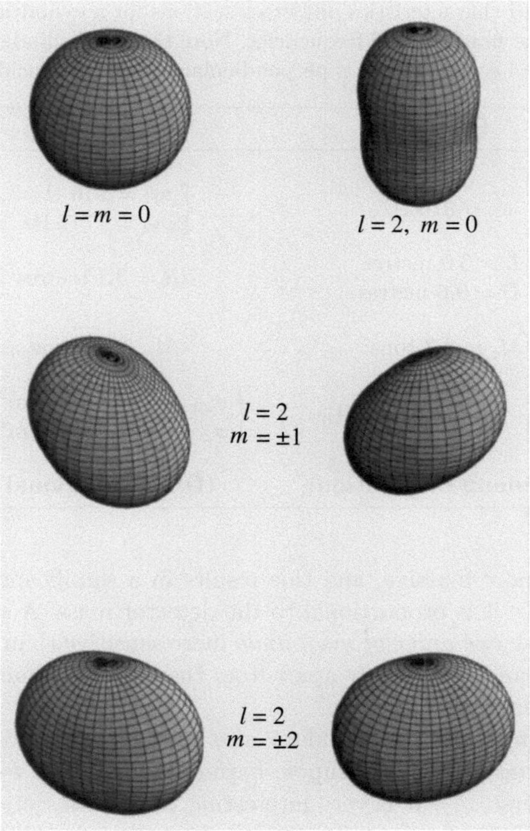

Fig. 6. The five quadrupole spheroidal oscillation modes of a solid elastic sphere, and the monopole mode. Note the latter (top left) is a spherically symmetric 'breathing' mode, while the rest have non-symmetric shapes. Because the eigenmodes (45) are actually *complex* (the spherical harmonics for $m \neq 0$ are complex) suitable combinations of them and their complex conjugates have been used to make plotting possible. These *shapes* are shared by *all* corresponding harmonics, i.e., quadrupole profiles for example are those shown no matter the harmonic number n of their frequency ω_{n2}: simply, they oscillate faster for higher n, but always keep the represented profile

with $\aleph$. For example, if the latter is General Relativity (GR) then

$$\aleph = \mathrm{GR} \Rightarrow \begin{cases} \mathcal{K}_0(\aleph) = 0\,, \\ \mathcal{K}_2(\aleph) = \dfrac{16\,\pi^2}{15}\,, \end{cases} \tag{50}$$

while if it is e.g. Brans-Dicke [17] then these expressions get slightly more complicated [42,43], etc.

Sticking to GR, a few illustrative figures are in order. They are shown in Table 1, where a material of aluminum Al5056 alloy has been chosen. It appears that a sphere having the same fundamental frequency (ν_{12}) as a cylinder (ν_1) is

Table 1. Compared characteristics and cross sections for a cylindrical and a spherical GW detector of like fundamental frequencies. Note that the cylinder is assumed to be optimally oriented, i.e., with its axis perpendicular to the GW incidence direction.

Cylinder	Sphere
$\nu_1 = 910$ Hz	$\begin{cases} \nu_{12} = 910 \text{ Hz} \\ \nu_{22} = 1747 \text{ Hz} \end{cases}$
$\begin{cases} L = 3.0 \text{ metres} \\ D = 0.6 \text{ metres} \end{cases}$	$2R = 3.1$ metres
$M_c = 2.3$ tons	$M_s = 42$ tons
$\sigma_1 = 4.3\times10^{-21}$ cm^2 Hz	$\begin{cases} \sigma_{12} = 9.2\times10^{-20} \text{ cm}^2 \text{ Hz} \\ \sigma_{22} = 3.5\times10^{-20} \text{ cm}^2 \text{ Hz} \end{cases}$
(**Optimum orientation**)	(**Omnidirectional**)

about 20 times more massive, and this results in a significant improvement in cross section, since it is proportional to the detector mass. A spherical detector is therefore almost *one order of magnitude* more sensitive than a cylinder in the same frequency band – obviously apart from the important fact that the sphere has *isotropic* sensitivity.

But there is more to this. Table 1 also refers to the cross section of the sphere in its *second* higher quadrupole harmonic frequency, ν_{22} – almost twice the value of the first, ν_{12}. It is very interesting that cross section at this second frequency is *only* 2.61 times smaller than that at the first [44] while, as stressed in Sect. 2.1, it is *zero* for the second mode of the cylinder. Figure 7 shows a plot of the cross sections *per unit mass* of a cylinder and a sphere of like fundamental frequencies. It graphically displays the numbers given in the table, but also shows that, even per unit mass, the sphere is a better detector than a cylinder – its cross section 'curve' stays above that of the cylinder. In particular, the *first* quadrupole resonance turns out to have a cross section which is 1.17 times that of the cylinder (per unit mass, let me stress again) [44], i.e., $\sim$17% better. This constitutes the *quantitative* assessment of the discussion in the paragraph immediately following (48).

5.3 The *Motion Sensing* Problem

In order to determine the actual GW induced motions of an elastic solid a *motion sensing* system must be set up. In the case of currently operating cylinders this is done by what is technically known as *resonant transducer* [45]. The idea of such device is to couple the *large* oscillating cylinder mass (a few tons) to a *small* resonator (less than 1 kg) whose characteristic frequency is accurately tuned to that of the cylinder. The *joint* dynamics of the resulting system {cylinder +

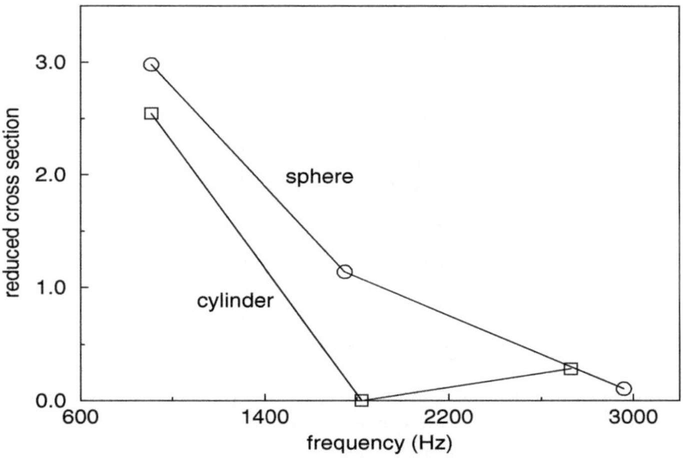

Fig. 7. Cross section *per unit mass* of a cylinder and a sphere of like fundamental frequencies, in units of Gv_s^2/c^3. Note the (slightly) better Fig. (17 %) for the lowest mode in the sphere, as well as the appreciable value in the second harmonic of the latter, in sharp contrast with the *null* coupling of this mode in the cylinder. Third harmonics show a considerable reduction in sensitivity

resonator} is a two-mode *beat* of nearby frequencies given by

$$\omega_{\pm} \simeq \omega_0 \left(1 \pm \frac{1}{2}\,\eta^{1/2}\right) , \tag{51}$$

where ω_0 is the frequency of either oscillator when uncoupled to the other, and $\eta \equiv M_{\rm resonator}/M_{\rm cylinder}$. The key concept of this device is the *resonant energy transfer*, which flows back and forth between cylinder and sensor with the period of the beat, i.e., $2\pi(\eta\omega_0)^{-1/2}$. This means that, because the mass of the sensor is very small compared to that of the cylinder, the amplitude of its oscillations is enhanced by a factor of $\eta^{-1/2}$ relative to those of the cylinder, whence a *mechanical amplification* factor is obtained *before* the sensor oscillations are converted to electrical signals, and further processed – see a more detailed account of these principles in [46].

The same principles can certainly be applied to make resonant motion sensors in a spherical antenna. In this case, however, a special bonus is there, associated to the *degeneracy* of the quadrupole frequencies: because all five quadrupole modes oscillate with the *same* frequency, it is possible to attach five (or more) identical resonators, tuned to a given quadrupole frequency, at suitable positions on the sphere surface, thus taking *multiple samples* of the motion of the sphere. This makes possible to retrieve the oscillation amplitudes of the five degenerate modes – Fig. 6 –, and thereby of the GW quadrupole amplitudes $g^{(l,m)}(t)$, since both are linearly related through (46).

A *single* spherical antenna can thus deconvolve completely the quadrupole GW signal, and do so with *isotropic sky coverage*. These characteristics are *unique* to the spherical detector, and they make it a *theoretically* superior system

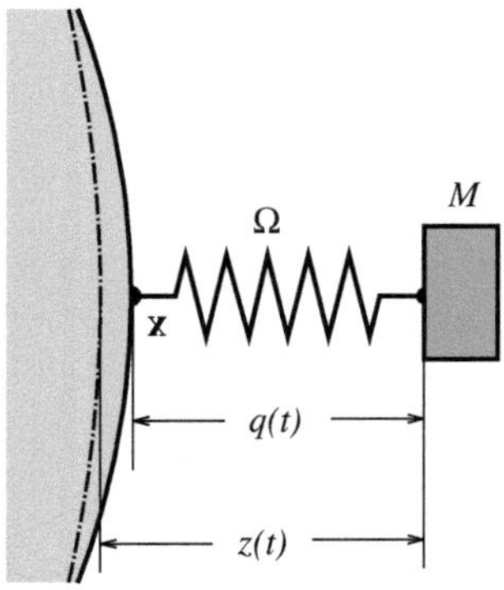

Fig. 8. Schematic diagramme of the coupling between a solid sphere and a resonator, modeled as a small mass linked to a spring attached to the surface of the sphere. The dashed-dotted arc line on the left indicates the position of the *undeformed* surface of the sphere, and the solid arc, its *actual* position

compared to either interferometers or Weber bars. In addition, a sphere can naturally measure the amplitude of the non-degenerate monopole mode, as it is conceptually simple to sense the amplitude of an isotropically breathing pattern.

The conceptual idea of a resonant sensor is shown in Fig. 8, and the equations of motion for such a system are [47]

$$\varrho\frac{\partial^2\boldsymbol{u}}{\partial t^2} = \mu\nabla^2\boldsymbol{u} + (\lambda+\mu)\,\nabla(\nabla\cdot\boldsymbol{u}) + \sum_{b=1}^{J} M_b\Omega_b^2\,\left[z_b(t) - u_b(t)\right]\,\delta^{(3)}(\boldsymbol{x}-\boldsymbol{x}_b)\,\boldsymbol{n}_b + \boldsymbol{f}_{\rm GW}(\boldsymbol{x},t)\ , \tag{52a}$$

$$\ddot{z}_a(t) = -\Omega_a^2\,\left[z_a(t) - u_a(t)\right] + \xi_a^{\rm GW}(t)\ , \qquad a = 1,\ldots,J\ , \tag{52b}$$

where M_a and Ω_a are the mass and characteristic frequency of the a-th resonator, $\boldsymbol{f}_{\rm GW}(\boldsymbol{x},t)$ is the GW tide on the sphere – see (37) –, and $\xi_a^{\rm GW}(t)$ is the GW induced tidal acceleration on the resonator itself, relative to the centre of the sphere; $\delta^{(3)}$ is the three dimensional Dirac density, i.e., point-like connections between sphere and sensors are assumed. The mathematical detail of the analysis of these equations is somewhat sophisticated. The interested reader will find complete information in [47]; the rest of this section will be devoted to a brief discussion of the main conclusions of that analysis.

First thing to stress is that (52a-b) cannot be solved analytically, they must instead be solved by a *perturbative* procedure. The small perturbation parameters are the ratios

$$\eta_a \equiv \frac{M_a}{M}\ , \qquad \eta_a \ll 1\ , \qquad a = 1,\ldots,J\ , \tag{53}$$

where M is the total mass of the sphere. Actually, the analysis assumes that the resonators are all *identical*, any deviations from this being eventually assessed by suitable methods [47,48]. The fundamental result links the spring deformations to the GW amplitudes $g^{(l,m)}(t)$ by the following formula, expressed in terms of *Laplace transforms* – noted with a caret (ˆ):

$$\hat{q}_a(s) = \eta^{-1/2} \sum_{l,m} \hat{\Lambda}_a^{(lm)}(s;\Omega)\, \hat{g}^{(l,m)}(s) \ , \qquad a = 1,\ldots,J \ , \tag{54}$$

where it is assumed that the frequency of the resonators Ω is tuned to either a monopole or a quadrupole harmonic of the sphere. The *transfer functions* $\hat{\Lambda}_a^{(lm)}(s;\Omega)$ naturally depend on whether a monopole or a quadrupole mode is selected for resonator tuning; I will quote here only the quadrupole case, as it is the most interesting one – see again full information in [47]:

$$\begin{aligned}\hat{\Lambda}_a^{(lm)}(s;\omega_{n2}) = (-1)^J \sqrt{\frac{4\pi}{5}}\, a_{n2} \sum_{b=1}^{J} \Bigg\{ \sum_{\zeta_c \neq 0} \frac{1}{2} \left[\left(s^2+\omega_{c+}^2\right)^{-1} - \left(s^2+\omega_{c-}^2\right)^{-1} \right] \\ \times \ \frac{v_a^{(c)} v_b^{(c)*}}{\zeta_c} \Bigg\}\, Y_{2m}(\boldsymbol{n}_b)\, \delta_{l2} \ ,\end{aligned} \tag{55}$$

where $v_a^{(c)}$ is the c-th normalised eigenvector of the matrix $P_2(\boldsymbol{n}_a\cdot\boldsymbol{n}_b)$, associated to its *non-zero* eigenvalue ζ_c^2, P_2 is a Legendre polynomial, and $\boldsymbol{n}_a$ is the position of the a-th resonator on the surface of the sphere. Finally,

$$\omega_{a\pm}^2 = \omega_{n2}^2 \left(1 \pm \sqrt{\frac{5}{4\pi}}\, |A_{n2}(R)|\, \zeta_a\, \eta^{1/2} \right) \ , \qquad a = 1,\ldots,J \ . \tag{56}$$

Equations (54)–(56) are the key to the GW signal deconvolution problem: they show that *beats* occur around the tuned frequency (ω_{n2} in this case), and that the resonators oscillate with amplitudes enhanced by a factor $\eta^{-1/2}$, as indeed expected. Note that these beats have frequencies which depend on the positions of the resonators $\boldsymbol{n}_a$, as shown by the presence of the ζ_a coefficient in (56).

The deconvolution problem consists in inferring the GW amplitudes $\hat{g}^{(l,m)}(s)$ from the readouts of the telescope $\hat{q}_a(s)$. Thus *at least* 5 sensors must be attached to the surface of the sphere if e.g. the 5 quadrupole amplitudes are looked for: once the corresponding five $\hat{q}_a(s)$ are *measured* the system (54) is solved for $\hat{g}^{(2,m)}(s)$, that's it.

Crucial at this point is to decide *where* to implant the resonators, as such decision bears fundamentally on the simplicity, or even the possibility of solving the posed problem. There are two major proposals in the literature for this, and they are displayed in Fig. 9. They permit the definition of so called *mode channels*, which are linear combinations of the system readouts $\hat{q}_a(s)$ which are directly proportional to the GW amplitudes $\hat{g}^{(2,m)}(s)$. They happen to be of the form [47]

$$\hat{y}^{(m)}(s) = \sum_{a=1}^{5 \text{ or } 6} v_a^{(m)*}\, \hat{q}_a(s) \ , \qquad m = -2,\ldots,2 \ , \tag{57}$$

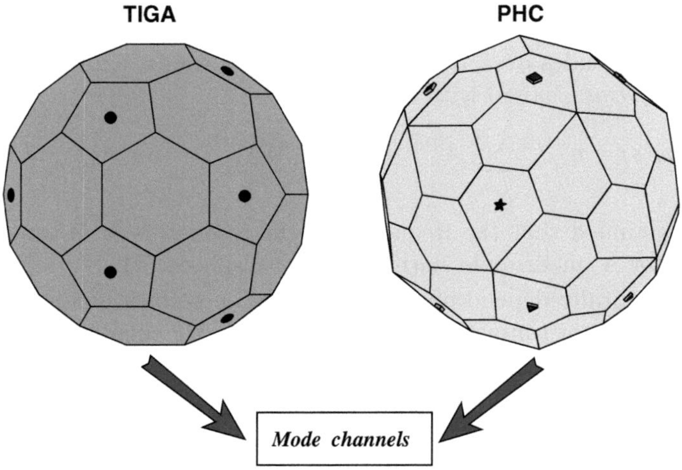

Fig. 9. The *TIGA* and *PHC* resonator distributions. In the former, *six* sensors are attached to the pentagonal faces of a truncated icosahedron, while in the latter there are *two* sets of *five* quadrupole sensors, respectively tuned to the first quadrupole frequency (squares) and to the second (triangles); there is an 11-th sensor (star) which is tuned to a monopole frequency. The relevant common characteristic of these layouts is that they enable the definition of *mode channels* – see text for details

both for *TIGA* and *PHC*. The actual result of these linear combinations is the following:

$$\hat{y}^{(m)}(s) = \eta^{-1/2}\, a_{n2}\, \frac{1}{2} \left[\left(s^2 + \omega_{m+}^2\right)^{-1} - \left(s^2 + \omega_{m-}^2\right)^{-1} \right] \hat{g}^{(2,m)}(s)\,, \tag{58}$$

i.e., they are *convolution products* of the signal and the system beats.

This formula appears to be very powerful, as it shows that suitable sensor systems enable a *single* spherical detector to fully deconvolve *all* the GW amplitudes. One should however be careful about this conclusion, for the formula *also* indicates that the relevant information *can only be obtained at the resonance frequencies* $\omega_{m\pm}$ – in an ideal, non-dissipative system. In a real system, resonant linewidths are never infinitely sharp, they have instead a certain breadth; this actually makes possible the observation across wider bandwidths, provided the amplifier noise can be kept sufficiently small – I shall briefly come to this below in Sect. 6.

5.4 *Hollow* and *Dual* Spheres

The real merit of the just described spherical GW detector comes from its *symmetry*. A *hollow* sphere does of course share the symmetry properties of a solid sphere, and one might therefore expect it to be an interesting alternative, too. A detailed study of the performance of a hollow spherical GW detector can be found in [49]. The added bonus of a hollow sphere is that there is one more structural parameter one can adjust to enhance this or that property, and this

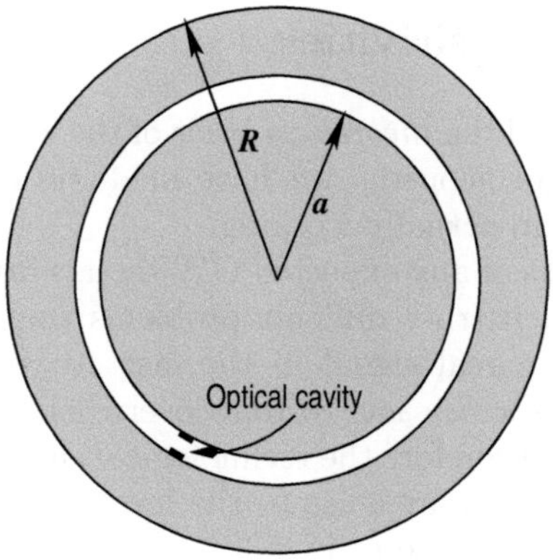

Fig. 10. Conceptual scheme of the *dual sphere*: the GW signal drives the facing surfaces between the nested spheres into oscillatory motions; when the signal frequency falls between the resonances of the spheres, oscillations happen with phase opposition, thereby enhancing the response of the device. Motion sensing can be accomplished e.g. by non-resonant optical transducers

is of course the *thickness* of the spherical shell. It appears that cross sections at e.g. different quadrupole harmonics have characteristic behaviours when plotted as functions of thickness; also, one can decide to attach the motion sensors to the inner *or* to the outer side of the shell, as their GW induced oscillations have different amplitudes.

To actually build and suspend a hollow sphere in the laboratory may be a difficult task from a technological point of view. Recently, though, a new concept spherical detector appeared in the literature [50]: this is called *dual sphere*, and consists in a solid sphere nested inside a hollow one, concentric with it and with a narrow gap between them – Fig. 10. An incoming GW drives *both* spheres into oscillation. Clearly, the inner solid sphere has a higher first (quadrupole) mode frequency than the outer hollow one, therefore an incoming GW with a frequency *between* those two will drive the oscillations of the spheres with opposite phase[11], thus enhancing the signal by a rough factor of 2.

The motion sensing in either hollow or dual spheres is conceptually analogous to that in the solid sphere, with the added flexibility that in the hollow piece one can choose to sense the displacements of either side of the shell. Actually, though, it appears that non-resonant detectors seem to constitute a better choice in dual spheres, for this enables a significant bandwidth enlargement [50].

Let me briefly discuss now, for completeness, a few essentials of GW detector sensitivity.

[11] As shown in textbooks on Mechanics, see e.g. [51], there is phase change in a response of the oscillator to a periodic excitation as the frequency of the latter shifts from below the natural frequency of the oscillator (Ω, say) to above it; the transition region has a width of order Ω/Q centred at Ω, where Q is the mechanical quality factor of the oscillator, about 10^7 or more in GW detectors.

6 GW Detector Sensitivities

So far I have only discussed the *theoretical* basis of the workings of GW detectors, whether interferometric or acoustic, yet have made no mention of the *practical* difficulties of getting them actually working...

The extreme weakness of any expected GW signals arriving in the Earth [52] is in fact a source of such truly difficult problems that it has prevented GW detectors from sighting a real signal in the last 40 years, since the times of J. Weber. Local *detector noise* has to date overwhelmed any signals possibly hitting the antennas, and therefore the technological challenge has been for years, and still is today, to reduce that noise to the level where it can be filtered out with a meaningful signal to noise ratio [53].

During the last decade or so, a number of people in different countries worldwide have managed to get important GW detection research projects funded which constitute a major step forward in detector technology. Their goal is to improve the sensitivity to the point where a significant *event rate* becomes available to the GW astronomer. We thus find such laboratories as *VIRGO* (a French-Italian collaboration), *LIGO* (USA alone), or *LISA* (a space mission, jointly funded by *NASA* and *ESA*, the European Space Agency). In addition to those, somewhat smaller experiments are *GEO-600* (a German-British venture) and *TAMA* (the Japanese project, currently making strong progress in both and stability).

So much for *interferometric* antennas. But endeavours have not declined in the *acoustic detector* arena, either. In fact, the five acoustic detectors of the Weber type (*NAUTILUS, EXPLORER, AURIGA, ALLEGRO and NIOBE*) constitute the *only working systems* in the world today. Unfortunately, though, they are only sensitive to catastrophic events happening in our galaxy, with a far too low occurrence rate. These detector systems are periodically upgraded, and their sensitivity has gone up a few orders of magnitude since their origins. As already stressed in Sect. 1, new generation *spherical* GW detectors are being programmed in Brazil, Holland and Italy, and these should suddenly improve over bars by at least one more order of magnitude.

Figure 11 contains a recent plot of GW sensitivity of various earth based detectors, therefore in the frequency range near 1 kHz. On the other hand, *LISA* will be sensitive at frequencies far away from the range plotted in Fig. 11 – see *LISA*'s web site at `http://sci.esa.int/home/lisa` and Fig. 12 below. What we see in ordinates in plots like the one in Fig. 11, which are standard in GW science, is a *spectral density* – or, rather, its *square root*. It is to be understood as follows.

Noise of whatever origin causes any detector to generate random outputs, *stochastic* time series, as this is technically known. This noise competes with any GW signals which may be present in the antenna readout, obscuring their detection. Noise of course does not come from the sky, it rather gets added to the GW signal at almost all the different stages of the detection process. For example, the GW induced oscillations of a given mass compete with thermal oscillations of that mass and its suspension systems; then there is noise in the conversion

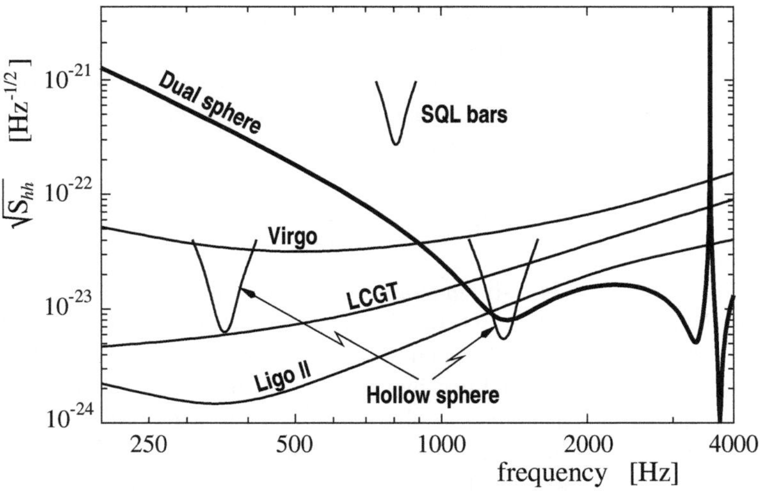

Fig. 11. Spectral strain sensitivities of various future generation GW detectors, including the "dual sphere"

from mechanical or optical GW signals to electrical signals; then there is noise introduced by the electronics in the amplifiers of the latter; and so forth...

In the end, the detector output is a certain physical magnitude (e.g. a voltage) which consists of a number of various superimposed sources of noise plus, possibly, a GW signal *converted* to volts by the detector hardware. One infers the value of the *actual* dimensionless GW amplitude $h(t)$, say, from the output voltage if one knows the precise physics of the transduction process – and this is obviously the case in any useful instrument.

Conversely, it is also possible to *back-convert* all the sources of noise voltage picked up across the various detector stages to an equivalent dimensionless "GW noise", which gets directly added to real GW signals, and travels through an ideally noiseless detector. This artifact is expedient because, for a fixed antenna, it enables a quick assessment of the detectability of a given GW signal, normally calculated by methods of gravitation theory, by direct comparison with suitably constructed detector characteristic curves – such as those in Fig. 11.

Let us then call $x(t)$ any one of the readout channels of the detector, *back-converted* to a GW amplitude by the above described procedure. We split this up into a GW signal proper $h(t)$ plus a noise term $n(t)$:

$$x(t) = h(t) + n(t) \ . \tag{59}$$

For stationary Gaussian noise, the statistical properties of $n(t)$ are encoded in its *spectral density* function, $S_h(\omega)$, which is the Fourier transform of the *autocorrelation function*

$$R(\tau) \equiv \langle x(t)\, x(t+\tau)\rangle \ , \qquad S_h(\omega) \equiv \int_{-\infty}^{\infty} R(\tau)\, \mathrm{e}^{\mathrm{i}\omega\tau}\, \mathrm{d}\tau \ , \tag{60}$$

where $\langle - \rangle$ stands for *ensemble average* [54]. It can be shown [53] that the *optimum filter* to extract the signal $h(t)$ from the system readout $x(t)$ is the so called *matched filter*, whose transfer function is the Fourier transform of the signal $\tilde{h}(\omega)$ divided by the spectral density $S_h(\omega)$, and the *detection threshold* can be set from the integrated signal to noise ratio:

$$SNR = \int_{-\infty}^{\infty} \frac{|\tilde{h}(\omega)|^2}{S_h(\omega)} \frac{d\omega}{2\pi} \,. \tag{61}$$

The GW signal $h(t)$ is a dimensionless quantity, as it measures a perturbation of the Minkowski metric, see (3). Therefore the spectral density $S_h(\omega)$ has dimensions of time, or inverse frequency, Hz^{-1}, according to the definitions (60). Now, signal to noise ratio as defined by the integral (61) is made up of the contributions of the ratio between the signal power $|\tilde{h}(\omega)|^2$ to the noise power $S_h(\omega)$ at all frequencies; the idea of the graphical representation in Fig. 11 is thus to show which is the level of noise at different frequencies by means of an rms quantity, such as the square root of the noise spectral density is. The appropriate units for this representation are accordingly $\mathrm{Hz}^{-1/2}$, as indicated.

Different interesting sources of GWs are being considered by other authors in this volume, so I will not go into such matters here. It is nevertheless instructive to present an example graph of a few signals on top of the sensitivity curves of various detectors in order to get a picture of the actual possibilities of each instrument, and also to grasp what are the spectral orders of magnitude of different GW signals. One such plot is presented in Fig. 12. This is, let me insist, the standard way to assess the detectability of the different GW sources.

7 Concluding Remarks

This article is a brief review on the nature of the interactions of GWs with *test* particles and *test* electromagnetic fields, as they specifically happen in currently conceived detection devices. While the fundamental principles are not new in themselves, their application in actual systems is still in many cases subject of research, as we have seen. A thorough understanding of these matters is absolutely essential for an adequate interpretation of the antenna readouts, the more so if one considers the extreme *weakness* of any signals reaching us from even the most intense sources.

I have omitted any detailed discussion of the practical problems faced by real detector building. This is a major research field in itself, of an intrinsically sophisticated and multidisciplinary nature – involving such fundamental issues as quantum measurement limits and techniques, or Quantum Optics [56]. But it is of course not directly related to Astrophysics or Relativity... I have however considered appropriate to summarily brief the interested reader on how detector sensitivities are defined, and on which are the detection thresholds in current state of the art GW detectors: this is a key issue for an astrophysicist/relativist wishing to assess the detectability of a given GW signal, whether by existing instruments or future planned.

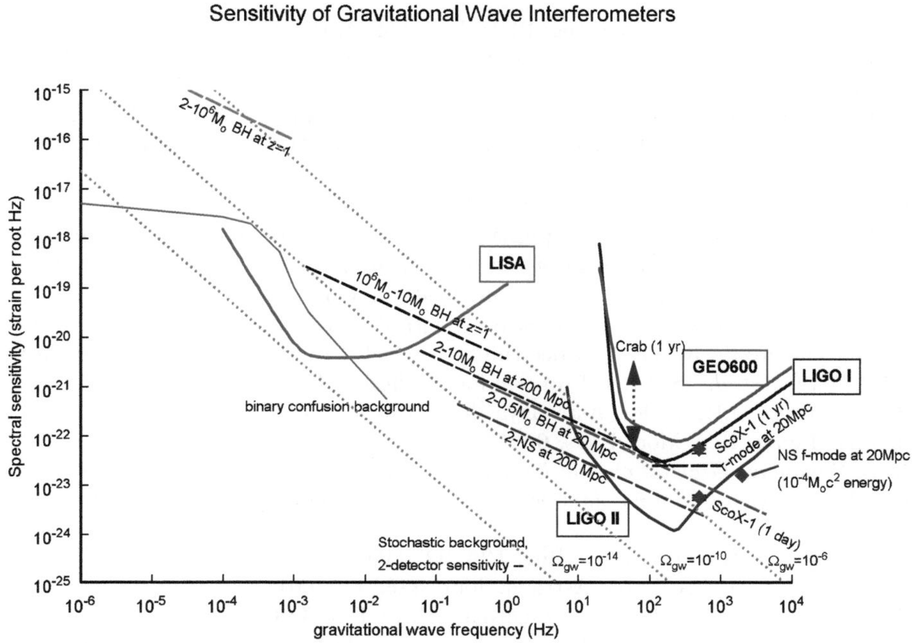

Fig. 12. Root mean square noise spectral densities, referred to GW amplitudes, for some of the upcoming interferometric GW detectors, together with the spectral intensities of various signals. The latter are estimated by numerical calculation, while the noise is modeled on the basis of its origin and instrumental characteristics. See [55]

GW detection endeavours have been the subject of intensive research during the last 4 decades, though the ultimate goal of sighting GW events has still not been accomplished. While this may look like a major failure, one may not forget that detector sensitivities have gone up by a remarkable *six orders of magnitude* (in energy, three in GW amplitude) since the very first telescopes constructed by J. Weber. Also, a look at the trend in this development indicates that we are getting closer to the objective.

Prospects look now better than ever yet the real challenge is still there...

References

1. A. Einstein: Sitz. Ber. Kön. Preus. Ak. Wiss., p. 688 (1916), and p. 154 (1918)
2. C.W. Misner, K.S. Thorne, J.A. Wheeler: *Gravitation* (Freeman, San Francisco 1973)
3. C.M. Will: *Theory and experiment in Gravitational Physics* (Cambridge University Press, Cambridge 1993)
4. J. Weber: Phys. Rev. Lett. **22**, 1320 (1969); **24**, 276 (1970); **25**, 180 (1970)
5. J. Weber: 'Supernova 1987 A Rome and Maryland GW radiation antenna observations'. In: *Gravitational wave experiments*, Proceedings of the 1st Edoardo Amaldi Conference. ed. by E. Coccia, G. Pizzella, F. Ronga (World Scientific, Singapore 1995), p. 416

6. G.W. Gibbons, S.W. Hawking: Phys. Rev. **D4**, 2191 (1971)
7. R. Weiss: Quarterly Progress Report, Research Laboratory of Electronics **105**, 54 (MIT 1972)
8. R. Drever: 'Interferometric detectors for gravitational radiation'. In: *Gravitational radiation*. ed. by N. Deruelle and T. Piran (North Holland, Amsterdam 1983), p. 321
9. G.A. Prodi et al.: Int. J. Mod. Phys. **D9**, 237 (2000)
10. O.D. Aguiar *et al.*: 'The first phase of the Brazilian Graviton project'. In: *Gravitational waves*, Proceedings of the 3rd Edoardo Amaldi Conference. ed. by S. Meshkov (World Scientific, Singapore 2000), p. 413
11. G. Frossati: '*MINIGRAIL*, a sensitive spherical gravitational wave antenna for frequencies around 3.6 kHz', Projectruimte Normale Programma FWO (Leiden 2000)
12. P. Astone *et al.*: '*SFERA*: Proposal for a spherical GW detector' (Roma 1997)
13. R.A. Hulse, J.H. Taylor: Astrophys. Jour. **195**, L51 (1975)
14. J.H. Taylor: Rev. Mod. Phys. **66**, 711 (1994)
15. S. Weinberg: *Gravitation and Cosmology* (Wiley & Sons, New York 1972)
16. J.B. Griffiths: *Colliding plane waves in General Relativity* (Clarendon Press, Oxford 1991)
17. C. Brans, R.H. Dicke: Phys. Rev. **124**, 925 (1961)
18. E. Gaztañaga, J.A. Lobo: Astroph. Jour. **548**, 47 (2001)
19. D.M. Eardley, D.L. Lee, A.P. Lightman: Phys. Rev. **D8**, 3308 (1973)
20. S. Chandrasekhar: *The mathematical theory of Black Holes* (Pergamon Press, Oxford 1983)
21. B.J. Meers: Phys. Rev. **D38**, 2317 (1988)
22. J.A. Lobo: Class. Quantum Grav. **9**, 1385 (1992)
23. F.I. Cooperstock, V. Faraoni: Class. Quantum Grav. **10**, 1189 (1993)
24. G. Pizzella: *Fisica Sperimentale del Campo Gravitazionale* (Nuova Italia Scientifica, Roma 1993)
25. L.D. Landau, E.M. Lifshitz: *Theory of Elasticity* (Pergamon Press, Oxford 1970)
26. J.A. Lobo: Phys. Rev. **D52**, 591 (1995)
27. R.G. Hier, S.N. Rasband: Astroph. Jour. **195**, 507 (1975).
28. S.N. Rasband: Jour. Acous. Soc. Am. **57**, 899 (1975)
29. P. Astone *et al.*: Phys. Rev. **D47**, 362 (1993)
30. P. Astone *et al.*: Astropart. Phys. **7**, 231 (1997)
31. S.V. Dhurandhar, M. Tinto: Mon. Not. Roy. Astr. Soc. **234**, 663 (1988), and **236**, 621 (1989)
32. Z.A. Allen *et al.*: Phys. Rev. Lett. **85**, 5046 (2000)
33. R. Forward: Gen. Rel. Grav. **2**, 149 (1971)
34. N. Ashby, J. Dreitlein: Phys. Rev. **D12**, 336 (1975).
35. R.V. Wagoner, H.J. Paik: 'Multimode detection of gravitational waves by a sphere'. In: *Experimental Gravitation*, Proceedings of the Pavia International Symposium, Accad. Naz. dei Lincei (Roma 1977)
36. W.W. Johnson, S.M. Merkowitz: Phys. Rev. Lett. **70**, 2367 (1993)
37. S.M. Merkowitz: Truncated Icosahedral Gravitational Wave Antenna. PhD Thesis Memoir, Louisiana State University (Baton Rouge 1995)
38. E. Coccia: private communication
39. M. Bianchi, E. Coccia, C.N. Colacino, V. Fafone, F. Fucito: Class. Quantum Grav. **13**, 2865 (1996)
40. A.R. Edmonds: *Angular Momentum in Quantum Mechanics* (Princeton University Press, Princeton 1960)

41. J.A. Ortega: Spherical gravitational wave detectors. PhD Thesis Memoir, University of Barcelona (Barcelona 1997)
42. M. Bianchi, M. Brunetti, E. Coccia, F. Fucito, J.A. Lobo: Phys. Rev. **D57**, 4525 (1998)
43. E. Coccia, F. Fucito, J.A. Lobo, M. Salvino: Phys. Rev. **D62**, 044019-1 (2000)
44. E. Coccia, J.A. Lobo, J.A. Ortega: Phys. Rev. **D52**, 3735 (1995)
45. H.J. Paik: 'Electromechanical transducers and bandwidth of resonant mass GW detectors'. In: *Gravitational wave experiments*, Proceedings of the 1st Edoardo Amaldi Conference. ed. by E. Coccia, G. Pizzella, F. Ronga (World Scientific, Singapore 1995), p. 201
46. J.A. Lobo: 'Spherical gravitational wave detectors and geometry'. In: *Gravitational waves*, Proceedings of the 2nd Edoardo Amaldi Conference. ed. by E. Coccia, G. Pizzella, G. Veneziano (World Scientific, Singapore 1998), p. 168
47. J.A. Lobo: Mon. Not. Roy. Astr. Soc. **316**, 173 (2000) gr-qc/0006055
48. S.M. Merkowitz, W.W. Johnson: Phys. Rev. **D56**, 7513 (1997)
49. E. Coccia, V. Fafone, G. Frossati, J.A. Lobo, J.A. Ortega: Phys. Rev. **D57**, 2051 (1998)
50. M. Cerdonio, L. Conti, J.A. Lobo, A. Ortolan, L. Taffarello, J.P. Zendri: Phys. Rev. Lett. **87**, 031101 (2001)
51. K.R. Symon: *Mechanics*, 2nd edition, Chap. 2 (Addison-Wesley, Reading 1960)
52. K.S. Thorne: 'Gravitational radiation'. In: *300 Years of Gravitation.* ed. by S.W. Hawking and W. Israel (Cambridge University Press, Cambridge 1988), p. 330. See also more recent reviews by K.S. Thorne, e.g. gr-qc/9704042, or visit the website `http://fermi.phys.ualberta.ca/~ccgrra/thorne/index.html`
53. A. Królak, J.A. Lobo, B.J. Meers: Phys. Rev. **D48**, 3451 (1993)
54. S.M. Kay: *Modern spectral estimation: theory and application* (Prentice Hall, New Jersey 1988)
55. B.F. Schutz: Class. Quantum Grav. **16**, A131 (1999)
56. V.B. Braginsky, F.Ya. Khalili: *Quantum measurement* (Cambridge University Press, Cambridge 1995)

Lecture Notes in Physics

For information about Vols. 1–589
please contact your bookseller or Springer-Verlag

Vol.590: D. Benest, C. Froeschlé (Eds.), Singularities in Gravitational Systems. Applications to Chaotic Transport in the Solar System.
link.springer.de/link/service/series/2669/tocs/t2590.htm

Vol.591: M. Beyer (Ed.), CP Violation in Particle, Nuclear and Astrophysics.
link.springer.de/link/service/series/2669/tocs/t2591.htm

Vol.592: S. Cotsakis, L. Papantonopoulos (Eds.), Cosmological Crossroads. An Advanced Course in Mathematical, Physical and String Cosmology.
link.springer.de/link/service/series/2669/tocs/t2592.htm

Vol.593: D. Shi, B. Aktaş, L. Pust, F. Mikhailov (Eds.), Nanostructured Magnetic Materials and Their Applications.
link.springer.de/link/service/series/2669/tocs/t2593.htm

Vol.594: S. Odenbach (Ed.),Ferrofluids. Magnetical Controllable Fluids and Their Applications.
link.springer.de/link/service/series/2669/tocs/t2594.htm

Vol.595: C. Berthier, L. P. Lévy, G. Martinez (Eds.), High Magnetic Fields. Applications in Condensed Matter Physics and Spectroscopy.
link.springer.de/link/service/series/2669/tocs/t2595.htm

Vol.596: F. Scheck, H. Upmeier, W. Werner (Eds.), Noncommutative Geometry and the Standard Model of Elememtary Particle Physics.
link.springer.de/link/service/series/2669/tocs/t2596.htm

Vol.597: P. Garbaczewski, R. Olkiewicz (Eds.), Dynamics of Dissipation.
link.springer.de/link/service/series/2669/tocs/t2597.htm

Vol.598: K. Weiler (Ed.), Supernovae and Gamma-Ray Bursters.
Online version forthcoming

Vol.599: J.P. Rozelot (Ed.), The Sun's Surface and Subsurface. Investigating Shape and Irradiance.
Online version forthcoming

Vol.600: K. Mecke, D. Stoyan (Eds.), Morphology of Condensed Matter. Physcis and Geometry of Spatial Complex Systems.
link.springer.de/link/service/series/2669/tocs/t2600.htm

Vol.601: F. Mezei, C. Pappas, T. Gutberlet (Eds.), Neutron Spin Echo Spectroscopy. Basics, Trends and Applications.
link.springer.de/link/service/series/2669/tocs/t2601.htm

Vol.602: T. Dauxois, S. Ruffo, E. Arimondo (Eds.), Dynamics and Thermodynamics of Systems with Long Range Interactions.
link.springer.de/link/service/series/2669/tocs/t2602.htm

Vol.603: C. Noce, A. Vecchione, M. Cuoco, A. Romano (Eds.), Ruthenate and Rutheno-Cuprate Materials. Superconductivity, Magnetism and Quantum Phase.
link.springer.de/link/service/series/2669/tocs/t2603.htm

Vol.604: J. Frauendiener, H. Friedrich (Eds.), The Conformal Structure of Space-Time: Geometry, Analysis, Numerics.
link.springer.de/link/service/series/2669/tocs/t2604.htm

Vol.605: G. Ciccotti, M. Mareschal, P. Nielaba (Eds.), Bridging Time Scales: Molecular Simulations for the Next Decade.
link.springer.de/link/service/series/2669/tocs/t2605.htm

Vol.606: J.-U. Sommer, G. Reiter (Eds.), Polymer Crystallization. Obervations, Concepts and Interpretations.
Online version forthcoming

Vol.607: R. Guzzi (Ed.), Exploring the Atmosphere by Remote Sensing Techniques.
link.springer.de/link/service/series/5304/tocs/t3607.htm

Vol.608: F. Courbin, D. Minniti (Eds.), Gravitational Lensing:An Astrophysical Tool.
link.springer.de/link/service/series/2669/tocs/t2608.htm

Vol.609: T. Henning (Ed.), Astromineralogy.
Online version forthcoming

Vol.610: M. Ristig, K. Gernoth (Eds.), Particle Scattering, X-Ray Diffraction, and Microstructure of Solids and Liquids.
link.springer.de/link/service/series/5304/tocs/t3610.htm

Vol.611: A. Buchleitner, K. Hornberger (Eds.), Coherent Evolution in Noisy Environments.
link.springer.de/link/service/series/2669/tocs/t2611.htm

Vol.612 L. Klein, (Ed.), Energy Conversion and Particle Acceleration in the Solar Corona.
Online version forthcoming

Vol.613 K. Porsezian, V.C. Kuriakose, (Eds.), Optical Solitons. Theoretical and Experimental Challenges.
link.springer.de/link/service/series/2669/tocs/t3613.htm

Vol.614 E. Falgarone, T. Passot (Eds.), Turbulence and Magnetic Fields in Astrophysics.
link.springer.de/link/service/series/5304/tocs/t3614.htm

Vol.615 J. Büchner, C.T. Dum, M. Scholer (Eds.), Space Plasma Simulation.
Online version forthcoming

Vol.616 J. Trampetic, J. Wess (Eds.), Particle Physics in the New Millenium.
Online version forthcoming

Vol.617 L. Fernández-Jambrina, L. M. González-Romero (Eds.), Current Trends in Relativistic Astrophysics, Theoretical, Numerical, Observational
Online version forthcoming

Vol.618 M.D. Esposti, S. Graffi (Eds.), The Mathematical Aspects of Quantum Maps
Online version forthcoming

Vol.619 H.M. Antia, A. Bhatnagar, P. Ulmschneider (Eds.), Lectures on Solar Physics
Online version forthcoming

Vol.620 C. Fiolhais, F. Nogueira, M. Marques (Eds.), A Primer in Density Functional Theory
Online version forthcoming

Monographs

For information about Vols. 1–30 please contact your bookseller or Springer-Verlag

Vol. m 31 (Corr. Second Printing): P. Busch, M. Grabowski, P.J. Lahti, Operational Quantum Physics. XII, 230 pages. 1997.

Vol. m 32: L. de Broglie, Diverses questions de mécanique et de thermodynamique classiques et relativistes. XII, 198 pages. 1995.

Vol. m 33: R. Alkofer, H. Reinhardt, Chiral Quark Dynamics. VIII, 115 pages. 1995.

Vol. m 34: R. Jost, Das Märchen vom Elfenbeinernen Turm. VIII, 286 pages. 1995.

Vol. m 35: E. Elizalde, Ten Physical Applications of Spectral Zeta Functions. XIV, 224 pages. 1995.

Vol. m 36: G. Dunne, Self-Dual Chern-Simons Theories. X, 217 pages. 1995.

Vol. m 37: S. Childress, A.D. Gilbert, Stretch, Twist, Fold: The Fast Dynamo. XI, 406 pages. 1995.

Vol. m 38: J. González, M. A. Martín-Delgado, G. Sierra, A. H. Vozmediano, Quantum Electron Liquids and High-Tc Superconductivity. X, 299 pages. 1995.

Vol. m 39: L. Pittner, Algebraic Foundations of Non-Com-mutative Differential Geometry and Quantum Groups. XII, 469 pages. 1996.

Vol. m 40: H.-J. Borchers, Translation Group and Particle Representations in Quantum Field Theory. VII, 131 pages. 1996.

Vol. m 41: B. K. Chakrabarti, A. Dutta, P. Sen, Quantum Ising Phases and Transitions in Transverse Ising Models. X, 204 pages. 1996.

Vol. m 42: P. Bouwknegt, J. McCarthy, K. Pilch, The W3 Algebra. Modules, Semi-infinite Cohomology and BV Algebras. XI, 204 pages. 1996.

Vol. m 43: M. Schottenloher, A Mathematical Introduction to Conformal Field Theory. VIII, 142 pages. 1997.

Vol. m 44: A. Bach, Indistinguishable Classical Particles. VIII, 157 pages. 1997.

Vol. m 45: M. Ferrari, V. T. Granik, A. Imam, J. C. Nadeau (Eds.), Advances in Doublet Mechanics. XVI, 214 pages. 1997.

Vol. m 46: M. Camenzind, Les noyaux actifs de galaxies. XVIII, 218 pages. 1997.

Vol. m 47: L. M. Zubov, Nonlinear Theory of Dislocations and Disclinations in Elastic Body. VI, 205 pages. 1997.

Vol. m 48: P. Kopietz, Bosonization of Interacting Fermions in Arbitrary Dimensions. XII, 259 pages. 1997.

Vol. m 49: M. Zak, J. B. Zbilut, R. E. Meyers, From Instability to Intelligence. Complexity and Predictability in Nonlinear Dynamics. XIV, 552 pages. 1997.

Vol. m 50: J. Ambjørn, M. Carfora, A. Marzuoli, The Geometry of Dynamical Triangulations. VI, 197 pages. 1997.

Vol. m 51: G. Landi, An Introduction to Noncommutative Spaces and Their Geometries. XI, 200 pages. 1997.

Vol. m 52: M. Hénon, Generating Families in the Restricted Three-Body Problem. XI, 278 pages. 1997.

Vol. m 53: M. Gad-el-Hak, A. Pollard, J.-P. Bonnet (Eds.), Flow Control. Fundamentals and Practices. XII, 527 pages. 1998.

Vol. m 54: Y. Suzuki, K. Varga, Stochastic Variational Approach to Quantum-Mechanical Few-Body Problems. XIV, 324 pages. 1998.

Vol. m 55: F. Busse, S. C. Müller, Evolution of Spontaneous Structures in Dissipative Continuous Systems. X, 559 pages. 1998.

Vol. m 56: R. Haussmann, Self-consistent Quantum Field Theory and Bosonization for Strongly Correlated Electron Systems. VIII, 173 pages. 1999.

Vol. m 57: G. Cicogna, G. Gaeta, Symmetry and Perturbation Theory in Nonlinear Dynamics. XI, 208 pages. 1999.

Vol. m 58: J. Daillant, A. Gibaud (Eds.), X-Ray and Neutron Reflectivity: Principles and Applications. XVIII, 331 pages. 1999.

Vol. m 59: M. Kriele, Spacetime. Foundations of General Relativity and Differential Geometry. XV, 432 pages. 1999.

Vol. m 60: J. T. Londergan, J. P. Carini, D. P. Murdock, Binding and Scattering in Two-Dimensional Systems. Applications to Quantum Wires, Waveguides and Photonic Crystals. X, 222 pages. 1999.

Vol. m 61: V. Perlick, Ray Optics, Fermat's Principle, and Applications to General Relativity. X, 220 pages. 2000.

Vol. m 62: J. Berger, J. Rubinstein, Connectivity and Superconductivity. XI, 246 pages. 2000.

Vol. m 63: R. J. Szabo, Ray Optics, Equivariant Cohomology and Localization of Path Integrals. XII, 315 pages. 2000.

Vol. m 64: I. G. Avramidi, Heat Kernel and Quantum Gravity. X, 143 pages. 2000.

Vol. m 65: M. Hénon, Generating Families in the Restricted Three-Body Problem. Quantitative Study of Bifurcations. XII, 301 pages. 2001.

Vol. m 66: F. Calogero, Classical Many-Body Problems Amenable to Exact Treatments. XIX, 749 pages. 2001.

Vol. m 67: A. S. Holevo, Statistical Structure of Quantum Theory. IX, 159 pages. 2001.

Vol. m 68: N. Polonsky, Supersymmetry: Structure and Phenomena. Extensions of the Standard Model. XV, 169 pages. 2001.

Vol. m 69: W. Staude, Laser-Strophometry. High-Resolution Techniques for Velocity Gradient Measurements in Fluid Flows. XV, 178 pages. 2001.

Vol. m 70: P. T. Chruściel, J. Jezierski, J. Kijowski, Hamiltonian Field Theory in the Radiating Regime. VI, 172 pages. 2002.

Vol. m 71: S. Odenbach, Magnetoviscous Effects in Ferrofluids. X, 151 pages. 2002.

Vol. m 72: J. G. Muga, R. Sala Mayato, I. L. Egusquiza (Eds.), Time in Quantum Mechanics. XII, 419 pages. 2002.

Vol. m 73: H. Emmerich, The Diffuse Interface Approach in Materials Science. VIII, 178 pages. 2003

MIX
Papier aus verantwortungsvollen Quellen
Paper from responsible sources
FSC® C105338

If you have any concerns about our products,
you can contact us on
ProductSafety@springernature.com

In case Publisher is established outside the EU,
the EU authorized representative is:
Springer Nature Customer Service Center GmbH
Europaplatz 3, 69115 Heidelberg, Germany

Printed by Libri Plureos GmbH
in Hamburg, Germany